Fortschritte der Chemie organischer Naturstoffe

Progress in the Chemistry of Organic Natural Products

53

Founded by L. Zechmeister
Edited by W. Herz, H. Grisebach, G.W. Kirby, and Ch. Tamm

Authors:
L. F. Alves, A. Chimiak, M. J. Milewska, T. Nomura

Springer-Verlag
Wien New York 1988

Dr. W. Herz, Professor of Chemistry, Department of Chemistry,
The Florida State University, Tallahassee, Florida, U.S.A.

Prof. Dr. H. Grisebach, Biologisches Institut II, Lehrstuhl für Biochemie der Pflanzen,
Albert-Ludwigs-Universität, Freiburg i.Br., Federal Republic of Germany

G.W. Kirby, Sc. D., Regius Professor of Chemistry, Chemistry Department,
The University, Glasgow, Scotland

Prof. Dr. Ch. Tamm, Institut für Organische Chemie der Universität Basel,
Basel, Switzerland

With 72 Figures

© 1988 by Springer-Verlag/Wien
Softcover reprint of the hardcover 1st edition 1988

Library of Congress Catalog Card Number AC 39-1015

ISSN 0071-7886

ISBN-13:978-3-7091-8989-4   e-ISBN-13:978-3-7091-8987-0
DOI: 10.1007/978-3-7091-8987-0

Contents

List of Contributors

ALVES, Dr. L.F., Rua das Marrecas 38, sala 302, CEP 20031 Lapa, Rio de Janeiro, Brazil.

CHIMIAK, Prof. A., Institute of Organic and Food Chemistry and Technology, Technical University, Majakowskiego 11/12, 80-952 Gdańsk, Poland.

MILEWSKA, Dr. M.J., Institute of Organic and Food Chemistry and Technology, Technical University, Majakowskiego 11/12, 80-952 Gdańsk, Poland.

NOMURA, Prof. T., Ph.D., Faculty of Pharmaceutical Sciences, Toho University, Miyama 2-2-1, Funabashi, Chiba, 274, Japan.

Chemical Ecology and the Social Behavior
of Animals

By L.F. ALVES, Rio de Janeiro, Brasil

With 2 Figures

Contents

1. Introduction

The last fifteen or twenty years have witnessed an increasing amount of research on the role of chemical signals in the social behavior of animals. The present article attempts to present a survey of the information available at this time. General aspects have been discussed in a number of books and reviews (1–16). In this review only behavioral aspects of semiochemicals (for definition see next section) will be discussed. Purely chemical aspects such as isolation, identification and synthesis will be omitted. For reviews of the techniques and results used in chemical investigations see (17–37).

Semiochemicals are well documented in mammals (38–56), arthropods (57–85), fishes and other aquatic organisms (86–96), invertebrates except insects (97–106) and humans (107–118).

BRAND et al. (17) stated in 1979 that about 10000 papers on insect chemistry alone had been published by that time and ROPARTZ (119) argued in 1977 that it would be impossible to compile an exhaustive list of articles on chemical communication among rodents.

Odors are involved in several aspects of the life cycles of animals: scent marking, kin, caste, age, conspecifics, individual, social status, physiological status and sexual recognition, territorial marking, mating, sexual attraction, alarm, predator-prey interaction, defense, migration, reproductive isolation and feeding.

According to ROPARTZ (119) almost all social information in rodents is transmitted through olfactory channels and GARRY (120) has suggested that all activities within a bee hive are potentially mediated by pheromones.

2. Terminology

Over twenty-five years ago, KARLSON and BUTENANDT (1) and KARLSON and LÜSCHER (121) proposed the term *pheromone* – from the Greek *pherein* (to carry) and *hormon* (to stimulate) – to designate substances which are secreted by an animal to the outside and then cause a specific reaction in an individual of the same species, such as the release of certain behavior or the induction of certain physiological responses.

WILSON (3) and WILSON and BOSSERT (4) divided *pheromones* into *release pheromones* (if the stimulus triggers an immediate behavioral response in the receiving organism) and *primer pheromones* (if the response is a long lasting one). Substances involved in alarm, territorial marking, trail, social status, sexual attraction and reproductive isolation are *release pheromones* while the odors involved in puberty acceleration, pregnancy block, estrus synchronization or suppression are *primer pheromones*.

Many synonymous and overlapping terms have been suggested as categories of chemical releasing stimuli; the topic has been reviewed by NORDLUNG (*122*).

NORDLUNG and LEWIS (*123*) broadened the definition of a pheromone to include chemicals elaborated by plants. Indeed, interactions within the plant kingdon can also be attributed to semiochemicals (*124–132a*).

In 1971, WHITTAKER and FEENY (*133*) proposed the term *allelochemic* for a substance involved in interspecific communication. BROWN *et al.* (*134*) subdivided *allelochemics* into a) *allomones*, defined as those chemical substances produced or acquired by an organism which, when in contact with an individual of another species in the natural context, evoke in the recipient organism a behavioral or physiological reaction adaptively favorable to the emitter and b) *kairomones* which are chemical messengers the adaptive benefit of which falls on the receiver rather than the emitter.

In 1971, LAW and REGNIER (*6*) proposed the term *semiochemical* (Gr. *semion* = mark or signal) to describe a chemical that delivers any message to any organism.

In 1976 NORDLUNG and LEWIS (*123*) proposed two new terms: *synomones* (Gr. *syn* = with or jointly) are substances which evoke a response adaptively favorable to both emitter and receiver, e.g. a floral scent, and *apneumones* (Gr. *a-pneum* = lifeless or breathless) are substances emitted by non-living material which evoke a behavioral or physiological reaction adaptively favorable to a receiving organism, but are detrimental to an organism of another species that may be found in or on the non-living material. For example, the ichneumonid parasite *Venturia canescens* is attracted to the odor of its host's food, i.e. fresh oatmeal and the braconid *Alysia manducator* and the chalcid *Nasonia vitripenis* are attracted to meat even though it never contained the dipteran hosts.

Controversy on the use of the terms *pheromone, allomone* and *kairomone* has been a problem in chemical ecology (*122, 123, 135–145*). When KARLSON and LÜSCHER (*121*) proposed the term pheromone, they stated that "strict species-specific activity is not required; certain overlaps between closely related species may occur". Indeed, HASKINS *et al.* (*146*) reported that formic acid, a potent alarm pheromone and the defensive secretion (thus an allomone) from *Camponotus* sp. is also an effective kairomone for its predator *Myrmecia gulosa*, because it induces more vigorous attacks by *M. gulosa* on *Camponotus*.

However, overlap of the type suggested by KARLSON and LÜSCHER has been detected not only between related species, but also across insect orders and across classes. Thus, HUMMEL and METCALF (*147*)

reported that the sex pheromone Z-7-dodecenyl acetate (1) of the female cabbage looper (*Trichoplusia ni*, Lepidoptera) also attracts males
of *Scirtes orbiculatus* (Coleoptera) and mentioned the earlier observation that the click beetle *Melanotus depressus* (Coleoptera) responds
to the sex pheromone E-11-tetradecenyl acetate (2) and E-11-tetrade-
cenol (3) of the female tufted apple budworm, *Platynota idaeusalis*.

(1)

(2)

(3)

(4)

(5)

(6)

(7)

(8)

The interaction between *pheromone-allomone-kairomone* is also well
established. For example, *Anagasta kuhniella* moths emit from the mandibular gland an epideitic (or a dispersal) pheromone which also induces parasitization by *Venturia canescens* (148). STERNLICHT (149) reported that *Aphytes mellinus* and *A. coheni*, the two parasitic wasps
of the coccid *Aonidiella aurantii*, were attracted by the sex pheromone

of the host and LEWIS *et al.* (*150*) found that a blend of hexadecanal (**4**), Z-7-hexadecenal (**5**), Z-9-hexadecenal (**6**) and Z-11-hexadecenal (**7**), the sex pheromone of female *Heliothis zea*, increases the rates of parasitization of *Heliothis* eggs by *Trichogramma pretiosus*, while according to EBERHARD (*151*) the mature female bolas spider *Mastophora* sp. (class Arachnida) attracts its prey, the fall armyworm *Spodoptera frugiperda*, with a substance which apparently mimics the female moth's pheromone, Z-9-dodecen-1-ol acetate (**8**).

Recently, PAYNE (*152*) reported that frontalin (**9**), the sex pheromone of the bark beetle *Dendroctonus frontalis*, is also a kairomone which attracts its predator *Thanasimus dubius*.

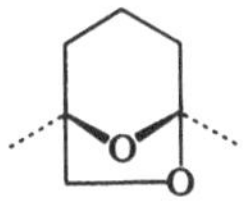

(**9**) 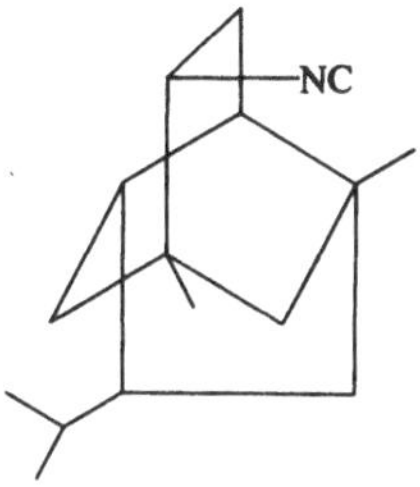   (**10**)

According to SCHEUER (*88*), pupukeanane (**10**), the defensive chemical secretion of the nudibranch *Phyllidia varicosa*, must be regarded as a kairomone as well as an allomone, since the mollusks obtain it through association with a sponge tentatively identified as *Hymenacion* sp. on which the nudibranch feeds.

Since distinguishing between allomone and kairomone is difficult or even impossible, WILSON (*144*) has suggested that the term kairomone be dropped and that the term allomone be used in a broader sense. BLUM (*137*) has argued that kairomones are nothing more than pheromones or allomones which in a few instances have evolutionarily boomeranged and stated "it is totally premature as well as evolutionarily unsound to regard these compounds as maladaptive", as argued elsewhere (*134*) rejects (*118*). PASTEELS (*140*) also reject the appropriateness of the term kairomone and makes a caricatural analogy between mosquitoes, fleas and human blood: "Since mosquitoes and fleas like human blood, blood is non-adaptive or mal-adaptive to humans". However, WELDON (*143*) reaffirms that the term kairomone is valid and useful.

Homeochemics is a new term suggested by MARTIN (*139*) for chemicals released by an individual which convey information to one or more individuals of the same species, but SMITH (*142*) has advocated use of a more euphonious word.

3. Origin

Many types of glands are involved in the production and/or emission of semiochemicals: the labial, mandibular, Dufour, hindgut, poison, pygidial, sternal, anal, supraanal, cephalic, tibial, Pavan, epidermal, salivary, and cibarial glands in insects and the tarsal, inguinal, chin, orbital, infraorbital, frontal and caudal glands in mammals.

THIESSEN has pointed out that thermoregulatory processes can be involved in the production and utilization of pheromones (153–155). His data (156, 157) suggest that chemical communication among Mongolian gerbils *Meriones unguiculatus* is associated with both thermoregulation and osmoregulation. The pheromone is emitted under heat stress conditions and is associated with thermoregulatory spreading of saliva.

On the other hand, insect pheromones may be obtained from plants on which the insects feeds, biosynthesized by the insects themselves or acquired through symbiosis with microorganisms. For example, CULVENOR and coworkers (158–162) noted that male butterflies obtain the precursors of their attractants by visiting plants containing 1,2-dehydropyrrolizidine alkaloids.

However, there exists a controversy on how the female oak leaf roller *Archips semiferanus* acquires its sex attractants. Evidence to support the theory that numerous insect semiochemicals are of plant origin has been presented by HENDRY et al. (163–166). HENDRY (163) identified 21 isomers of tetradecenyl acetate in the active sexual attraction fraction of the female moth, reported that females frequently attempt to copulate with the host leaves that had been damaged by larval feeding and suggested that the pheromone might be present in the plant. This was supported by the finding that all species of oak leaves tested contained tetradecenyl acetates (164). In conflict with this were subsequent reports (167, 168) that pheromone blends from females kept either on a semisynthetic diet or fed on leaves from three oak species did not vary significantly.

In addition, SCHNEIDER and colleagues (169) reported that in males of *Creatonotos gangis* and *C. transiens*, hydroxydanaidal (the major volatile component of the scent organ) is derived from pyrrolizidine alkaloids ingested by the larva. They also observed that dietary alkaloids also influence the morphology of the coremata (the scent organ).

Arthropod defensive secretions may also be sequestered from food plants (68). Thus FERGUSON and METCALF (170) found that four species of *Diabrotica* beetles sequester cucurbitacins as their defense compounds.

Environmental and physiological factors controlling the production and/or cmission of insect sex pheromones were reviewed by SHOREY (8). Light intensity, temperature, host habitat, and air velocity are the main environmental variables. Physiological factors involved are time of day, age, previous experience, population density, diet, hormones and seasonal rhythms.

That microorganisms in the gut of an insect are capable of converting substances from food plants into chemicals which can be used as pheromones has been demonstrated by several authors. Thus while PEARCE et al. (171) reported that alpha-cubebene (11) one of the components in the aggregation pheromone released by Scolytus multistriatus, is produced by the host, BRAND et al. (172) found that a bacterium isolated from the gut of Ips paraconfusus converts alpha-pinene (12) to cis- and trans-verbenol (13, 14) and myrtenol (15).

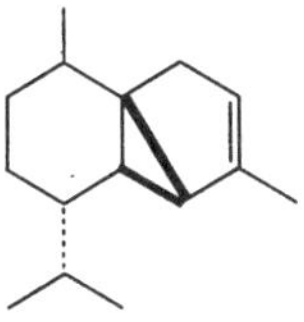

(11)

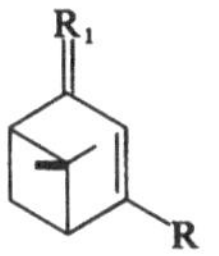

(12) R=Me;   R₁=H₂
(13) R=Me;   R₁=cis-OH, H
(14) R=Me;   R₁=trans-OH, H
(15) R=CH₂OH;   R₁=H₂
(16) R=CH₃;   R₁=O

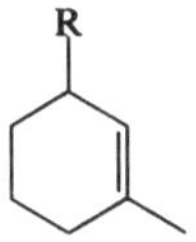

(17) R=OH
(18) R=O

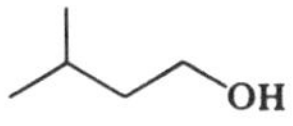

(19)

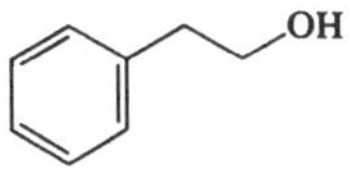

(20)

Similarly, SBJ-113, a filamentous fungus present as the yeast phase in the mycangium of Dendroctonus frontalis females, is capable of transforming trans-verbenol (14) to verbenone (16) and 3-methyl-2-cyclohexen-1-ol (17) into the corresponding ketone. Verbenone (16) and 3-methyl-2-cyclo-hexen-1-one (18) may be involved in nullifying the attractive properties of some substances in D. frontalis, while cis- and trans-verbenol (13, 14) are oxidation products of alpha-pinene (12) which may be acquired from the host (173). Three yeasts isolated from D. frontalis, Hansenula holstii, Pichia pinus and P. bovis produce four main metabolites, isoamyl alcohol (19), 2-phenylethanol (20) and the corresponding acetates. The esters greatly enhanced the attractiveness of a mixture of frontalin-trans-verbenol-turpentine to bark beetles (174).

WIYGUL and coworkers (*175, 176*) noted that in the boll weevil *Anthonomus grandis* the emission of sex pheromone is affected by diet, chemosterilants, bacterial contaminants as well by age, while WIYGUL and SIKOROWSKI (*177*) demonstrated that enterotoxin B produced by *Staphylococcus aureus* decreases male sex pheromone production in fat bodies isolated from male boll weevils.

Temperature affects the production of and response to sex pheromone in the American cockroach, *Periplaneta americana* (*178*), and modifies the expression of female calling in *Platynota stultana* (*179*). In the female cockroach, sex pheromone production is directly related to acclimatization temperature and rapidly decreases one week after acclimatization to lower temperature (*178*).

There are in addition, other factors which affect the biosynthesis of an insect pheromone.

The influence of juvenile hormone (JH) or its analogues in *Musca domestica*, *A. grandis*, *D. frontalis* and *P. stultana* sex pheromone biosynthesis as well as its role in enhanced production of aggregation pheromone of *Oryzaephilus surinamensis*, *O. mercator*, *Cryptolestes ferrugineus* and *Trilobium castaneum* has also been discussed (180–185). WEBSTER and CARDÉ (*183*) applied JH analogues and JH I, II, III exogenously to *P. stultana* virgin females and observed that they elicited oviposition comparable to that in mated females and that JH analogues appear to block pheromone production in virgin females.

As was mentioned earlier, pheromones potentially mediate all activities within a bee hive (*120*). In this sense, JH also plays a role in this insect society, since ROBINSON's data (*185*) support the hypothesis that this hormone regulates temporal division of labor in the honey bee colony.

HUGHES and RENWICK (*186*) observed that JH induces the synthesis of three pheromones ipsenol (**21**), ipsdienol (**22**) and 2-phenylethanol (**20**) by the male bark beetle *I. paraconfusus* and enhances the conversion of *alpha*-pinene (**12**) to *cis*-verbenol (**13**). According to HEDIN (*187*), the biosynthesis of grandlure (**23**) by boll weevil males is greater in the summer months, and somewhat less grandlure is produced by males in the presence of females than by males which are isolated. The beetles are capables of oxidizing myrcene (**24**) and limonene (**25**) to allylic alcohols, which suggests that pheromone precursors may be at least in part inhaled rather than ingested. The hemolymph then transports the precursor to the gut or to some alternative site where allylic oxidation occurs.

On the other hand, WIYGUL and SIKOROWSKI recently added glucose and ATP to fat bodies from adult boll weevil males and observed an increase in sex pheromone production. This supports their hypothe-

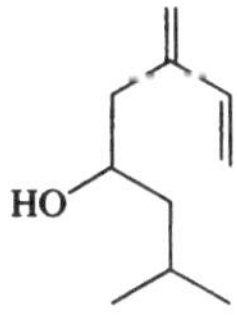

(21) 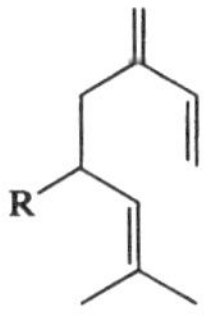   (22) R=OH   (25)
       (24) R=H

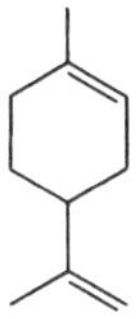

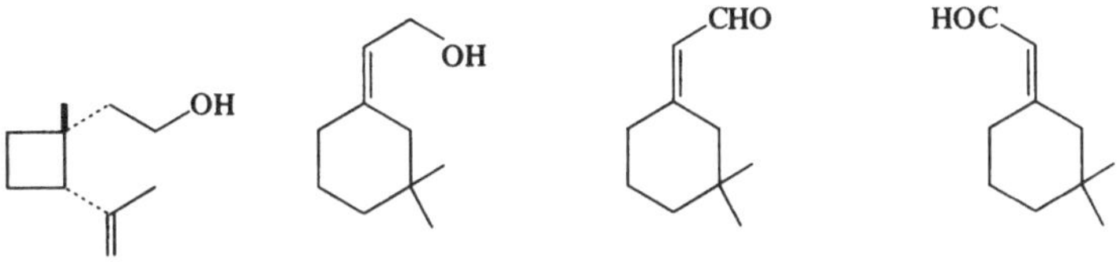

Ga   Gb   Gc   Gd

Ga—Gd=(23)

sis that added energy sources can cause increased pheromone production in incubating male boll weevil fat bodies (*188–189*).

In vertebrates, the sources of semiochemicals vary widely (even urine and feces may be involved), and microorganisms also play an important role in the production of mammalian odors (*190–192*).

Some scents probably result from incomplete substrate associated with anaerobic processes. Several mammals possess anatomical structures which are adequate for such processes. Examples have been observed in the anal sac of *Vulpes vulpes* and *Panthera leo* (*190–192*), the anal pocket of the Indian mongoose *Herpetes auropunctatus*, the vaginal secretion of the rhesus monkey *Macaca mulata* (*191*) and in the interdigital gland of black-tailed deer *Odocoileus hemionis columbianus* (*193*).

Volatile secretions emanating from the vaginal and prepuce of mammals are of microbial origin (*191*). That microorganisms are involved in the production of the wolf's anal sac secretion was deduced on treating the anal sac with antibiotics (*194*).

The role of enzymatic activity in the biosynthesis of the sex pheromone *trans*-11-tetradecenyl acetate (2) of the eastern spruce budworm *Choristoneura fumirans* was investigated by MORSE and MEIGHEN (*195*).

4. Pheromones as a Multicomponent System

In 1964, WRIGHT (*196*) argued that a multicomponent chemical communication system would be more advantageous than one including only a single substance. It is now well established that his assumption was correct (*197*).

It has been documented that an insect pheromone system may contain more than 30 compounds. For example, FRANCKE *et al.* (*198*) demonstrated that the complex volatile secretion from the mandibular gland of the bee *Andrena haemorrhoa* consists of 40 substances, composed primarily acid of three groups of compounds: spiroacetals, straight chain fatty acids derivatives (methyl ketones, primary and secondary alcohols, acetates and hydrocarbons) and isoprenoids. Five different types of spiroacetals were identified: four 1,6-dioxaspiro-[4.4]-nonanes (**26, 26a, 27, 27a**), four 1,6-dioxaspiro-[4.5]-decanes (**28–31**) four 1,7-dioxaspiro-[5.5]-undecanes (**32, 32a, 33, 33a**), one 1,6-dioxaspiro-[4.6]-undecane (**34**) and one 1,7-dioxaspiro-[5.6]-dodecane (**35**). Later, BERGSTRÖM *et al.* (*199*) isolated 50 volatile compounds (isoprenoids, spiroacetals and unbranched cyclic compounds) from the mandibular gland secretion of *Andrena wilkella*, *A. ovulata*, and *A. ocreata* males and females. The 19 spiroacetals isolated by them belonged to the four of the five types which were obtained by FRANCKE *et al.* (*198*) lacking only the 1,7-dioxaspiro-[5.6]-dodecane.

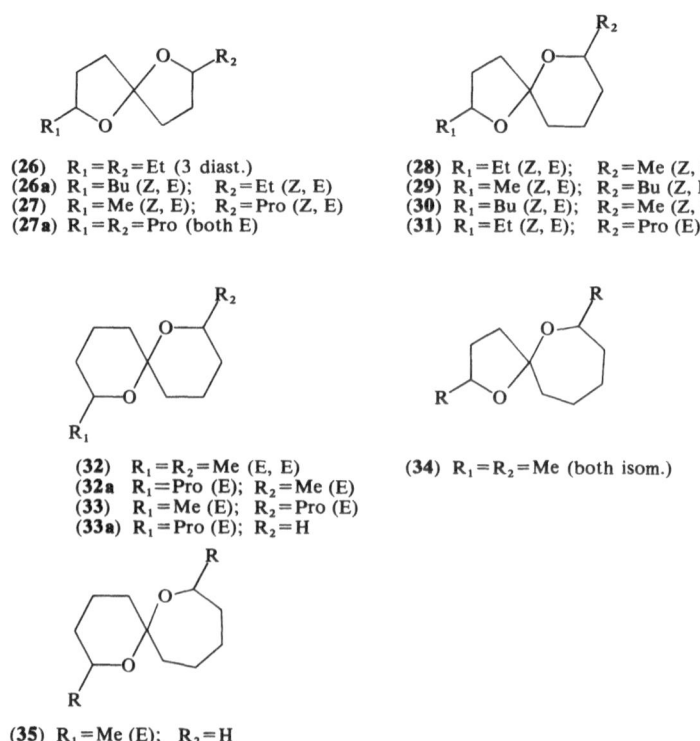

(**26**) $R_1 = R_2 = Et$ (3 diast.)
(**26a**) $R_1 = Bu$ (Z, E); $R_2 = Et$ (Z, E)
(**27**) $R_1 = Me$ (Z, E); $R_2 = Pro$ (Z, E)
(**27a**) $R_1 = R_2 = Pro$ (both E)

(**28**) $R_1 = Et$ (Z, E); $R_2 = Me$ (Z, E)
(**29**) $R_1 = Me$ (Z, E); $R_2 = Bu$ (Z, E)
(**30**) $R_1 = Bu$ (Z, E); $R_2 = Me$ (Z, E)
(**31**) $R_1 = Et$ (Z, E); $R_2 = Pro$ (E)

(**32**) $R_1 = R_2 = Me$ (E, E)
(**32a**) $R_1 = Pro$ (E); $R_2 = Me$ (E)
(**33**) $R_1 = Me$ (E); $R_2 = Pro$ (E)
(**33a**) $R_1 = Pro$ (E); $R_2 = H$

(**34**) $R_1 = R_2 = Me$ (both isom.)

(**35**) $R_1 = Me$ (E); $R_2 = H$

DUFFIELD et al. (*200*) found that Dufour's gland secretion of *Svastra obliqua* contains a series of 32 aliphatic esters, while according to BRAD-SHAW the mandibular gland secretion of the weaver ant *Oecophylla longinoda* consists of 33 chemical entities (*201, 202*).

Recently, FRANCKE et al. (*203*) identified for the first time several alkyl-3,4-dihydro-2H-pyrans (**36–41**) and alkyl-2,3-4H-pyran-4-ones (**42–45**) from both solitary and social bees (*Apis mellifera, Scaptotrigona bipunctata, Nannotrigona testaceicornia, Plebeia droriana, Partamona cupira* and *Tetragona clavipes*) and from male *Hepialus hecta* moth.

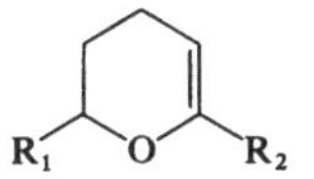

(**36**) $R_1 = R_2 = Me$
(**37**) $R_1 = Me$; $R_2 = Pro$
(**38**) $R_1 = Me$; $R_2 = Pentyl-$
(**39**) $R_1 = Me$; $R_2 = Heptyl-$
(**40**) $R_1 = Me$; $R_2 = Nonyl-$
(**41**) $R_1 = H$; $R_2 = Bu$

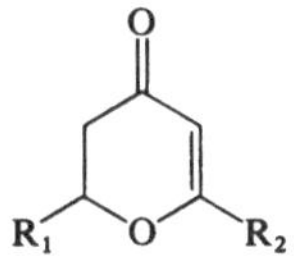

(**42**) $R_1 = Me$; $R_2 = Et$
(**43**) $R_1 = R_2 = Et$
(**44**) $R_1 = Et$; $R_2 = Me$
(**45**) $R_1 = Me$; $R_2 = Pentyl-$

The cephalic gland secretion from both sexes of *Dufourea inermis* and *D. minuta* also contains a series of multicomponent pheromones. The secretion is composed of sex species-specific blends methyl-carbinols and the corresponding long chain carboxylic esters (*204*). Multicomponent insect pheromones have also been identified in trail (*205*) and sex pheromones (*206, 207*).

BERGSTRÖM (*208*), has proposed a reason for the complexity of insect chemical secretions which he terms "Biochemical-Evolutionary-Readyness". According to this hypothesis, if several compounds are produced by a gland of an insect, they may form a ready made basis for further adaptation to play upon. The adaptation can go in the direction of higher specificity.

5. Perception of Pheromones by Insects and Vertebrates

The pheromone communication system is very primitive and chemical communication among protozoan cells was an evolutionary step preliminary to the formation of metazoans (*4, 8, 209, 210*). It is suggestive that according to recent behavioral studies fetal rats may be sensitive to the odor quality of aminiotic fluid (*211*).

A mathematical model for the transmission of a chemical signal has been developed by BOSSERT and WILSON (2). According to this model, the substance is transmitted through the air in accordance with simple laws of diffusion when the air is still and the substance cannot adsorb on or react with other substances. At any point away from the emitter, pheromone concentration is a function of 1) the rate of molecular emission, 2) the diffusion coefficient of the substance, 3) the distance from the source and 4) the time from the initiation of the emission.

The perception of pheromones by insects and vertebrates is a matter of several reviews and discussions (212–239). Hypotheses attempting to explain changes in the odor of organic molecules with changes in molecular structure and with spatial arrangement of the atoms, and to explain the effect of chirality on odor perception have been advanced by several authors (212, 216, 219, 223, 238). Other theories are based on vibrational energy levels of the molecules, intermolecular interactions and molecular shape and size (219, 220, 222, 233, 239).

Insects detect pheromones in surprisingly small amounts. Thus, male *Bombyx mori* begin to react when air contains as little as 100 molecules of the female sex pheromone per cm^3 (210). The trail pheromone of the leaf cutting ant *Atta texana*, methyl-4-methyl-pyrrole-2-carboxylate (46) is perceived at a concentration of 0,08 pg/cm^3, that is 3.48 molecules/cm^3 (240), about the same as that recorded for the trail pheromone *cis*-3-*cis*-6-*trans*-8-dodecatrien-1-ol of *Reticulitermes virginicus* termite workers (47) (241).

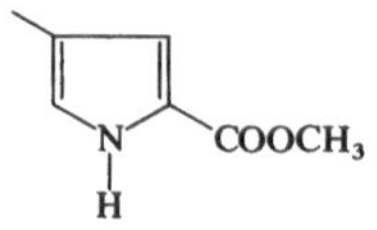

(46) 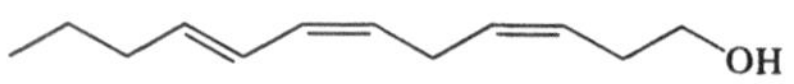   (47)

It has long been known that pheromones in insects are detected by means of organs located in the antennae, but chemoreceptors have also been found in the tips of maxillary and labial palps, in the buccal cavity, in the epipharynx and at the inner face of the clipeolabrum (217).

Insect antennae possess characteristics such as shape, size and abundance, type and location of the sensilla that are critical to pheromone perception (214). The sensilla are hairs perforated by large numbers of pores through which the pheromone molecules diffuse after being adsorbed onto the surface. The binding of these molecules to the acceptor sites relies on the stereochemical shape of the pheromone and the receptor (218).

References, pp. 56–85

SCHNEIDER and STEINBRECHT (in *214*) after electrophysiological recordings considered the following types of sensilla to be olfactory:

1. sensilla trichodea (in Lepidoptera)
2. sensilla basiconica (in Lepidoptera, Diptera, Coleoptera and Hymenoptera)
3. sensilla placodea (in Hymenoptera)
4. sensilla coelonica (in Orthoptera)

PRIESNER (*242, 243*) found specific receptor cells in the male antennae sensilla trichodea for the sex pheromones of two moth species, Z-9-tetradecenyl acetate (**48**) in *Adoxyphies orana* and Z-11-tetradecenyl acetate (**49**) and Z-9-tetradecenyl acetate (**48**) in *Eulia ministrana*. Z-11-tetradecenyl acetate (**49**) and Z-9-tetradecenol (**50**) were also isolated from *A. orana*.

$$\text{(structure)}$$

(**48**) R=OAc
(**50**) R=OH

$$\text{(structure)}$$

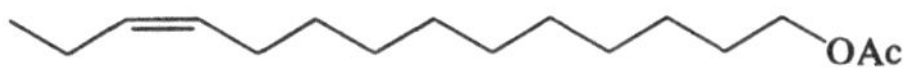

(**49**)

VOGT and RIDDIFORD (*228, 229*) and VOGT *et al.* (*234*), reported identification of three proteins involved in sex pheromone reception by the male silk moth *Antheraea plyphemus*. The three proteins which interact with the pheromone of the wild moth are a binding protein of molecular weight of 15000 daltons, which is present in the sensilla in high concentration, and two esterases which degrade the pheromone to prevent its accumulation and hence sensory adaptation. Their report suggest a molecular model for pheromone reception in which a pheromone binding protein act as a pheromone carrier and an enzyme acts as a rapid pheromone inactivator, thus maintaining a low stimulus noise level within the sensory hairs.

The antennae of male cabbage looper moth *Trichoplusia ni* have two classes of pheromone-sensitive sensilla. The sensilla from each class contain two receptor neurons, one producing the larger impulse and the other producing the smaller. The first of these is excited by low doses of Z-7-dodecenyl acetate (**1**), whereas the other is excited by low doses of the corresponding alcohol (**51**) (*244*).

As stated above, the shape of the molecule is an important factor in insect olfaction. A series of experiments were undertaken by KOSTELC and collaborators (*223*) to elucidate the possible role of chemical shape

(51)

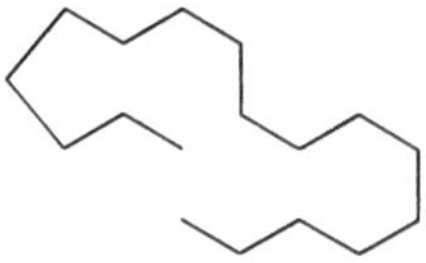

(52)

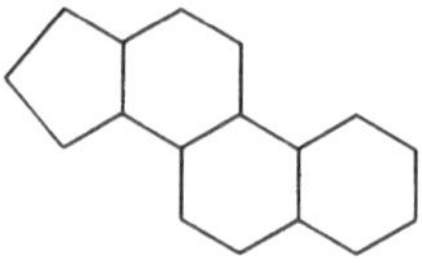

(53)

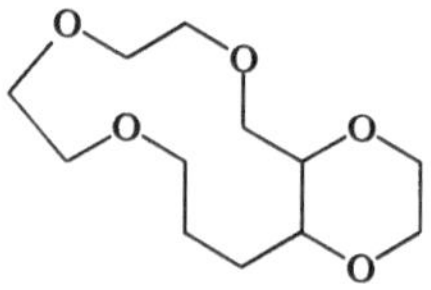

(54)

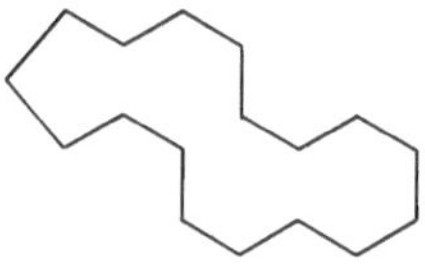

(55)

in the pheromone system of the sciarid fly *Lycoriella mali*. The major sex attractant identified in this fly was heptadecane (52) which when appropriately folded may adopt the shape of perhydrocyclopentane-phenanthrene (53). Other sex pheromones identified were tetradecenyl acetates and ambrolides which when in appropriate conformation can also assume a steroidal shape.

KOSTELC tested the steroidal hypothesis by examining the response of sciarid males to a spectrum of molecules. Among the substances tested, crown ethers were particularly interesting since they fit several important criteria: 1) their occurrence in nature is unknown; 2) they can assume a steroidal shape and their polarity and functionality are different from that of the natural pheromone. The most active sub-stances among the fifteen compounds tested were 17-crown-5 (54) and cycloheptadecane (55). Since the natural pheromone heptadecane (52) and the crown ether (54) are two entirely unrelated molecules, the re-sults are in accordance with the proposed hypothesis.

In related work CHAPMAN *et al.* (*219*) defined the conformation of Z-11-tetradecenyl acetate (49), the sex attractant of the European corn borer and the red banded leaf roller, in the pheromone chemore-ceptor system, of these two species.

In mammals semiochemicals stimulate several intranasal chemosen-sory structures, including olfactory receptors, free nerve endings from the trigeminal nerves and specialized epithelial receptors from the vo-meronasal organ of Jacobson and the septal organ of Masera (*55*). The available data suggest that the various intranasal chemosensory organs develop at different rates and thus may play different roles in early life (*55*). Indeed there is some evidence that chemical substances are detected prenatally by fetal rats (*55*, *211*).

6. Chemical Communication in Vertebrates

Among vertebrates, semiochemicals have been reported to occur in primates, mice, rabbits, wolves, deer, reindeer, hyaenas, foxes, dogs, cats and fishes (*1–16*), but chemical signals have also been found in tigers (*245–246*), capybaras (*247*), camels (*248*), elephants (*249*), galago (*250*, *251*), antelopes (*252*) and snakes (*253*). In such animals an odor may identify sex physiological status, dominant from subordinate individuals, conspecifics, population, groups, home or territoriality and also play a role in fish migration.

Chemical discrimination between species, subspecies, groups, and individuals was observed in reindeer (*254*, *255*), guinea pig (*256*), rabbits (*257*), lizards (*258*), lemurs (*259*, *260*), fishes (*261*, *262*), chimpanzees (*269*) and marmosets (*270*). However, the recognition of sex in conspecifics by individual red swamp crayfish *Procambarus clarkii* is a matter of discussion (*263–268a*).

Volatile constituents associated with female urine and/or vaginal discharge often increase male and female sexual activity (*271–279*). Urinary marking of the environment is also an important feature of rodent and primate behavior (*270*, *280*, *281*). The compounds found in the urine of males and females generally vary with the breeding season or with the physiological status of the females; this has been observed in the ram (*274*), red fox (*282*), the goat (*283*, *284*), and the wolf (*285*).

Flehmen (or lipcurl) is a response of male mammals primarily to female urine. It is seen primarily during the reproductive season and might function as a means of transport of chemicals to the vomeronasal organ (*286*). To elucidate the components of female urine which release the flehmen, CRUMP *et al.* (*277*) fractionated the urine of black-tailed deer *Odocoileus hemionus columbianus* and observed that behavioral responses disappeared completely before fractionation had reached the level of a single component. Similar results were found with beaver (*287*).

In a few cases these compounds have been identified. Dimethyl sulfide (**56**) and methylthiol butyrate (**57**) (MTB) were identified in the vaginal secretion of female hamsters (*288*, *289*). Methylthiol butyrate is readily converted to methyl disulfide (**58**) (DMS) by hydrolysis and even more readily by aminolysis followed by air oxidation of the released methanethiol (59). DMS is responsible for most of the male attraction and MTB might act as its precursor (288–290).

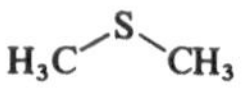

(56) 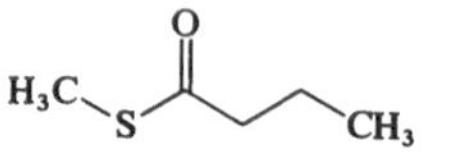   (57) 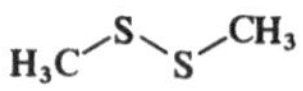   (58)   (59)

A series of volatile acids termed copulins which includ acetic, propionic, isobutyric, isovaleric, and isocaproic acids was isolated from the vaginal secretion of female rhesus monkeys *Macaca mulata* (273, 291, 292). Application of copulins to the sexual skin of female rhesus increased the number of ejaculations, mounting attempts and mounting received by females (273).

JORGENSON (282) isolated the following substances from red fox urine: isopentenyl methyl sulfide (60), 2-phenylethyl methyl sulfide (61), 6-methyl-5-hepten-2-one (62) and *trans*-geranylacetone (63) from both sexes and additionally 2-methylquinoline (64) from urine of males.

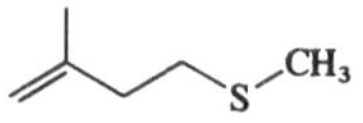

(60) 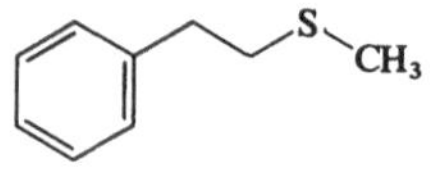   (61) 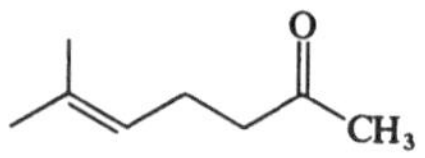   (62)

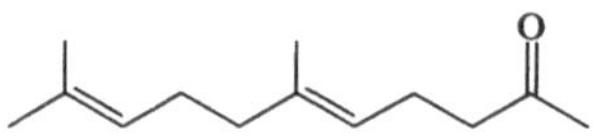

(63) 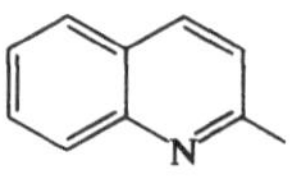   (64)

The urine of the mouse is the source of estrus synchronization, pregnancy block, group and individual recognition, sexual attraction, aggregation, histocompatibility-related mating preference, fear and stress (293).

NOVOTNY et al. (293) identified a volatile substance in male mouse (*Mus musculus*) urine, whose concentration depends on the level of testosterone, as 7-*exo*-ethyl-5-methyl-6,8-oxabicyclo-[3.2.1]-3-octene (65) or *exo*-3,4-dehydrobrevicomin, and suggested that it might be a pheromone or pheromone adjuvant. The proposed structure is identical, except for the 3,4 double bond, with that of the pine bark beetle *Dendroctonus frontalis* pheromone *exo*-brevicomin (66).

In the same article reference is also made to a dissertation which describes identification in the urine of 2-(*sec*-butyl)-dihydrothiazole

(**67**). The concentration of (**65**) and (**67**) is greater in normal than in castrated males but testosterone injection restores the normal level of at least *exo*-3,4-dehydrobrevicomin (**65**) and both of these compounds are synergistically active in inducing inter-male agonistic behavior (*294*).

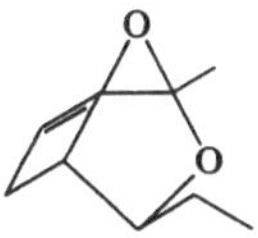

(**65**) 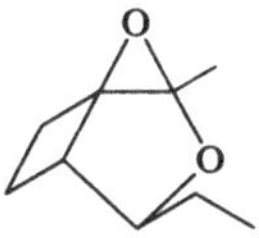   (**66**) 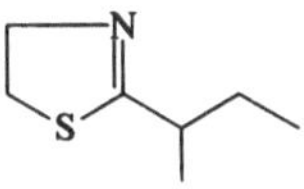   (**67**)

Recently SCHULTZ and coworkers (*295*) isolated and identified methyl butyl sulfide, propyl sulfide, trimethylamine and acetone among the volatile constituents of female beagle urine.

Black-tailed deer possess three glands which along with urine play a role in social chemical communication among their conspecifics. The tarsal scents acts in individual recognition, the metatarsal odor functions as an alarm pheromone and the substances from the forehead are employed for marking (*193*).

The tarsal gland is carried on a tuft of hair located in the medial site of the tarsal joint. When in a group, free or in captivity, the deer sniff and lick each other's tarsal gland, more intensively in the dark or when a strange female or male is introduced into the group. These observations suggest that the tarsal organ is involved in individual recognition (*193*).

From the tarsal tuft, BROWNLEE *et al.* (*296*) isolated a substance identified as 4-hydroxy-6-dodecenoic acid lactone (**68**) or deer lactone. Deer can discriminate between the geometric isomers, responding to the Z-isomer only (*297*). Urine is the source of the lactone which is deposited on the tarsal tuft by "rub-urination" (*298*).

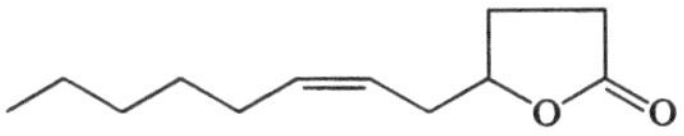

(**68**)

The response to metatarsal odor consists of increased alertness. Only females show a response and they do so to both male and female metatarsal scent (*254*), but *Odocoileus virginicus*, the white tailed deer, does not exhibit this alarm behavior system to the same degree (*299*). It has also been observed that two subspecies of blacktailed deer, *O. hemionus columbianus* and *O.h. hemionus*, can distinguish each other in this manner. Such discrimination may be important in sexual isolation of the species (*254*).

Reindeer, *Rangifer tarandus tarandus*, possess several glands (caudal, interdigital, preorbital and tarsal) which release chemical substances employed for different purposes (*300*). The volatile ketones released from the interdigital gland exhibit seasonal variation (*301*). In sexual behavior, males examine the anogenital areas of females and their urine both by olfaction and taste while in general behavior, reindeer are scented at a distance rather than through physical contact. However, when two individuals meet, mainly after isolation, they sniff each other's tail, nose and general body (*300*). In maternal and sexual behavior, the reindeer sniff the tail gland of the other individual, but the caudal gland is also involved in sexual behavior, alarm and social encounters between previously separated individuals. During sexual behavior a male keeps away from his harem and sniffs his cow's tail from time to time (*302*).

Like cervids and other mammals, rodents can also discriminate their conspecifics. SKEEN and THIESSEN's data suggest that young Mongolian gerbils *Meriones unguiculatus* recognize the odor of other gerbils and that they prefer those associated with the diet on which they and their parents were raised (*303*). In male Mongolian gerbils, Harderian glands may also be a source of olfacients (*304–305*).

GOODWIN *et al.* (*271*) have found methyl *p*-hydroxybenzoate (**69**) as a constituent of the vaginal secretion from female dogs and proposed that it acts as a male attractant, but KRUSE and HOWARD have challenged this conclusion (*306*).

Hamster *Mesocricetus auratus* vaginal odor influences the ability of males to discriminate between females on different days of the estrus cycle (*276*), whereas features of the complex mixture from the vaginal discharge of female chimpanzees, *Pan troglodytes*, distinguish these apes from other anthropoids and may therefore, be involved in the odor signal that identifies chimpanzees (*269*).

Saliva can also be important in mammalian reproduction. Sexually aroused boar produce from the submaxillary gland a copious, sticky saliva containing 3-α-hydroxy-5α-androst-16-ene and 5-α-androst-16-ene-3-one (**70**, **71**) which act as sexual pheromones for the male (*49, 307, 308*).

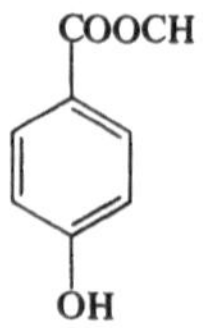

(69)

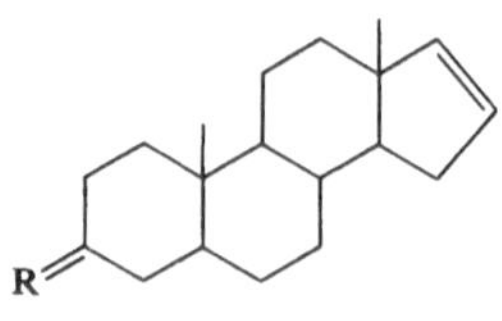

(70) R = H, OH
(71) R = O

The presence of these glands is essential for normal behavior pattern in both males and females. PERRY et al. (309) excised them from the young boars and observed that disappearance of the white, frothy saliva which normally contains and disperses these androstenes not only reduces the boar's sexual ability, but also reduce its libido.

The above discussion shows that both urine and vaginal discharges are closely associated with steroidal hormones as will be discussed in the next section. However, vertebrate semiochemistry is not restricted to reproductive behavior nor to mammals. Territoriality, defense and odor recognition may also depend on emission of chemical signals.

Marking territorial behavior (with urine or feces) is well documented in the beaver (310, 311), in primates (42, 44, 46, 270), gerbils (157, 312–314) and salamanders (315).

The function of allomones in invertebrate defense is well established, but there is some evidence that such chemicals also occur in vertebrates, mainly in mustelids (316–325), but also in turtles *Sternotherus odoratus* from which EISNER et al. (326) isolated four ω-phenyl alkanoic acids identified as phenylacetic, 3-phenylpropionic, 5-phenylpentanoic and 7-phenylheptanoic acid.

Sulfur compounds (72–92) have been identified as malodorous components of the anal sac secretion from the striped skunk *Mephites mephites* (316), the polecat *Mustela putorius* (317–319), the stoat *Mustela erminea* (317, 320), the mink *Mustela vison* (317, 321), the ferret *Mustela putorius* forma *furo* (318, 322) and the weasel *Mustela nivalis*

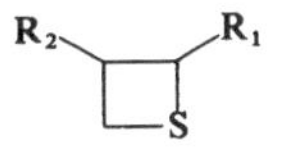

(72) R_1 = Me; R_2 = H
(73) R_1 = Et; R_2 = H
(74) R_1 = Pro; R_2 = H
(75) R_1 = Isoprop; R_2 = H
(76) R_1 = Pentyl; R_2 = H
(77) R_1 = R_2 = trans-Me
(78) R_1 = R_2 = cis-Me
(79) R_1 = Et; R_2 = Me

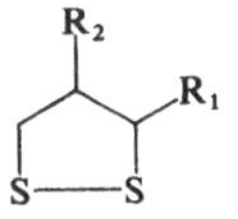

(80) R_1 = Et; R_2 = H
(81) R_1 = Prop; R_2 = H
(82) R_1 = Me; R_2 = H
(83) R_1 = Me, Me; R_2 = H
(84) R_1 = R_2 = trans-Me
(85) R_1 = R_2 = cis-Me

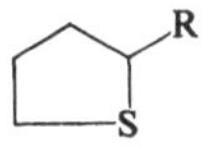

(86) R = Me

(87)  (88) 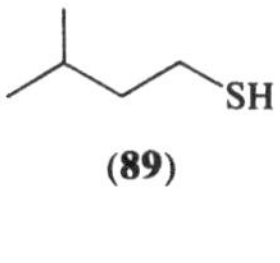 (89)

(90)  (91)  (92)

(*318*). However, the malodorous substances of the beach marten *Martes foina*, of the pine marte *Martes martes* (*318*) of the raccoon *Procyon lotor* (*324*) and the subcaudal gland of the badger *Meles meles* are fatty acids rather than sulfur compounds (*323*).

Although pheromones have been most investigated among mammals, there is some evidence that reptiles and amphibians may also use chemical signals for several purposes. GARSTKA and CREWS (*253*) observed that serum and extracts of tissues from the female garter snake *Thamnophis sirtalis parietalis* act as a pheromone and elicit male courtship behavior when applied to the back of another male. On the other hand, the lizard *Eumeces laticeps* produces a pheromone which may function for species identification and may serve as a reproductive isolation mechanism (*258*).

7. Endocrine Hormone-Pheromone Interaction

The role of the endocrine system in pheromone perception and production is now unequivocally demonstrated (*280, 312, 313, 327–354*).

The primer pheromone (for definition see page 2) of sexually mature females increases the plasma testosterone level of male mice (*329*), raises the basal levels of plasma corticosteroids in individually housed females as well as in virgin female mice and influences the attractiveness of female rats to sexually experienced males (*280*).

Production of a female attractant pheromone by the preputial glands of the male mouse is androgen dependent, since it is suppressed by injection of the anti-androgen ciprotene acetate, but is restored by treatment with testosterone. This releaser pheromone is absent from the preputial glands of castrated males (*346*).

SCHWENDE and collaborators (*354*) reported that female mice that had been made to go into estrus through implantation showed an increase in the levels of *n*-pentyl acetate, *cis*-2-penten-1-yl acetate, *p*-toluidine and 2-heptanone excreted in urine.

In an experiment to determine whether the pituitary gland controls the production of those chemical signals present in female mouse urine which are responsible for LH release in males and the attractiveness of this urine to males, JOHNSTON and BRONSON (*278*) concluded that there are at least two distinct odors in female urine. The first is dependent on the pituitary and provokes LH release in males, while the second is dependent on the ovary and attracts males.

There is some evidence that primer pheromones play a role in agon istic and hierarchical behavior among rodents, pigs, primates and other mammals as well as fishes (*90, 119, 215, 355–363*). In the house mouse *Mus musculus* this effect is not influenced by the genotype (*364*).

The urine of a dominant male rat or mouse contains an aversive substance which is absent in the urine of subordinate animal (*119*). Castrated male rat urine also lacks the aversive substance which may be restored after treatment with androgen (*365*).

The concentration of steroidal hormone is also influenced by primer pheromones. Testosterone levels of isolated *Microcebus marinus* males decrease after submitting them to the urinary odor of a dominant con-specific. However, the level is restored when the subordinate male is olfactorily isolated from urine odors of dominant males (*366*). Indeed mammalian scent marking, closely associated with aggressive behavior and social dominance, is apparently testosterone dependent (*363, 367*). Thus the data of TAYLOR *et al.* (*361*) suggest that differences in andro-gen level influence agonistic behavior among male rats on exposure to females and that size and secretory activity of odor-producing glands have a relation to sex, reproductive status, age and social status whereas the odor of an attractive female talapoin monkey *Miopithecus talapoin* raised testosterone plasma levels in dominant, but not in subordinate males (*368*).

Dominant fishes *Ictalurus natalis* also use chemical signals to main-tain their social status (*90*).

However, new evidence from GOLDFOOT (*369*) has shown that fe-males given estradiol are more attractive to males in part because the females allow males to mount much more frequently only when the females have initiated the sexual interaction. It appears that aliphatic acids, the copulins, may not be the exclusive component of vaginal cues to which male normally responds, although they may play a role in such communication. Only sexually experienced males paid attention to vaginal products when these are placed on environmental surfaces. GOLDFOOT (*369*) concluded that male rhesus monkeys may utilize the chemical signals from the female during sexual encounters, but these cues are neither necessary nor sufficient for mating.

Both social and environmental factors may affect sexual maturation in female house mice (*370*), female house mice puberty (*352, 374*), fertility in cotton top tamarins *Saguinus oedipus oedipus* (*371*), and the endocrine response of *Macaca fascicularis* males to receptive fe-males (*372*).

The puberty of juvenile female mice may be modified by primer pheromones originating from urine of both sexes. Treatment of juvenile females with urine from normal males provokes puberty acceleration,

while treatment of these females with urine from intact females that
had been grouped together, but not with urine of single caged females,
substantially delays puberty (330, 339, 352, 374). This effect is observed
in ovariectomized females but not if the adult female have been adrenalectomized (339).

NOVOTNY et al. (352) isolated from the female house mouse urine
six components, whose concentrations were depressed after adrenalectomy. These components were identified as 2-heptanone, trans-5-hep-
ten-2-one, n-pentyl acetate, cis-2-penten-1-yl acetate and 2,5-methyl
pyrazine. When these substances were added to previously inactive
urine from adrenalectomized females, the biological activity was fully
restored.

On the other hand, NISHIMURA and colleagues (374) obtained isobu-
tyl amine and isoamyl amine from male mouse urine responsible for
female puberty acceleration.

The physiological effects of primer pheromones are well known
and are named after their discoverers (333, 334, 337, 338, 340, 344,
347–351).

7.1. Bruce Effect

When a newly impregnated female mouse is confined with a strange
male, that is a male which has an odor different from that of her
mate, implantation failure and a rapid return to estrus is observed.
This pregnancy-blocking is not observed if the female is exposed to
a castrated male, but treatment of the male with testosterone restores
his production of pregnancy-blocking pheromone (342).

GANGRADE and DOMINIC (341) observed that this effect is high if
a female is placed and housed below correlated males, but is reduced
when the female is housed above the males and concluded that the
pheromone involved in the Bruce effect is apparently nonvolatile and
acts on females through contact.

MARCHLEWSKA-KOJ (330) fractionated male mouse urine and found
that the pregnancy-blocking pheromone is a peptide or a substance
bound with the peptide fraction.

The rate of pregnancy-blocking is also reduced when the strange
male belongs to an inbreeding strain or on administration of alpha-
methyl dopa suggesting that dopaminergic nervous system are involved
in the Bruce effect (338).

A sociobiological approach (144) suggests that the Bruce effect has
evolved to promote heterogeneity in the population since the female
is available for reinsemination. However, KEVERNE and DE LA RIVA

(332) have argued that the presence of a strange male in a mouse colony is an infrequent event and have suggested another explanation. Their experiments show that the effect exists not so much to ensure that a strange male will block pregnancy, but to ensure that her own mate male will not do so.

Pregnancy inhibition experiments indicated that there were significant strain differences in the Bruce effect and that rates of pregnancy block were low when the strange male belonged to an inbred strain. Thus genetic factors appeared to be responsible for strain differences in male pheromone-induced pregnancy block (334).

7.2. Whitten Effect

The male mouse produces a scent in his urine which induces and accelerates the estrus cycles of the female. This effect will be greater in females in which estrus was suppressed by aggregation. According to NICHOLS and CHEVIN (373), male urine is less effective than male presence in inducing the estrus cycle, while GANGRADE and DOMINIC's (341) studies suggested that in contrast to the Bruce effect, the pheromone involved in the Whitten effect is a volatile substance.

Recently JEMIOLO et al. (351), observed that two volatile constituents of male mouse urine, 2-(sec-butyl)-4,5-dihydrothiazole (67) and 3,4-dehydro-exo-brevicomin (65), which is also involved in promoting aggressive behavior in males, when added in appropriate concentration were found to be as effective as normal urine in inducing the Whitten effect.

GANGRADE and DOMINIC (340) also observed that adrenalectomized females respond to the odor of male urine by exhibiting estrus acceleration and synchronization. Thus, adrenals are not involved in estrus cycle irregularities observed in the Whitten effect.

7.3. Lee-Boot Effect

A female's estrus cycle is suppressed concomitant with simultaneous pseudopregnancy development when four or more females are put together in a absence of a male. PANDEY and PANDEY (343, 347) concluded that the clitoral gland does not play any role in the production of the pheromone which suppresses the estrus cycles in other females and that the pheromone is thus associated directly or indirectly with ovarian hormones.

Recently McCLINTOCK (375) showed that preovulatory pheromone from female rats shortened or phase advanced the ovarian cycles, whereas ovulatory odors lengthened or phase-delayed the cycle of females.

7.4. Ropartz Effect

The odor of other mice causes the adrenal glands of individual mice to grow heavier and to increase their production of corticosteroids which results in a decrease in the reproductive capacity of the animal.

LAWTON and WHITSETT (376), observed that application of urine from male deer mice *Peromyscus maniculatus bardii*, when applied to the nose of young males retarded seminal vesicle growth but removal of olfactory bulbs of males at 3 weeks of age blocked the inhibitory response.

8. Maternal Pheromones

Between 14 and 21 days of lactation female mice excrete a pheromone in their feces which strongly attracts the young. When the young cease to respond to the emission of pheromone at age 27 days, the female stops releasing it. Males injected with bile from females which have been lactating for 21 days also release maternal pheromone (377). High circulating levels of prolactin are necessary for the pheromone emission (378).

Young mice approach and consume feces which contain a high level of deoxycholic acid (93), a compound in which they are deficient and which promotes brain growth and neurobehavior maturation (378–383).

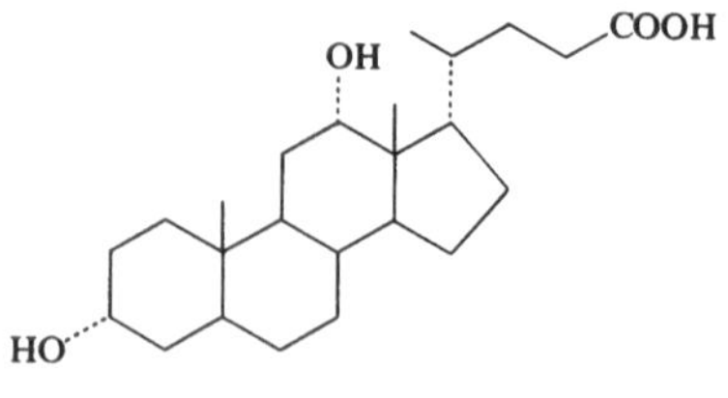

(93)

9. Chemical Communication in Marine Organisms

In contrast to research on chemical communication of terrestrial organisms, research in marine ecological chemistry is still in its beginning stages. SCHEUER (88) mentions two main reasons for this discrepancy – the greater economic, sociological and political importance of the former and the technical difficulties of monitoring and isolating minute amounts of biological activity substances secreted by marine organisms. In SCHEUER's words "seawater is probably the last solvent an organic chemist would choose as the solvent for an organic compound or as the vehicle for a reaction" (93). However, recent studies have indicated that chemical mesengers are important in communication among marine organisms (90–96).

RYAN (384) was the first to observe that premolt female crab *Portunus sanguinolentus* began to release sex pheromone 7–8 days prior to ecdysis and suggested that a pheromone was present in the urine of premolt female. However, according to DUNHAM (385) RYAN's suggestion that the moulting hormone crustecdysone (94) functions as a sex pheromone is inconclusive.

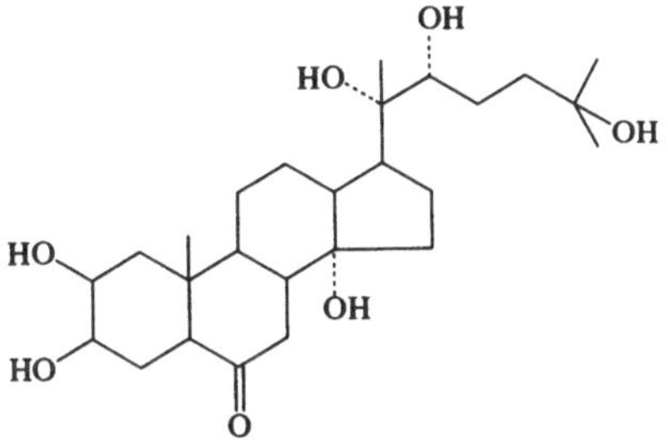

(94)

When mature *P. sanguinolentus* males were exposed to the female sex pheromone, a characteristic display was elicited but when the female's nephopore was blocked with paraffins this behavior could not be observed. When the paraffin was removed, normal behavior could again be obtained. The males recognize the premolt female and grasp and carry her until ecdysis (386). However, GLEASON et al. (387) concluded that crustecdysone (94) did not stimulate courtship and that their results did not support an evolutionary relation between the moulting hormone and sex pheromone communication in Crustacea.

Sex pheromones play a major important role in many teleosts. In females, pheromone production increases at the time of ovulation, in conjunction with onset of sexual behavior, whereas in males, pheromone production may be extended during the breeding season (95).

Some data suggest that female of *Ictalurus punctata*, *Petromyzon marinus*, *Salmo gairdineri*, *Blennius pavo*, *Gobius jozo* and *Brachdanio rerio* are attracted to the odor of mature conspecific males (90, 94).

Female pheromones that attract and stimulate conspecific males have been identified in the frillfin goby, *Bathygobius soparator*, the sea lamprey, *P. marinus*, the pond smelt, *Hypomerus olidus*, the loach,

Misgurnus anguillicaudatus, rainbow trout, *S. gairdineri*, the ayu, *Plecoglossus altivelis* and the zebrafish, *B. rerio* (*94*).

Testicular or ovarian androgen in conjuction form may be responsible for sex attractant behavior in several species of teleosts (*90, 94*). The sex attractant secreted by female zebrafish is a mixture of estradiol and testosterone glucuronides (*388*); the female guppy *Poecilia reticulata* is also under hormonal control (*389*).

Introduction of water from a tank containing a male *B. rerio* induced ovulation in an isolated female (*94*). The chemical substances affect both group and spawning behavior (*390*).

Allelochemical interactions in marine organisms are not restricted to sex pheromones. Predator-prey interaction, feeding behavior and species, sex and individual recognition are also involved.

Certain marine mollusks of the order Nudibranch which are softbodied and physically unprotected produce various chemical substances as a defense strategy. The mollusks sequester small organic molecules, which afford protection from natural enemies, from the diet (*391–401*).

It was demonstrated that the sponge *Hymenacion* sp. supplies the allomone pupukeanane (**10**) to its predator, the nudibranch *Phyllidia varicosa* (*395*). Furodysin (**95**), isolated from the mollusk *Catlina luteomarginata*, is also of dietary origin and shows antifeedant properties against the goldfish *Carassius auratus* (*401*). Similarly, two mollusks *Hypselodoris godeffroyana* and *Chromodoris maridadilus* accumulate nakafuran-8 (**96**) and nakafuran-9 (**97**) from the sponge *Dysidea fragilis* on which they feed (*397*), while CARTE and FAULKNER (*400*) isolated tambjamines A–D (**98–101**) from the carnivorous nudibranch *Roboastra tigris* which preys upon two other mollusks *Tambja abdere* and *T. eliora* which in turn feed upon the bryozoan *Sessibugula translucens*. All of them contain the four fish feeding inhibitors tambjamines.

Other defense allomones isolated from marine mollusks include panacene (**102**) from *Aplysia brasiliana* (*391*), onchidal (**103**) from *Onchidela binneyi* (*392*), navanones A–C (**104–106**) from *Navanax inermis* (*393, 394*), pu'lenal (**107**) from *Chromodoris albonotato* (*398*), spiniferin-2 (**108**) from *Hypselodoris daniellae* (*398*) and 8-hydroxy-4-quinolone (**109**) from the octopus *Octopus dofleini* (*401*).

Several of the defense allomones from marine invertebrates are structurally related to plant sesquiterpene antifeedants against insects (*398, 402*). The chemical defensive substances of alcyonaceans have been reviewed by TURSCH (*402*).

In spite of their physical defensive barrier, sponges are also protected from predation by allomones (*403*).

An interesting interaction between two marine organisms is that which involves the sea anemone *Anthopleura elegantissima* and its pre-

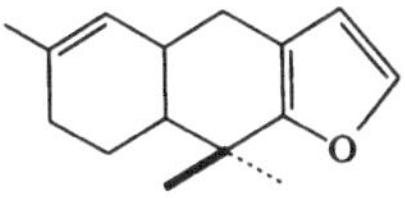

(95) 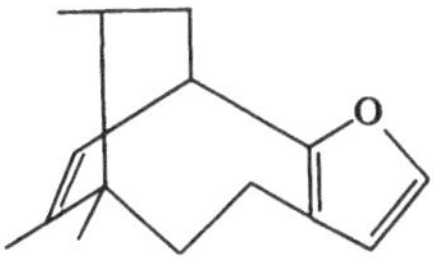   (96) 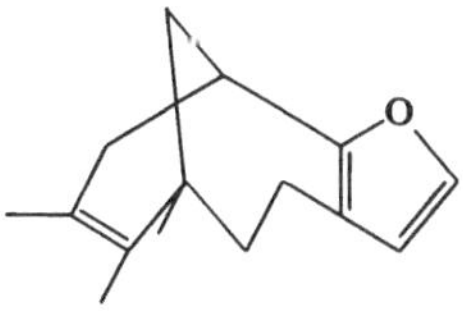   (97)

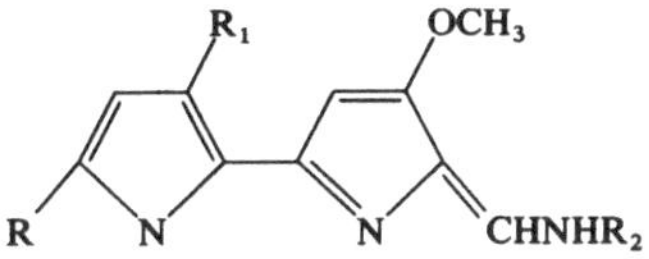

(98) R=R₁=R₂=H
(99) R=Br; R₁=R₂=H
(100) R=R₁=H,   R₂=CH₂CHMe₂
(101) R=H; R₁=Br;   R₂=CH₂CHMe₂

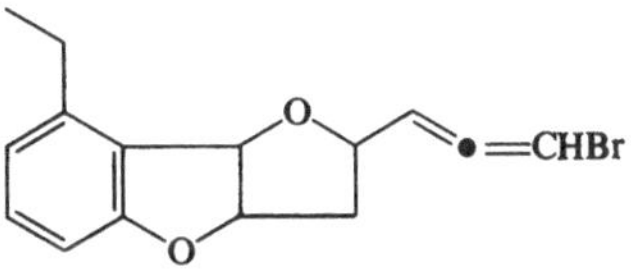

(102)

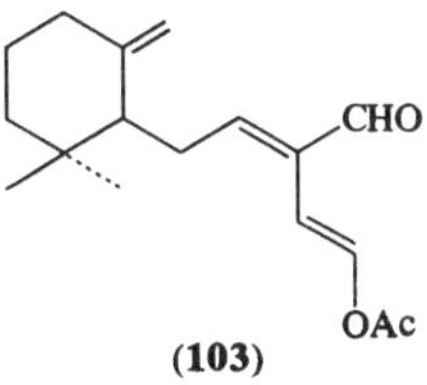

(103) 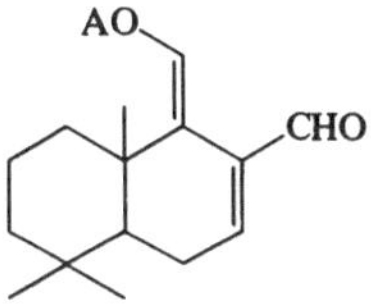   (107)

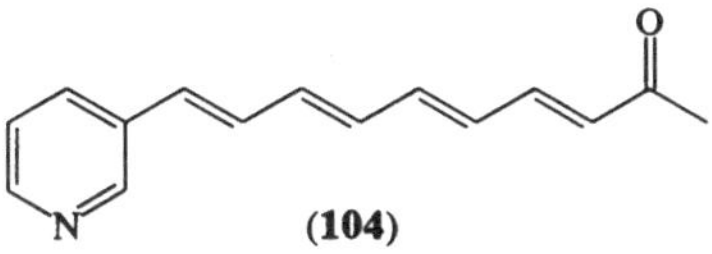

(104) 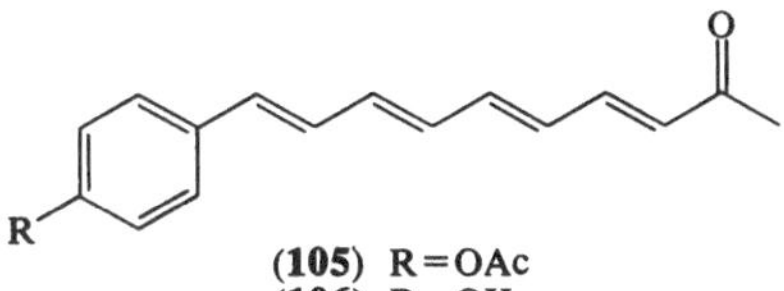   (105) R=OAc
                         (106) R=OH

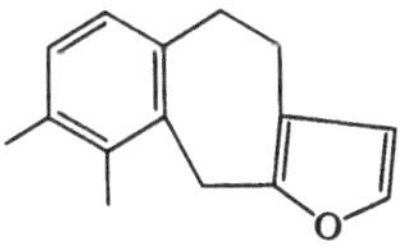

(108) 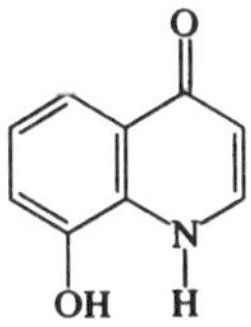   (109)

dator, the mollusk *Aeolidia papilosa*. Anthopleurine (**110**) is a betaine released by an injured member of the colonial *A. elegantissima* which evokes in other members of the colony an alarm response consisting of a rapid bending and shortening of the tentacles. *A. papilosa* is an anemone-eating nudibranch regularly associated with *A. elegantissima*. After the nudibranch eats the sea anemone, the anthopleurine (**110**) persists in the tissue of the mollusk for at least five days after ingestion. During this period the gastropods are capable of evoking an alarm response in anemones without touching them, presumably by releasing anthopleurine into the water. This suggests that the predator helps in transmission of the alarm pheromone which may reduce the predation of the sea anemone (*404*).

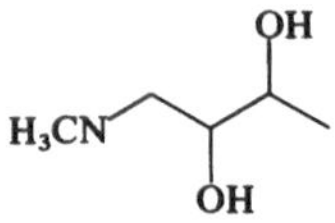

(**110**)

An association has also been described between young of the fish *Abudefduf leucogaster* and the alcyonarian *Lithophyton viridis*. The latter secretes a toxic chemical substance rendering it unpalatable to large predators. The toxin affects *A. leucogaster* more slowly than other fishes; they can thus remain near the soft coral and benefit indirectly from its chemical protection (*405*).

An unusual behavior observed in mollusks is the so-called Exploratory-Feeding Behavior. The mollusks secrete a chemical substance, frequently an amino acid, which attracts its prey. Two of these fish attracting principles are arcamine (**111**) and strombine (**112**) from *Arca zebra* and *Strombus gigas* respectively (*406*).

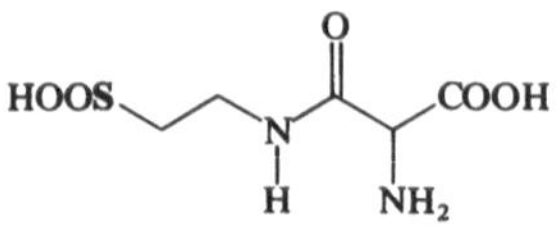

(**111**) 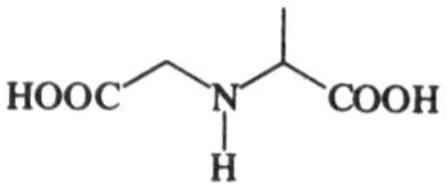   (**112**)

Chemical signals are also involved in fish migration. Fish which, like eels, spawn in the sea and swim up rivers to feed and mature are called *catadromous*, whereas those, like salmons, which perform the reverse movement are called *anadromous*.

Salmon *Salmo salar* and *S. alpinus* migration was observed more than 400 years ago. The sensory basis for homing in salmonids is partially understood. The so-called olfactory hypothesis has as its cornerstone the suggestion that juvenile salmon become imprinted with the unique chemical odor of their natal stream during the smolt stage and subsequently use this cue to locate their stream during the spawning migration (*406 a*).

Mature anadromous salmons also use specific odorant cues to discriminate between chemical signals produced by their own population. Intestinal contents were reported to include the more olfactory substances (*407–411*).

10. The Effect of Anosmia

In the preceeding chapters it was discussed that removal or block of the olfactory bulbs may change the expected stereotyped response. However, the effect of anosmia, i.e., the loss of the sense of smell, on the social behavior of vertebrates is not well established and is a matter of controversy (*119, 412–416*). While MICHAEL and KEVERNE (*412*) found that anosmic male rhesus monkeys show no interest in females receiving oestrogen until their olfaction was restored GOLDFOOT et al. (*413*) observed that such males display typical cycles of copulation.

ROPARTZ (*119*) suggested that the results depend upon the method used. He assumed that bilateral removal of the olfactory bulbs provokes a decrease in the social behavior of mice, while intranasal application of ZnSO₄ solution does not produce the same effect. However, GANGRADE and DOMINIC (*414*) showed that intranasal irrigation of ZnSO₄ induces anosmia in female mice, rendering these females unable to detect the pheromone originating from the males. Contrary to findings in some rodents and in primates, prepubertal bulbectomy had no significant effect on mating or aggressive behavior in male boar (*415*).

In male golden hamster, injection of testosterone restored normal sexual behavior in castrated males, but had no effect on the bilaterally bulbectomized animal (*416*). In gerbils *Meriones unguiculatus*, olfactory bulb extirpation led to a reduction in territorial marking (*312*).

11. Chemical Communication in Invertebrates

Semiochemicals have been identified most extensively in insects. However, the literature suggests that chemical communication may also be used by other invertebrates (*98–106, 384–386, 391–401*). Primer and releaser pheromones were detected both in Platyhelminthes and Aschelminthes (*99, 100*).

According to PERKINS and FRIED (*102*), free sterols and fatty acids may have pheromonal activity in planaria such as *Dugesia tigrina* and *D. dorotocephala*.

BONE (*99*) argued that the present knowledge of pheromone mediated behavior in helminthes is similar to that of insect communication over 20 years ago. Thus, insect semiochemistry will be discussed here in greater detail.

12. Caste and Kin Recognition in Social Insects

The chemical organization among social insects is unique. All activities within colonies of social insects seem to depend to some degree on the emission of chemical signals. Thus chemicals released by social insects are involved in sex attraction, reproductive isolation, recruitment, nest defense, nest building, predation, territoriality, kin recognition and many other aspects of their life.

According to WILSON (*417*), truly social insects or eusocial insects can be distinguished as a group from other insects by common possession of three traits:

1) Individuals of the same species cooperate in caring for the young.
2) There is a reproductive division of labor, with more or less sterile individuals working on behalf of fertile individuals.
3) There is an overlap of at least two generations in life stages capable of contributing to colony labor, so that offspring assist their parents during some period of life.

The truly social insects include ants, all termites and the more organized bees and wasps.

The following adjectives are useful in defining the social behavior of insects (*417*):

1. solitary – denotes insects which show none of the traits observed in eusocial insects.
2. subsocial – denotes insects where adults care for their own nymphs or larvae during certain periods.
3. communal – denotes insects where members of the same generation use the same composite nest but do not cooperate in brood care.
4. quasisocial – denotes communal insects where in addition individuals cooperate in brood care.
5. semisocial – denotes quasisocial insects where in addition there is a reproductive division of labor.

6. eusocial – denotes insects which possess the three traits discussed above.

Ants and termites have evolved the same three castes – reproductive, workers and soldiers. Workers and termites soldiers may be male or female, but among ants they are female only, while in the honeybee (*Apis* sp.) and stingless bees (*Melipona* and *Trigona*) the larger queens and workers differ from each other but there is no special soldier class (*418*). For discussion concerning the involvement of pheromones in caste formation and evolution see BLUM (*419*) and JAFFÉ (*420*).

Among the ants the males constitute an additional "caste" since no certain case of true caste polymorphism within the male sex has yet been reported. The contribution of the male to the labor of the colony is virtually nil (*417*).

Ants of different sexes and castes may produce different odoriferous compounds. BRAND *et al.* (*421*), reported isolation of methyl-6-methyl-salicylate (**113**), 2,4-dimethyl-2-hexenoic acid (**114**) and methyl anthranilate (**115**) from the mandibular glands of *Camponotus noveboracensis*, *C. nearticus*, *C. rasilis* and *C. subbarbatus*. These compounds have been found in alate females or in workers. This secretion is of crucial importance in initiating the activity of the females before swarming and is used to scent areas near the nest entrance and to induce the female to swarm from the nest when the male flight is at a maximum.

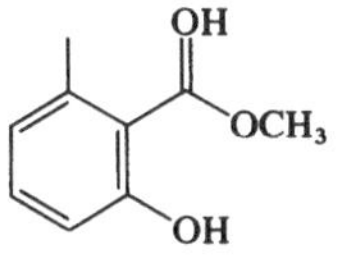

(**113**) 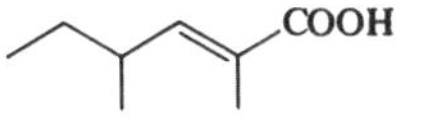   (**114**) 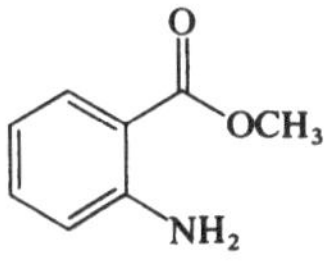   (**115**)

Differences in the secretions released by workers, females and queen ants have also been found. From the heads of *Tetramorium caspitum*, PASTEELS and coworkers (*422*) isolated 4-methyl-3-hexanol and the corresponding ketone. The heads of alate females contained both compounds, whereas only the alcohol was detected in workers. Both the alcohol and the ketone were attractants for workers (*422*). LÖF-QVIST and BERGSTRÖM (*423*) examined the Dufour's gland secretion of virgin females and queens of *Formica polyctena* and concluded that undecane was present in concentrations up to 74% in virgin females, while the queen gland contained less than 1% of this hydrocarbon. In addition they detected decane and nonane in virgin females but not in queens.

The pheromone(s) emitted by a termite queen are also important for the life of the colony. BRUINSMA and PRESTWICH (in *548*) found that a substance released by the queen abdomen of the subterranean grass-feeding termite *Macrotermes subhyalinus* induced replacement of queen cells. The pheromone was identified as *Z*-9-hexadecenoic acid **(116)**. Termite workers detect this substance in a concentration as small as 100 nanograms at a distance of 2–3 cm and build new cells in response to it. The *E*-isomer and other analogues were not effective.

(116)

HOWARD *et al.* (*424*), demonstrated that all caste forms of *Reticulitermes virginicus* possess the same *n*-methyl alkanes, dimethyl alkanes and dienes, but they showed caste specific proportions.

Nestmate and kin recognition against aliens is the rule in social insects. Among carpenter ants (*Camponotus* sp.) odor discriminators originate from the queen and are distributed among and learned by adult colony members (*425*). Colony founders of the social wasp *Polistes fuscatus* discriminate between kin and non-kin, accepting the combs of sisters but destroying young brood in the combs of less related females (*426*). Recent studies indicate that recognition odors have both a heritable and an environmental component. Young gynes (potential queens) deprived of the opportunity to learn recognition odors from their natal nests treat all gynes as nestmates, suggesting that the ontogeny of nestmate recognition ability involves intolerance to unfamiliar pheromones rather than the development of tolerance to familiar odors (*433*).

Guard bees of *Lasioglossum zephyrum* discriminate between residents and non-residents on the basis of individual odors, allowing residents to enter, but rejecting non-residents (*427*). It was originally assumed that the guards either learn or become habituated to the odors of each colony member, but BUCKLE and GREENBERG (*428*) showed that guards learn the odor of their nestmates. Later, GREENBERG (*429*), demonstrated that there exist both a genetic odor mechanism and a learned component by which guard bees distinguish odors of kin and non-kin. On the evidence guards, by noting their own odor as reference for recognition of nestmates, allow individuals with different odors to live together and enhance kin selection by providing a more efficient basis for discriminating between relatives and non-relatives.

HEFETZ and collaborators established the composition of the Dufour's gland secretion of four sympatric halicitine bees *Evylaeus malacharum*, *E. carneiventre*, *E. politum* and *Halictus resurgens*, as consisting of macrocyclic lactones, isopentenyl esters and hydrocarbons and concluded that these secretions are responsible, at least in part, for nestmate recognition (430) Similarly, the Dufour's gland secretion of *Eucera palestinae* which is composed of hydrocarbons and the methyl esters of unsaturated and saturated fatty acids also provides the nest with an individual odor and helps the bees to locate their own nest (431).

The variation in the ratio of the two major alarm pheromone components in individuals of *Crematogaster castanea* and *C. liengmei* was examined by BRAND and PRETORIUS (432), who suggested that these two semiochemicals (3-octanone and 3-octanol) may have roles in species or nestmate recognition.

In termites intracolony, intraindividual and intraspecific variation may be provided by the secretions of the soldiers. The multiple functions of these semiochemicals in defense, recognition and reproductive isolation is an unusual means of chemical communication and will be discussed later.

13. Insect Sex Pheromones

Insect sex pheromones have been studied extensively because of their potential and actual use in pest management (17, 434–436).

Sex pheromones may act not only as attractants but also as repellents or inhibitors. That of the male moth *Pseudaletia unipunctata* has been thought to act at close range as a sexual stimulant for the female, but its real effect is exerted upon other males by decreasing their tendency to approach sexually receptive pheromone-releasing females. It was suggested that the adaptive significance of this male pheromone may be related to the increased reproductive efficiency that results if multiple males are prevented from competing for a single female (437).

It is now well known that males of a number of insect species transfer an antiaphrodisiac pheromone during mating, thus making the female unreceptive to further mating (439, 440). SCHLEIN and collaborators observed a similar strategy for preventing homosexual activity in *Glossina morsitans morsitans* and *Musca domestica* (441).

Intrasexual competition is also a very important factor influencing mating behavior; insects have developed a series of strategies in competing for a potential mate and the necessary food resources. Young rove beetle males, *Aleochara curtula*, which are not sexually active yet, but need access to food sources avoid aggressive encounters and expulsion by mimicking the female sex pheromone (438).

Investigation of the post-mating courtship inhibition in *Drosophila melanogaster* females has led to the suggestion that a pheromone may be involved (442–447). JALLON et al. (442, 445), observed that male D. melanogaster transfer cis-11-octadecen-1-ol acetate (117) to females during copulation and proposed that this substance inhibits further mating for a period of hours. Although normal males produce a volatile substance which inhibits courtship, this chemical cue is not identical with the pheromone secreted by mated female flies (443). TOMPKINS and HALL (444) observed that mutant males are inhibited from mating by mated females but are unaffected by volatile pheromone from males, and concluded that the two inhibitors are not the same substance. After copulation, females produce less sex attractant and also begin to produce a mating inhibitory pheromone (444).

ZAWISTOWSKI and RICHMOND (446) suggested that cis-vaccenyl acetate (117) is hydrolyzed to cis-vaccenol by the action of a carboxylesterase, esterase 6, which is also present in male ejaculate. The alcohol then acts as a synergist to enhance the action of the acetate. However, in a further and conflicting report, VAN DER MEER et al. (447) argued that 1) the acetate is not converted to alcohol; 2) the amount transferred to female during copulation is below the level required to be effective for inhibition of courtship; 3) the substance is not transferred from the female reproductive tract to the abdominal cuticle; 4) esterase 6 has no effect on the rate of cis-vaccenyl acetate loss from the reproductive tracts of mated female; and 5) mutant males that do not reduce the postmating sexual attractiveness of females contain and transfer normal amounts of this acetate. Thus, the mechanism by which courtship in D. melanogaster is inhibited remains to be elucidated.

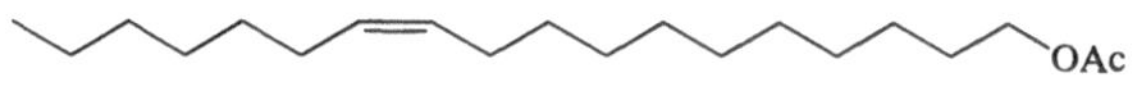

(117)

According to BELLES et al. (448), the structures of the sex pheromones of some species of Noctuidae and Tortricidae moths have taxonomic value, at least at the subfamily level.

Insect sex pheromones have very simple structures like butyric acid, the aphrodisiac pheromone of the male moth Bapta temerata and valeric acid from Limonius californicus or more complex ones like periplanone (118) one of the components of the sexual attractant of the American cockroach Periplaneta americana.

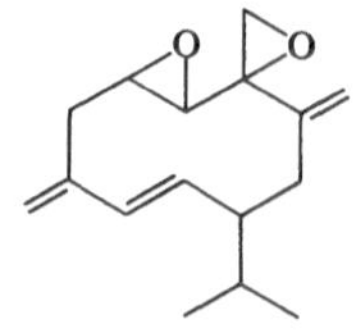

(118)

Male butterflies of the subfamily Danainae possess an extrudible organ called a hairpencil which releases aphrodisiac secretions during courtship. These secretions contain derivatives of dihydropyrrolizine alkaloids which adults sequester from their food plants and convert into pheromones (*158–162, 449–455*). BOPPRÉ *et al.* examined sex pheromone production by males of *Danaus chrysippus* and demonstrated that contact between the abdominal hairpencil and the pockets on the hindwings is a prerequisite for pheromone production, since pheromones are found in the hairpencil only after the males have ingested pyrrolizidine alkaloids and after the hairpencil has been dipped into the wing pockets (*456*).

14. Role of Pheromones in Reproductive Isolation of Insects

Most insects release sex attractants as a primary stimulus during mating behavior. In many cases, these insects utilize the chemicals as important barriers to interspecific mating. Reproductive isolation can be achieved because insects emit different pheromones, different blends of the same or even trace chemicals which prevent interspecific mating. Very similar pheromones but with a structural modification such as a change in position of the double bond, in chain length, in the number of unsaturations, in the configuration (*Z* or *E*) of the double bond or an alteration in the functional group, are likely to possess quite different activities. For example, in the family Tortricidae (Lepidoptera), two closely species *Grapholita prunivora* and *G. packardii* have essentially the same life cycle, but they are not cross-attracted since *G. prunivora* releases *cis*-8-dodecenyl acetate (**119**) while *G. packardii* produces the *trans* isomer (**120**).

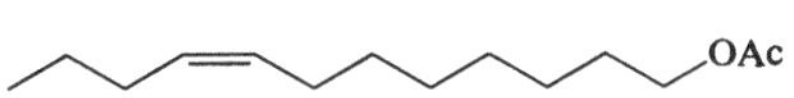

(**119**) 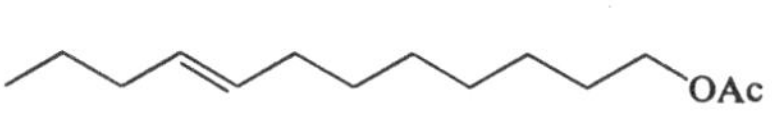   (**120**)

If more than one species employs the same attractant, the rate of pheromone release from the female may be species-specific. The concentration of pheromone emitted may also be important in eliciting a behavioral response.

Male discrimination among various release rates has been reported by several authors. Reproductive isolation among four sympatric ermine moths *Yponomeuta evonymellus*, *Y. cagnagellus*, *Y. padellus* and

Y. virgintipunctatus was analyzed (*458*). The species were found to share
11-Z-tetradecenyl acetate (**49**) in combination with varying proportions
of the *E* isomer as the primary components. *Y. cagnagellus* differs by
having a small amount of *E* isomer, *Y. padellus* also produces 11-Z-
hexadecenyl acetate (**121**) while this substance appeared to reduce trap
catches of *Y. evonymellus* and *Y. virgintipunctatus* when added to the
pheromone blend.

Different circadian mating rhythyms provide another means by
which insects achieve reproductive isolation. When an insect releases
its sex attractant, it is receptive to mating only during a brief period
of time. *Porthetria dispar* and *Lymanthria monacha*, two moths which
co-occur widely from Europe to Japan, are lured by the same attractant,
cis-7,8-epoxy-2-methylcyclooctadecane (**122**), but while *L. monacha*'s
mating period in Europe occurs from about 6 p.m. until 1 a.m. that of
P. dispar is diurnal. In North America, *P. dispar* has evolved a lengthened mating period in response to lessened communication interference
in the absence of *L. monacha* (*457*).

The importance of trace chemicals in reproductive isolation was
demonstrated by KLUN and coworkers (*459–461*) for *Heliothis zea* and
H. virescens. Both species produce relatively large amounts of 11-Z-
hexadecenal (**7**) with traces of 9-Z-hexadecenal (**6**), 7-Z-hexadecenal (**5**)
and hexadecanal (**4**). *H. virescens* females differ from *H. zea* because
they also release tetradecanal (**123**), 9-Z-tetradecenal (**124**) and 11-Z-
hexadecenol (**125**) as trace compounds. In both *Heliothis* species, trace

(**121**)

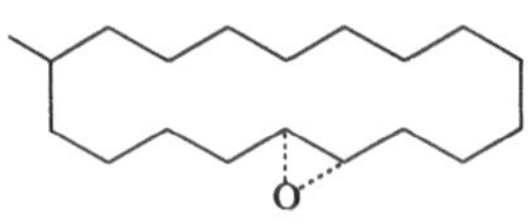

(**122**)

(**123**)

(**124**)

(**125**)

chcmicals arc fundamental both as to pheromone activity and specifici
ty of chemical signals.

An attractant-inhibitor interaction was observed by STECK and collaborators (462). Several species of noctuid moths such as *Mamestra
configurata, Scotogramma trifolii, Eurois occulta, Peridona saucia, Leu-
cania commoides* and *Crymodes devastator* utilize 11-Z-hexadecenyl acetate (121) as sex attractant for males. In addition certain olefins are
use for effective species-specific interaction. In nearly all cases at least
one of the co-attractants, i.e., the olefinic components, for each species
functions as a strong inhibitor for one of or more of the other species
in the group. For example, the female *S. trifolii* emits 11-Z-hexade-
cenol (125) which inhibits the attraction of *C. devastator* and *E. occulta*
males to sex lures.

Examples of reproductive isolation by means of pheromones in
insects other than Lepidoptera have also been observed.

LÖFQVIST and BERGSTRÖM reported such a case among Neuroptera.
Euroleon nostras, Grocus bore, and *Myrmeleon formicarius* are three
species of ant lion with short-lived adults which survive only for mating
and egg-laying. The species are sympatric in both space and time (463).
Males of the three species have large thoracic glands which secrete
mainly two nerol-derived compounds in each species. From *E. nostras*
males, the authors isolated nerol oxide (126) and another unidentified
monoterpene. The secretion obtained from *G. bore* consisted of two
monoterpenes. The first of these is also a nerol-derived compound,
whereas the other one was identical with the unidentified substance
from *E. nostras*. The secretion from *M. formicarius* was composed of
nerol (127) and an equal amount of another monoterpene. The glands
of the female are smaller than those of the males and secrete only
one substance in all three species, namely nerol oxide. The substances
obtained from these three species are all closely related to nerol and
the unique combination of two of them in males of each species forms
a species-specific secretion (463).

(126)   (127)

Attraction and mutual inhibition of several beetles of the genera
Ips, Trogoderma, and *Gnathotrichus* have been reported by many authors (464–469). Thus, *I. paraconfusus* and *I. pini* are two beetles which

colonize the same parts of *Pinus ponderosa* and *P. jeffreyi*. However, the species are not cross-attractive.

The aggregation pheromone of *I. paraconfusus* consists of a mixture of three components: *cis*-verbenone (16), ipsenol (21) and ips-dienol (22). Although the (S)-(−)-isomer of ipsenol acts as an attractant to *I. paraconfusus*, the (R)-(+) isomer acts as an inhibitor (469).

Ipsenol (21) and linalool (128), an isomer of ipsenol produced by male *I. pini* and female *I. paraconfusus*, had originally been implicated in the mutual inhibition of attractant pheromone response between these two species (464). However, this hypothesis was modified when BIRCH and LIGHT (466) reported that ipsenol inhibits attacks by *I. pini* on ponderosa pine logs baited with males of *I. pini* and suggested that the concentration of ipsenol (21) is critical for effective suppression of the attack. They therefore denied that linalool acts as inhibitors as was first thought.

(128)

FRANCKE *et al.* (469), described similar behavior in the *Ips amitinus-Ips typographus-Picea abies* interaction. The two beetles colonize different parts of this tree; this difference was attributed to the pheromone system of the beetles.

BIRCH *et al.* (470) also observed that chemically mediated behavioral interactions among four sympatric species of beetles, *Dendroctonus frontalis*, *Ips avulsus*, *I. calligraphus* and *I. grandicolis* appear to be a significant factor in delineating breeding areas within trees and in influencing the sequence of colonization.

According to BORDEN and MCLEAN (471), the attraction of *Gnathotrichus sulcatus* to logs infested with *G. retusus* suggests that these two beetles employ the same substance, sulcatol (129), as an aggregation pheromone. In addition, sulcatol functions as an efficient barrier in reproductive isolation, since the beetles respond differently to enan-

(129)

tiomers of sulcatol. *G. sulcatus* respond only when both enantiomers are present. They do not respond to optically active pure (*S*)-(+)-sulcatol, but even less than 1% of (*R*)-(−)- sulcatol, in an enantiomeric mixture, is sufficient for a response, while *G. retusus* respond to (*S*)-(+)-sulcatol, but not to the racemate (*465, 468*).

15. Arthropod Defensive Secretions

The literature on arthropod chemical defensive secretions is so thoroughly covered in reviews (*472–482*) that it will be discussed only briefly here. A recent and detailed book is that of BLUM (*68*).

The chemicals may be biosynthesized *de novo* or sequestered from dietary plants which contain compounds that may serve as precursors of the allomone or are toxic in themselves.

The qualitative and quantitative composition of arthropod defensive secretions may vary according to age, sex, physiological status, instar, caste and season.

There are two major types of defensive substances, those which are elaborated by special exocrine glands and those not strictly of glandular origin which are contained in the blood, gut or elsewhere in or on the body (*472*).

As regards their mode of action the glands are of different types, eversible as in *Papilio, Baronia* and *Euryades* butterflies which emit isobutyric acid and 2-methylbutyric acid in response to a disturbance; oozing as in the millipede *Narceus gordanus* which discharges a secretion containing benzoquinones such as 2-methoxy-3-methylbenzoquinone, in *Chauliognathus lecontei* in which the active principle is the acetylenic acid 8-*cis*-dihydromatricaria acid (**130**) and in *Apelloria corrugata* in which a cyanogenic secretion is emitted from a pair of prothoracic and abdominal glands; spraying glands as in the carabid beetle *Galerita janus* which sprays a secretion containing formic acid and controls the direction of the discharge by flexing the abdominal tip, reactor glands where the stored products are not the final constituents of the

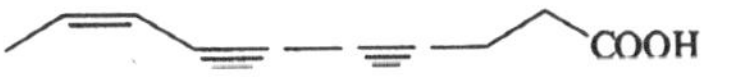

(**130**)

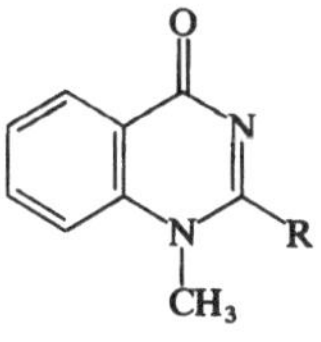

(**131**) R=Me
(**132**) R=Et

secretion but precursors of the defensive substance and finally tracheal glands (*472*).

The secretion may kill, as for example when HCN is discharged by *Geophitus vittatus* (*483*); it may repel as when isobutyric acid and 2-methylbutyric acid are emitted by *Papilio machaon* caterpillars (*472*) or it may imobilize predators as is the case of glomerin (**131**) and homoglomerin (**132**) which are given off by *Glomeria marginata* (*484*).

Certain defensive secretions emitted by arthropods are especially effective since they are emitted as a mixture rather than as a single compound. Thus, the carabid beetle *Helluomorphoides* sp. secretion consists of formic acid and nonyl acetate. The former is an irritant while the ester acts as a penetrating promoting agent (*472*).

An unusual means of defense is that exhibited by the aquatic beetle *Stenus comma*. The beetle lives on the water surface near river banks and in the case of danger it jets towards the banks and safety at a speed of 45–75 cm/s without using its legs. The beetle immerses the tip of its abdomen into the water and emits from two pairs of pygidial defensive glands a secretion that depresses the surface tension of the water, thus propelling itself forward over the water. The larger glands yielded stenusin (**133**) as the propelling agent (*485*).

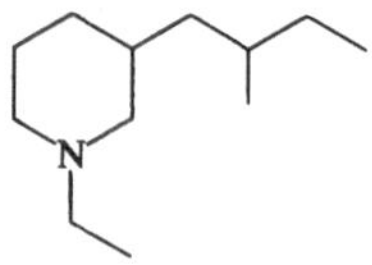

(133)

The evolutionary, ecological and taxonomic significance of chemical defensive substances in leaf beetles has been discussed in a recent article (*487*). Such substances may also affect sexual behavior of arthropods; for example it has been observed (*486*) that at high concentration the defensive tergal gland secretion of *Aleocharia curtula* beetle inhibits male copulatory response.

According to the so-called warning coloration hypothesis proposed in the second half of the last century by some naturalists, the bright colors exhibited by larvae and adult stages of some insects, the so-called aposematic insects, are a means of advertising to their vertebrate predators that they are unpalatable. Rather early the suggestion was made that this protection derived from storage of toxic substances obtained by the larvae from their food plants. A recent review (*68*) covers the subject adequately. However, PASTEELS and DALOZE (*488*, *489*), after investigating the cardenolide content in some chrysomelid beetles, concluded that in this case at least the cardenolides are synthesized by the beetles themselves rather than derived from their food plants.

The toxic compounds advertised by warning coloration include a great variety of compounds, although two classes predominate remarkably, pyrrolizidine alkaloids such as senecionine (134), seneciphylline (135), integerrimine (136), jacobine (137), jacozine (138) and jacoline (139) and cardiac glycosides such as nerigoside (140), oleandrin (141), calotoxin (142), calotropogenin (143), uscharidin (144) and uzarigenin (145).

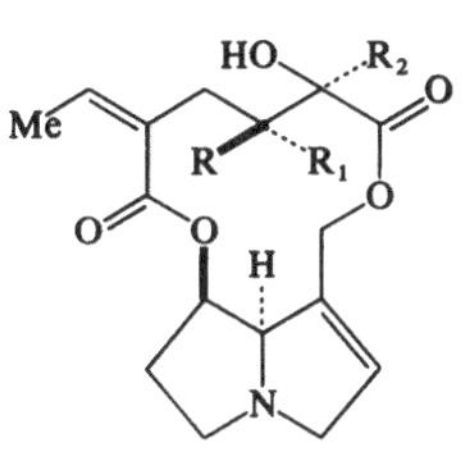

(134) R=H; R₁=R₂=Me   (136) R=H; R₁=R₂=Me 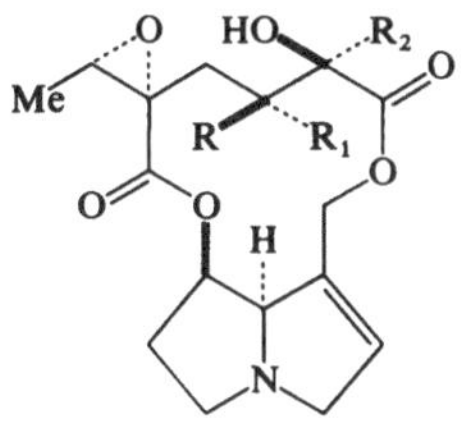   (137) R=H; R₁=R₂=Me
(135) R=R₁=CH₂; R₂=Me   (138) R=R₁=CH₂; R₂=M

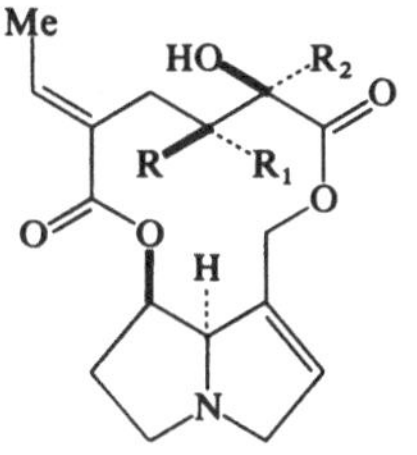

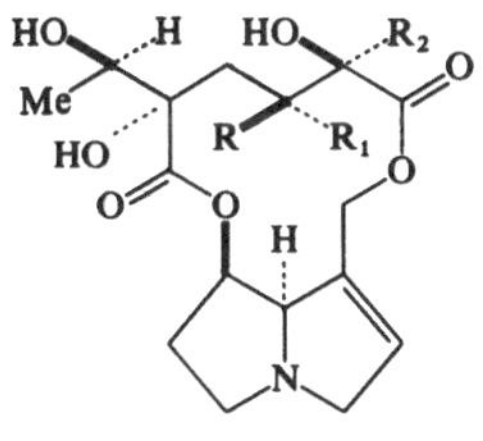

(139) R=OH; R₁=H; R₂=Me

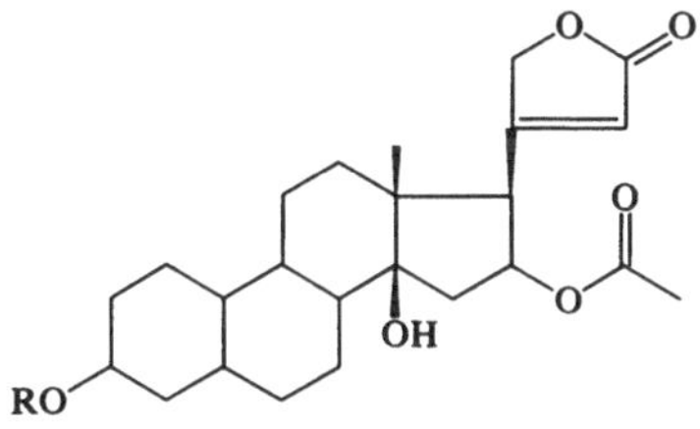

(140) R=D-diginose
(141) R=L-oleandrose

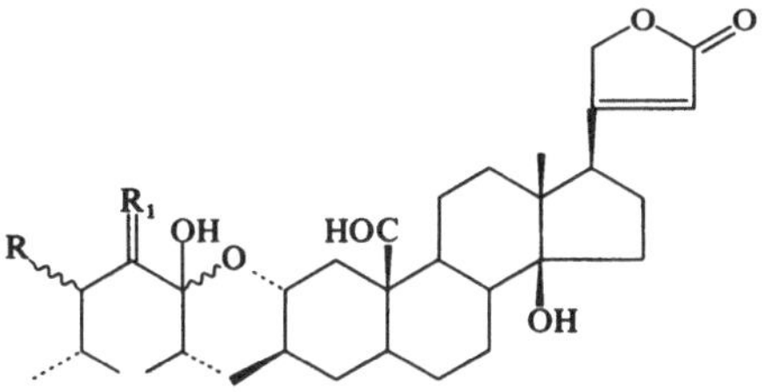

(142) R=R₁=H, OH
(144) R=H; R₁=O

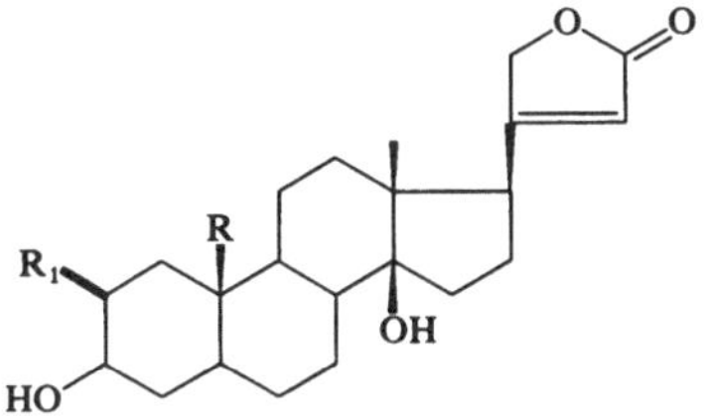

(143) R=CHO; R₁=OH
(145) R=Me; R₁=H

There are about 50 species belonging to seven orders in insects which are known to store cardiac glycosides which are found in about eleven families of angiosperms particularly in the Apocynaceae and Asclepiadaceae. These substances may be isolated from the roots, bulbs, rhizomes, wood, bark, stems, leaves, petals, fruits, pods and latex (475). Their concentrations depend on time of collection and provenance (*474, 476*). Recently it has been observed (*490, 491*) that cardenolide uptake by butterflies is a logarithmic function of their concentration in plants which serve as their source.

In spite of the protection provided by cardiac glycosides, for monarch butterflies, birds of several species have learned to penetrate the cardenolide-based chemical defense by discriminating among palatable and unpalatable parts (*492*). It also appears that predation is inversely proportional to colony size which may be an evolutionary explanation for the dense aggregation observed in these insects. Similarly, it has been reported (*493*) that tanagers, *Pipraeidea melanonota*, prey on *Mechanitis polymnia*, *M. lysimnia* and *Hypothyris daeta*, Ithomiinae butterflies, by accepting only their fatty abdominal contents and leaving the remainder of the bodies. Other recent articles dealing with aposematic insects are found in references (*494–496*).

16. Defense in Termites

The order Isoptera contains two superfamilies; lower termites and higher termites, both of which employ chemical substances as means of defense. These have not only interesting chemical structures but can be used in taxonomy and population recognition and provide support for a new evolutionary trend hypothesis (*497–547*).

QUENNEDEY (*503*) described three fundamental types of mechanisms by which termite soldiers defend their colonies (see Fig. 1).

1) Daubing which involves application of a secretion from the frontal gland through an elongated labrum (*Schedorhinotermes*).

2) Biting with simultaneous addition of a toxic substance from the frontal gland (*Amitermes, Cubitermes, Macrotermes, Odontotermes* and *Noditermes*).

3) Squirting which involves ejection by the soldiers of a viscous, sticky secretion from a specialized elongated rostrum called the nasus (*Nasutitermes, Trinervitermes*) and thus avoids physical contact between termite and enemy.

In termite soldiers the heads and bodies are so thoroughly modified into chemical and physical weapons that the soldiers can neither feed themselves nor reproduce. They are unique in exhibiting a diverse group of multipurpose allomones. Soldiers of the advanced genera have lost their mandibles through evolutionary processes, but have developed an ability to biosynthesize diterpenes which is unique in insects (*528, 531*).

Fig. 1. The drawings show three aspects of the chemical defenses that can be found in termites. In *Cubitermes*, the secretion from the frontal gland penetrates the wound made by the jaws. For *Schedorhinotermes*, the action of the jaws becomes secondary, since the secretion is directly applied against the adversary like a brush. In the case of *Trinervitermes*, the secretion is ejected from a distance, the nasute has no contact with the antagonist and the mandibles are withdrawn.
(Courtesy Andre Quennedey)

In the first category soldiers of some genera of the primitive mandibulate nasutes produce normal alkanes, simple mono- and sesquiterpenes and a homologous series of macrocyclic lactones with 22 to 36 carbon atoms (*531*).

Major and minor soldiers of *Schedorhinotermes lamanianus* produce a mixture of ten aliphatic ketones (**146–155**) when their colonies are disturbed. No wound is necessary for the toxic effect to appear which suggests that the long hydrocarbon chains of these defensive substances provide high lipid solubility and thus facilitate penetration on the cuticle of the enemy (*502, 548*).

Cubitermes umbratus is an example of a termite using the second type defense. Soldiers produce four diterpenes, biflora-4,10(19)-15-triene (**156**) (*522*), cembrane A (**157**) and 3-Z-cembrane A (**158**) (*521*) and cubitene (**159**). More recently, isolation and identification of cubegene (**160**), a diterpene with a novel carbon skeleton from *C. ugandensis* soldiers, has been reported (*539*). *Odontotermes badius* also bites with

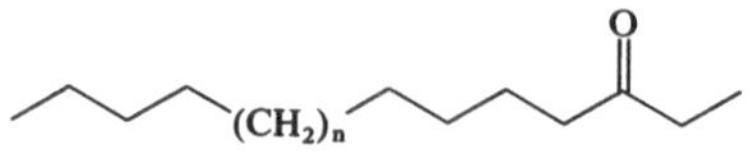

(146) n = 1
(147) n = 3
(148) n = 5

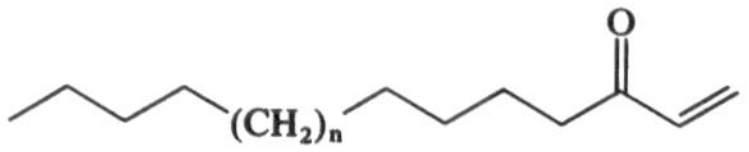

(149) n = 1
(150) n = 3
(151) n = 5

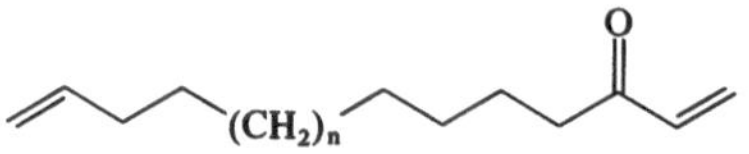

(152) n = 1
(153) n = 3
(154) n = 5

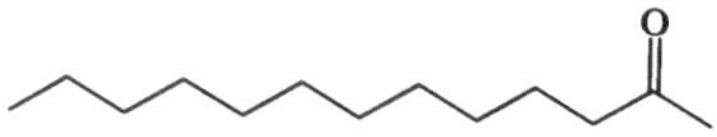

(155)

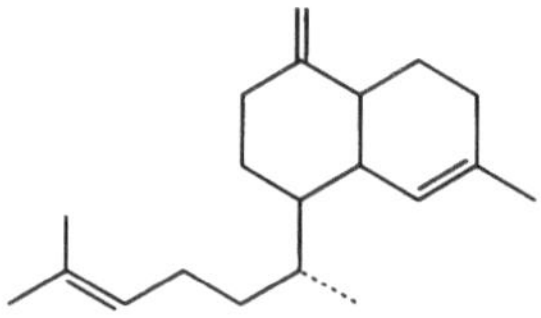

(156) 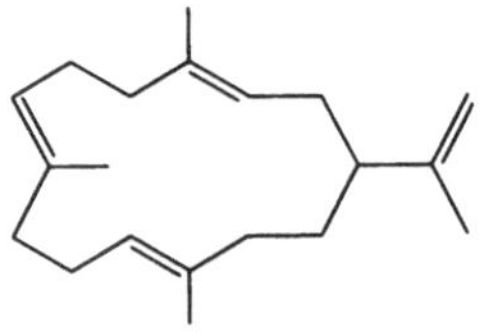   (157)

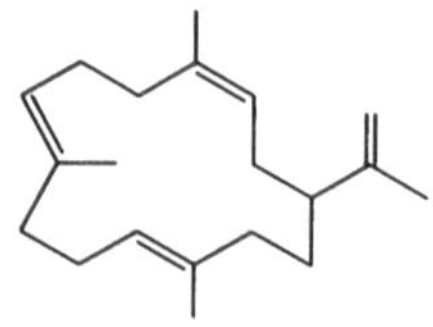

(158) 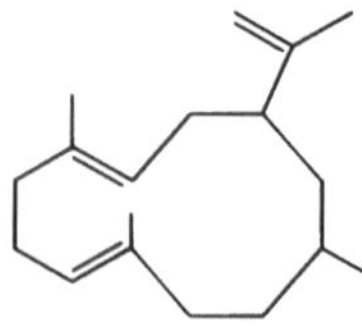   (159) 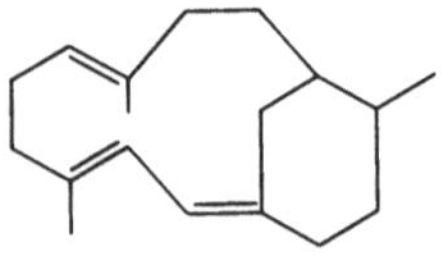   (160)

simultaneous emission of benzoquinone and protein from the salivary gland (504).

In the second category are also *Macrotermes subhyalinus* and *Amitermes* species. Soldiers of *M. subhyalinus* emit a secretion composed primarily of a mixture of long chain saturated and unsaturated hydrocarbons (161–166) (510). Chemical defensive secretions isolated from the soldiers of the genus *Amitermes* vary widely, making them of little value for interspecific systematic comparison (546). *A. unidentatus* produces a mixture of four 2-ketones with 12, 14, 16 and 18 carbon atoms (167–170) while *A. messinae* secretes limonene (25) and 4,11-epoxy-*cis*-eudesmene (171) (548). *A. evuncifer* also produces 4,11-

epoxy-*cis*-eudesmene while the secretion of *A. herbertensi* is 98% ter-
pinolene (172) (*500*).

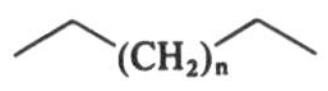

(**161**) n = 21
(**162**) n = 23
(**163**) n = 27 (with Me in 3 position)
(**164**) n = 27 (with Me in 5 position)

(**165**) n = 16
(**166**) n = 18

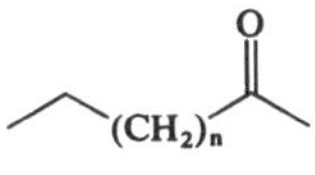

(**167**) n = 8
(**168**) n = 10
(**169**) n = 12
(**170**) n = 14

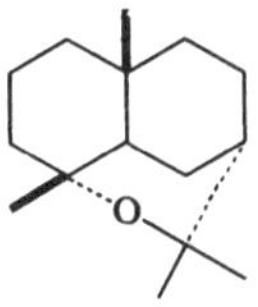

(**171**)

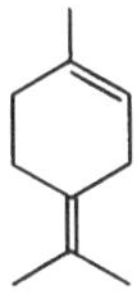

(**172**)

Detoxication mechanisms employed by the lower termites *Prorhino-termes simplex* and *Schedorhinotermes lamanianus* were investigated. Soldiers of these two species secrete nitroalkanes and vinyl ketones, respectively, in response to colony disturbance. Workers of these termites die when exposed to the volatiles of the other species, but survive when exposed to volatiles from conspecific soldiers (*529*). The workers have substrate-specific alkene reductases which, in the presence of a reduced cofactor, catalyze the reduction of the electron-deficient double bond of the unsaturated eletrophilic group. The alkanes are then recycled in vivo to acetate (see Fig. 2).

Termite chemical defensive secretions have been studied from the taxonomic and phylogenetic points of view (*517–519, 526, 531, 533, 535, 536, 546*). In chemical exocrine secretions PRESTWICH *et al.* have found intracolony, intraindividual and intraspecific variations (*513, 517–519, 523, 525–527, 531, 533–535, 543*). Thus PRESTWICH (*513*) on studying three allopatric population of *Trinervitermes gratiosus* soldiers encountered significant differences in the chemical composition of such secretion within a single species and even found that different population may produce isomeric compounds. Likewise, the secretions of three allopatric species of *T. bettonianus* exhibited intraspecific chemical variations which were genetically rather than nutritionally determined (*526*). Implications for systematics have been provided by the findings of PARTON *et al.* (*525*) who showed that secretions of termites belonging to the *Reticulitermes lucifugus* complex differed from those of *R. santonianus*. *R. santonianus* produces monoterpenes and specific sesquiterpene hydrocarbons, while monoterpenes are absent in *R. luci-*

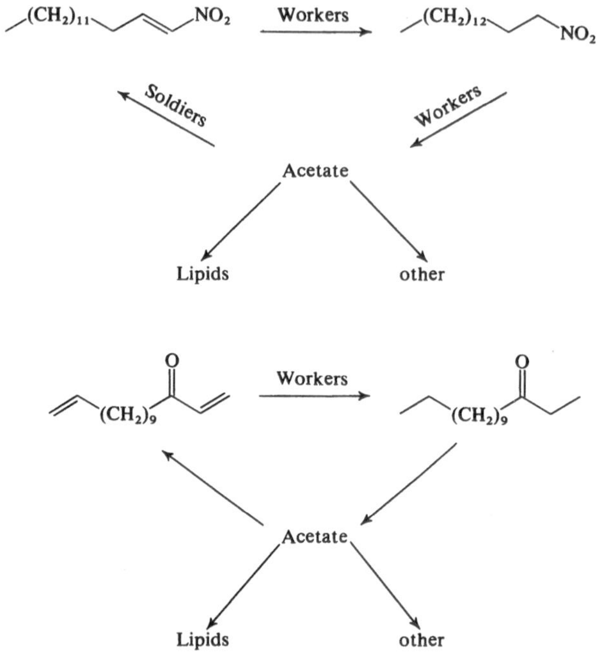

Fig. 2. Detoxication of soldier defense secretions. Termite workers of *Prorhinotermes simplex* and *Schedorhinotermes lamanianus* have substrate-specific alkene reductases which, in the presence of a reduced cofactor, catalyze the reduction of the electron-deficient double bond of the unsaturated electrophilic group. The alkanes are then recycled in vivo to acetate. (After Spanton, S.G. and G.D. Prestwich: Chemical Self-Defense by Termite Workers: Prevention of Autotoxication in two Rhinotermitids. Science *214*, 1363–1365 (1981))

fugus populations and sesquiterpenes when detected are different from these found in *R. santonianus*.

Four species of *Nasutitermes* soldiers, *N. columbicus, corniger, ephrates* and *nigriceps* from Costa Rica and Panama, can be distinguished by means of their chemical defense secretions (*543*). The authors of the study suggested that climate had an influence on the chemical composition of secretions from two Pacific Coast populations of *N. corniger* and postulated that the Atlantic Coast population of the species derived from a Pacific Coast population.

As to possible implication for phylogeny, the following provides an interesting example. In 1961, EMERSON (*497*) proposed that the defensive system of nasute soldiers, which eject a chemical secretion, is more effective against ants than the large mandibles of primitive mandibulate nasutes. As the nasus elongated (e.g. in the progression from *Cornitermes* via *Rhyncotermes* to *Nasutitermes*) fewer ants were repeled by the soldiers' bites and more by the frontal gland secretion; i.e., the regressive loss of mandibles led to a better protective adaptation.

PRESTWICH and COLLINS (527) showed that the defensive secretions of the advanced termites of the genera *Subulitermes* and *Nasutitermes* contained diterpenoids which are absent in the primitive mandibulate nasute genera *Cornitermes, Amitermes* and *Rhyncotermes*. On this basis, they questioned the hypothesis of diphyletic evolution of the Nasutitermitinae.

Nasutitermes exitiosus soldiers are polymorphic as both small and large soldiers are present in the same colony (499, 508). The behavior of small and large soldiers is very different, with only small soldiers defending the colony. While the small soldiers are attracted to a predator, the large non-combative soldiers retreat immediately after contacting an intruder (499). Although both types of soldiers produce a chemical secretion composed of *alpha*-pinene (12), *beta*-pinene and limonene (25), the non-combative soldiers do not employ their secretion in defense. According to KRISTON *et al.* (508) these soldiers act as messengers by raising alarm in the nest, while EISNER and collaborators (505) observed both soldiers and workers play an important function in the colony's defense, since the workers can bite ants and thus render them more vulnerable to the secretion emitted by the soldiers.

Termite chemical secretions also include oxygenated diterpenes (173–185) (506, 507, 509, 511, 513, 520), ancistrodial (186) and ancistrofuran (187) (512), longipenol (188) (532) and macrocyclic lactones (531).

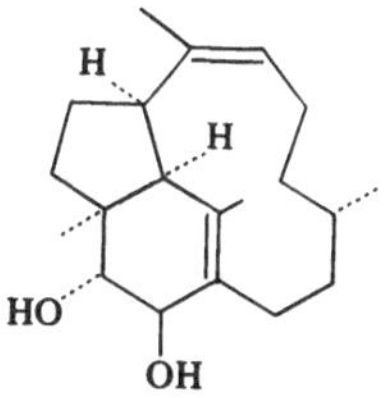

(173) 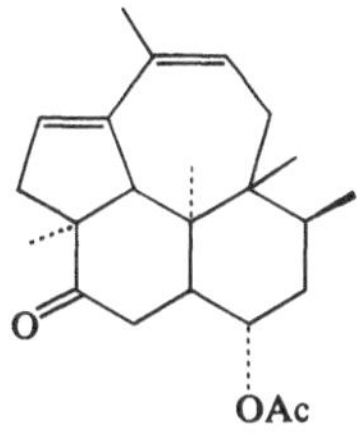   (174)

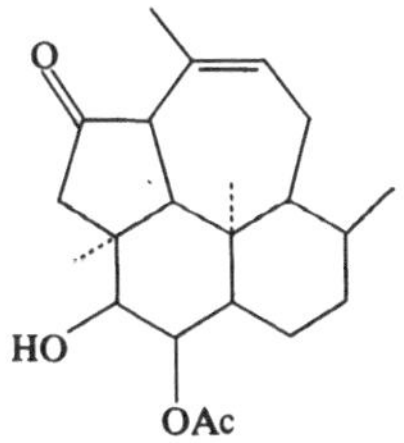

(175) 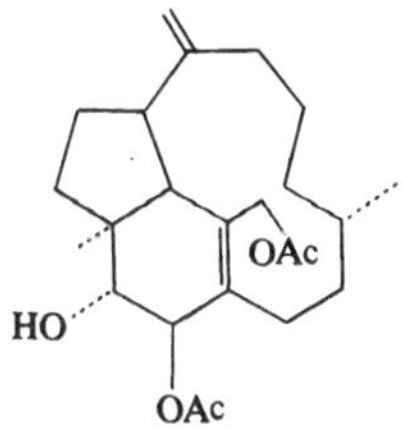   (176)

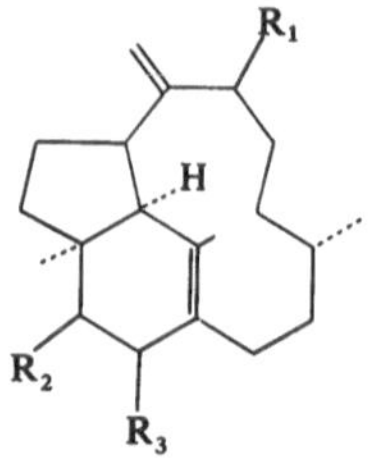

(177) R_1=t. OAc; R_2=t OH; R_3=c OH
(178) R_1=c OH; R_2=R_3=OH or OAc
(179) R_1=H; R_2=R_3=OH
(180) R_1=OH; R_2=R_3=OAc
(181) R_1=OH or OAc; R_2=R_3=H
(182) R_1=R_2=R_3=OAc
(183) R_1=H; R_2=R_3=OAc

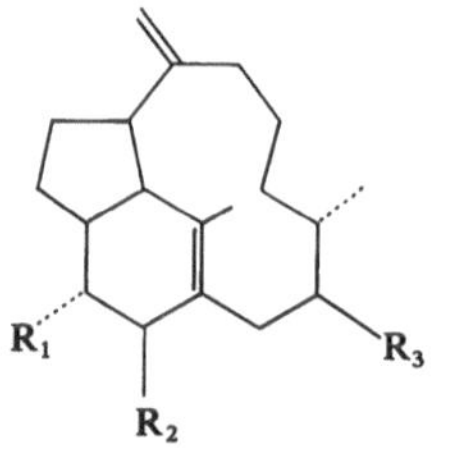

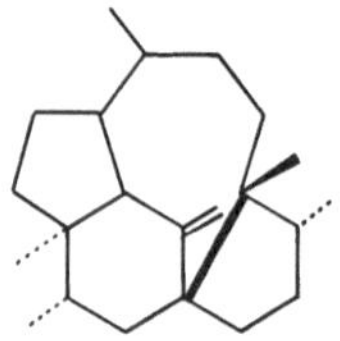

(188)

(184) R_1=R_2=R_3=H or OAc
(185) R_1=R_2=OAc; R_3=O

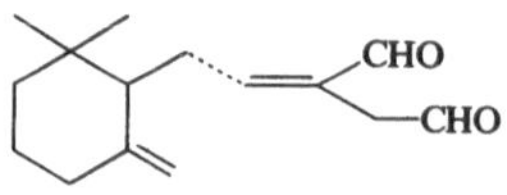

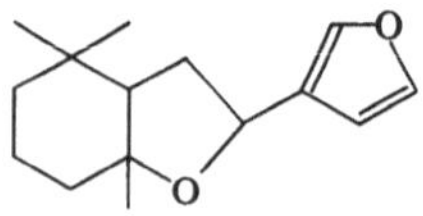

(186)   (187)

A hypothesis concerning the efficacy of such secretions has been advanced by PRESTWICH (519). On comparing the resin of Scotch pine (*Pinus sylvestris*) which contains *alpha* and *beta*-pinene with the secretion ejected by certain termite soldiers, he concluded that in both cases the diterpenes present in the mixture substantially retard the rates of evaporation of the monoterpenes, thus increasing the effect of the secretion as a sticky and irritant agent.

17. Predator-Prey Interactions

Emission of semiochemicals by insects is very closely associated with predator-prey interaction. Recent observations in connection with this topic will be discussed in the following paragraphs.

Ants have developed various ways to overcome their prey. Examples of slave-making ants, thief ants and crypsis in ants are now well known. Ants invade termite nests, forage on them and return to their own colony without being molested. How can this occur?

Analysing the chemical stimuli involved in a predator-prey interaction, LONGHURST *et al.* observed that specialized ants preying on termites emit non-repellent aliphatic alcohols as major components of the secretion from their mandibular glands whereas nonspecialized ants release aldehydes and ketones (*550*).

Decamorium uelense preys on termites. Scout workers search for foraging termites; the successful ones return to the nest and recruit a group of workers to the foraging area. The mandibular glands of these ants produce a non-repellent alcohol, 3-octanol, which prevent them from being detected by termites such as *Ancistrotermes cavithorax*, *Macrotermes bellicosus* and *Microtermes* sp. and thus permits them to move freely among the termites before attacking them. This behavior was also found in other ant species such as *Tetramorium termitobium*, which releases 2-undecanol from the mandibular gland, and certain *Crematogaster* species which give off 6-methyl-3-octanol and are therefore also able to prey on the above-mentioned termite species without being molested (*550*).

The mechanisms whereby parasitoids use kairomones to locate hosts are obviously of crucial importance in predator-prey interactions and may be divided in two categories depending on whether they are the result of long range or cole-range chemoreception. In some cases, olfaction can be influenced by the parasitoid's previous experience (*551*). A number of kairomones used by parasitoids as aids in host location have been identified. Hemolymph, cuticule, frass scale, mandibular gland and feces can be sources of such kairomones, and long-chain hydrocarbons are the main chemical stimuli responsible (*150, 552–569*).

Like *D. uelense* ants, *Solenopsis fugax* ants and *Nomada* bees have developed a way of predation which permits them to invade and steal brood from their prey. The European thief ants *S. fugax*, steal brood from nearby nests of other ant colonies by building a system of subterranean tunnels which lead to the nest of their prey. Scout ants lay down trail pheromone from the Dufour's gland to attract nest mates

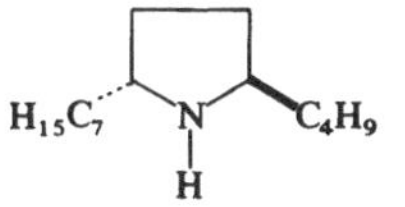

$$H_{15}C_7 \diagdown N \diagup C_4H_9$$
$$\mid$$
$$H$$

(**189**)

to the predation sites. When the thief ants arrive in the prey colony, they release from the poison gland a repellent identified as *trans*-2-butyl-5-heptylpyrrolidine (**189**), which prevents brood-tending ants from defending their own larvae (*570*).

Through cleptoparasitism females may appropriate food and resources from other females. An example is provided by female *Nomada* bees which prey on female *Andrena* bees; when a *Nomada* female finds an *Andrena* nest, she investigates the entrance and sits at the opening, possibly to guard against conspecifics (*571*). In the case of the interaction between *N. marshamella* and *A. sabulosa*, the communal nesting of *A. sabulosa* has been interpreted as a possible strategy to reduce territorial parasitism (*571*). Chemical similarities between *Nomada* bees and their prey have been investigated (*572–577*) and it has been suggested (*572*) that *Nomada* bees are scented by the males during copulation, thus making the predator more acceptable in *Andrena* nests.

All-*trans*-farnesyl hexanoate (**190**) was the main component of the female Dufour's gland secretion of eleven *Andrena* species, while geranyl octanoate (**191**) was the main component of the secretion from two other species. One or another of these two compounds is also the dominant component in the cephalic secretion of male *Nomada* bees. Apparently, *Nomada* bees prey only on a single *Andrena* host species. All-*trans*-farnesyl hexanoate (**190**) is the dominant component

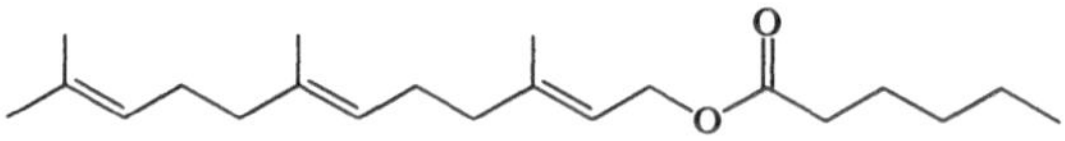

(**190**)

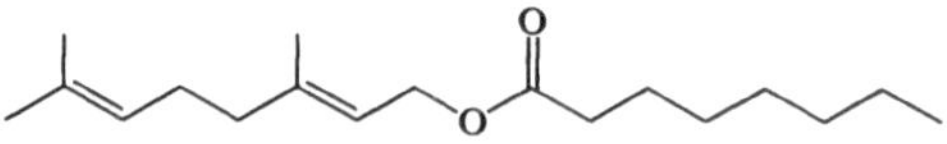

(**191**)

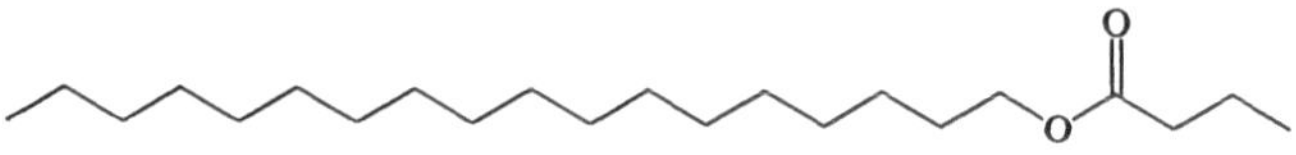

(**192**)

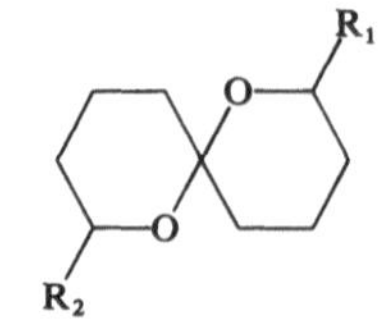

(**193**) R_1=Me; R_2=H
(**198**) R_1=H; R_2=Et

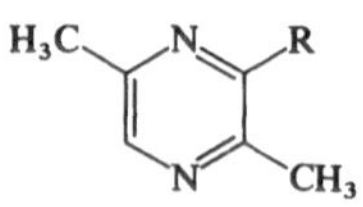

(**194**) R=Et
(**195**) R=Methylpropyl
(**196**) R=2-Methylbutyl
(**197**) R=3-Methylbutyl

in the *A. haemorrhoa-N. bifida* and *A. carantonica N. marshamella* pairs whereas geranyl octanoate (**191**) is the major compound in the *A. helvola-N. panzeri* and *A. clarkella-N. leucophthalma* pairs (*576, 577*).

The Dufour's gland secretions of *Melitta haemorrhoidalis* and *M. leporina* and the cephalic scent of *Nomada flavopicta* also exhibit chemical similarities. Octadecyl butyrate (**192**) has been isolated and identified as the dominant component in the gland of all three species (*575, 577*). Similar observations were made when the volatile secretions from females of four cleptoparasitic bees, *Epeolus cruciger*, *E. variegatus*, *Coelioxys quadrimata* and *C. mandibulata* were examined. The two *Epeolus* species elaborate spiroacetals (**32, 193**), pyrazines (**194–197**) and 2-alkanols, while the *Coelioxys* species produce spiroacetals (**29, 198**) alkanols and ketones (*578*).

On the other hand, the cephalic secretion of the cleptoparasitic bee *Holcopasites calliopsides* produces two main substances, 6-methyl-5-hepten-2-one (**62**) and geranyl acetone (**63**), whereas the mandibular gland secretion of its host *Caliopsis andreniformis* contains neral (**199**) and geranial (**200**). Thus in this case no evidence was found to support the hypothesis that *H. calliopsides* uses host mandibular gland secretion as a kairomone to locate *C. andreniformis* nests (*579*).

One of the most interesting, unusual and fully studied types of predator-prey interaction is that observed among slave-maker ants, an interaction sometimes referred to as dulosis. This kind of behavior was observed and described by the Swiss entomologist HÜBNER in 1810 among *Formica rufescens*.

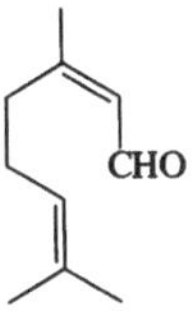

(**199**) 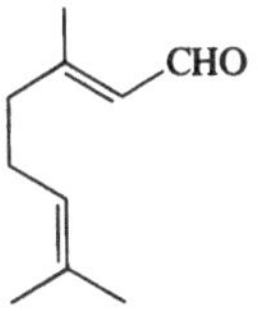   (**200**)

The slave-maker workers conduct raids on colonies of another related species by means of a trail pheromone which guides the worker columns to the target area, where they kill or repel defending workers, after which they penetrate into the nest and carry pupae and large larvae to their own colony. When the captured pupae develop into adult workers, they accept the slave-makers as nest mates and assist them in the tasks of the colony (*580, 581*).

The workers of species in early stages of dulotic evolution are relatively self-reliant assisting their slaves in common tasks and in some cases are able to survive alone. However, species of the most advanced form are completely dependent on their slaves. Fertile males and females of these species are capable only of conducting efficient slave raids (*581*).

DARWIN was fascinated with this behavior and proposed the first hypothesis for its origin. In "The Origin of Species", he described the experiment of HÜBNER who confined 30 individuals of *F. rufescens* without any slaves, but fully provided with food, and observed that the ants were unable to feed themselves. However, when a single slave (*Formica fusca*) was added *F. rufescens* fed and the survivors built new cells.

According to WILSON (*417*), there are at least 35 such ant species in six groups (*Leptothorax, Harpagoxenus* and *Strongylognathus* in the family Myrmecinae and Formica, *Rossomyrmex* and *Polyergus* in the family *Formicinae*). In some species such as those of the genera *Polyergus*, the workers are morphologically adapted to making raids because of the form of their mandibles, while others such as *Formica subintegra* and *F. pergandei*, members of the *F. sanguinea* group, possess hypertrophied Dufour's glands which contain large amounts of decyl, dodecyl and tetradecyl acetates. The acetates are discharged at defending workers during the slave raids and produce a very efficient long-lasting alarm signal which attracts the slave-makers but disperses the defenders (*417, 580, 581*).

Ants and termite colonies also share their nests with a large number of guests. They have developed the capacity to provide the correct chemical signal to these social insects and are thereby able to take advantage of the benefits of association (*62, 144, 417*). The guests are referred to as myrmecophiles (symbionts of ants) and termitophiles (symbionts of termites). WILSON presented an extensive list of such ectosymbionts (*417*). Both chemical and mechanical signals are involved in the interactions.

Reticulitermes virginicus and *R. flavipes* are two sympatric termite species with distinctly different cuticular hydrocarbons which function as species recognition cues. They also have different termitophilous cohorts which possess the same cuticular chemical substance as their host. Thus, *R. flavipes* is associated with *Tricopsenius frostii* while *R. virginicus* is associated with *T. depressus, Xenistusa hexagonalis* and *Philotermes howardii* (*582–586*).

18. Human Pheromones

There are several observations which suggest a connection between olfaction and human pheromones. Saliva, axillary sweat and vaginal odor appear to be the main stimuli involved (*107–118, 587–607*).

An extraordinary proportion of adults are anosmic for the various components of axillary and menstrual odor. AMOORE and FORRETIER (592) reported that 7% of human volunteers are anosmic to trimethylamine which has been isolated from menstrual blood. On the other hand, some individuals are able to distinguish 4000 different odors – well trained persons as many as 10000 – and can even identify other persons by smelling their hands only (118, 602). WYSOCKI and BEAUCHAMP (606) suggested that there is a genetic component in the variation in androstenone perception.

Psychoanalytic writers suggested the possibility of a special role for odor in the sexual development of children. According to this point of view, attraction to the odor of the opposite-sex parent may explain, at least in part, the development of the Oedipal complex (111).

The father of psychoanalysis, SIGMUND FREUD, speculated in 1930 in "Civilization and its Discontents" "The organic periodicity of the sexual process has persisted, it is true, but its effect on psychical sexual excitation has rather been reversed. This change seems most likely to be connected with the diminution of the olfactory stimuli by means of which the menstrual process produced an effect on the male psyche. Their role was taken over by visual excitation, which, in contrast to the intermitent olfactory stimuli, was able to maintain a permanent effect. The diminution of olfactory stimuli seems itself to be a consequence of man's raising himself from the ground, of his assumption of an upright gait; this made his genitals, which were previously concealed, visible and in need of protection (...)."

Substances which may act as human pheromones are secreted to the surface of the skin from three main sources, ecrine, apocrine and sebaceous glands.

The ultimate source of axillary odor are apocrine sweat glands. The secretion which appears initially on the surface is both sterile and odorless, but the characteristic odor is generated when resident microorganisms interact with it. LEYDEN et al. (600) and LABOWS et al. (602) found that the ability of bacteria to produce the pungent sweat odor is restricted to aerobic diphteroids.

It is well known that women detect and react to the boar's pheromone, 5-*alpha*-androst-16-en-3-one (71) (androstenone), more readily than men. This substance was also detected in human male axillary sweat (589, 593, 596, 600, 602), in the urine of both sexes (593) and in human saliva in very low concentration (603). In an interesting experiment KIRK-SMITH and BOOTH (597) sprayed androstenone (71) onto a seat in a position previously avoided by women in a dentist's waiting room. The incidence of women using the seat increased while as the dose increased, the incidence of men using the seat decreased.

DOTY *et al.* (*596*) provided data in support of the theory that 1) female and male responses to axillary odor are similar in magnitude and direction; 2) men's axillary odor is more intense and stronger than women's axillary odor and 3) the intensity and pleasantness of the odors are inversely related. They also suggested that caution is necessary in accepting the notion that human beings can readily determine gender from conspecific axillary odor.

The point that different human races possess their own characteristic odors is based on the fact that different ethnic groups have apocrine glands of different size. Thus, compared with the Europeans and Negroes, individuals of the Mongolian race possess only weakly developed apocrine and sebaceous glands, mainly in the axilla (*117*).

SCHLEIDT and coworkers (*601*) examined the ability of Japanese, Italians and Germans to distinguish between the odor of T-shirts worn by different individuals. 80% of the participants could differentiate such odors, whereas 50% identified correctly the person to whom the recognized odors belonged. Male odors were classified as more unpleasant than female odors. Italian and Japanese women classified, in contrast to German women, their own partner's odor as unpleasant.

Later, NIXON *et al.* (*605*) investigated the metabolism of testosterone, pregnenolone and 5-*alpha*-dehydrotestosterone by human axillary corynebacteria and demonstrated that pregnenolone was metabolized to 5-*alpha*-dehydroandrosterone and an unidentified compound, while testosterone was converted to 5-*alpha*-dehydrotestosterone and 17-*beta*-hydroxy-3-*beta*-androst-3-one.

Synchrony and suppression of the female cycle has been found in women living together in university dormitories which suggest that social interaction can have a strong effect on menstruation (*588, 595, 599*). GRAHAN and MCGREW (*595*) also demonstrated menstrual synchrony in females sharing a co-educational university environment, but not between neighbors or randomized pairs of girls. Individual variations in both odor production and perception exist and some data suggest that axillary odor from a donor female can also cause menstrual synchrony (*599*).

Recently in an interesting and intriguing article STODDART (*607*) examined the use of certain triterpenoid derivatives which occur in incense and have structures resembling steroids and suggested that they are used because their odors mimic those of steroid hormones.

Volatile fatty acids found in the vaginal discharge of female primates were also detected in human vaginal samples (*114, 291, 590*). However, although the data of DOTY *et al.* (*591*) indicated that secretions from preovulatory and ovulatory phases were slightly weaker and less unpleasant in odor than those from menstrual, early luteal and late luteal phase, they do not support the hypothesis that such odors are particularly attractive to humans in an *in vitro* test.

As stated above, odor may play a role in parent-child relationship. DARWIN had already suggested that babies rely on odor for identification of their own mothers. Indeed, babies who were born blind and deaf can differentiate familiar from unfamiliar persons on account of their smell. Similarly, a mother can also quickly learn to distinguish the odors of her own babies (*116, 594*).

RUSSELL conducted an experiment to test whether babies from two days to six weeks of age can distinguish the odor of their mother. At two days of age only one of the ten babies responded to any of the stimuli presented, while at six weeks only one of the infants responded to the strange mother. According to RUSSELL the initial identification of the mother may be due not to a response to her odor but rather to odors placed on her by the baby during earlier contacts (*594*). But MYKYTOWYCZ (*117*) argued that humans have historically had a very high maternal death rate, and suggested that hence it would be advantageous for a baby not to form too early an attachment to its mother.

19. Conclusion

The influence of chemical substances on the social behavior of animals has been demonstrated without a doubt. This interdisciplinary subject offers an almost unlimited opportunity for further work in chemistry, all branches of biology and even for the social sciences. Its practical applications are well demonstrated by the use of pheromones in pest and wildlife management and in artificial insemination.

Further study of allelochemicals from marine sources, further definition of the well established relationship between primer pheromones and steroidal hormones and further inquiry into the possible existence of human pheromones are certainly among the most interesting areas of study for chemical ecology.

Acknowledgements

I am indebted to Dr. Keith S. Brown Jr. (Departamento de Zoologia, Universidade de Campinas) for reading the manuscript and for valuable comments and suggestions. I am extremely grateful to Dr. Werner Herz (Department of Chemistry, The Florida State University) without whose perceptive and constructive comments this review would have been quite different and much inferior; I am also particularly grateful to Professor Anibal de Lima Pereira, Faculdade de Farmácia, Universidade Federal do Rio de Janeiro) for helpful and patient discussion.

Further thanks are owed to Drs. Gunnar Bergström (Department of Ecological Chemistry, Goteborg University), Leon Bone (Department of Physiology, Southern Illinois University), Murray S. Blum (Department of Entomology, University of Georgia), Robert Bonsall (Department of Psychiatry, Emory University School of Medicine), Michael Boppré (Max-Planck Institut für Verhaltensphysiologie), Franklin Bronson (Department of Zoology, University of Texas), Rémy Brossut (Laboratoire de Zoologie, Université de Dijon), Lincoln Brower (Department of Biology, Amherst College), Luigi Colombo (Instituto de Biologia Animale, Universitá de Padua), Richard Doty (Department of Otorhinolaryngology, University of Pennsylvania), J. Edgar (Division of Animal Health, CSIRO), John Ebling (Zoology Department, The University of Sheffield), John Eisenberg (Smithsonian Institution), Thomas Eisner (Department of Neurobiology, Cornell University), Gisela Epple (Biology Department, University of Pennsylvania), Bert Holldöbler (Department of Biology, Harvard University), Klaus Jaffé (Departamento de Biologia de Organismos, Universidad Simon Bolivar), John Labows (Monell Chemical Senses Center, University of Pennsylvania), Norman Liley (Department of Zoology, University of British Columbia), Richard Michael (Georgia Mental Health Institute), Dietland Müller-Schwartze (College of Environmental Science and Forestry, State University of New York), E. Morgan (Department of Chemistry, University of Keele), Jacques Pasteels (Laboratoire de Biologie Animale, Université Libre de Bruxelles), Thomas Pliske (Department of Biology, University of Miami), Glenn Prestwich (Department of Chemistry, State University of New York at Stony Brook), André Quennedey (Laboratoire de Zoologie, Université de Dijon), J.A.A. Renwick (Boyce Thompson Institute), Paul Scheuer (Department of Chemistry, University of Hawaii), W. Seabrook (Department of Biology, University of New Brunswick), Alan Singer (The Rockfeller University), Dietrich Schneider (Max Planck Institut für Verhaltensphysiologie), Jack Baillet (Laboratoire de Physio-Hormono-Recepterologie), Norman Stacey (Department of Zoology, University of Alberta), Margret Schleidt (Max Planck Institut für Verhaltensphysiologie), Michael Stoddart (Department of Zoology, University of Tasmania), Jan Tengö (Ecological Station of Uppsala University), Delbert Thiessen (Department of Psychology, University of Texas), and J. Vité (Forstzoologisches Institut der Universität Freiburg). The articles and reviews they sent to me (sometimes in a manuscript form) were essential in the preparation of this paper.

References

1. KARLSON, P., and A. BUTENANDT: Pheromones (Ectohormones) in Insects. Ann. Rev. Entomol. **4**, 39–58 (1959).

2. BOSSERT, W.H., and E.O. WILSON: The Analysis of Olfactory Communication Among Animals. J. Theor. Biol. **5**, 443–469 (1963).

3. WILSON, E.O.: Pheromones. Sci. Amer. **208**, 100–114 (1963).

4. – and W.H. BOSSERT: Chemical Communication Among Animals. Rec. Prog. Horm. Res. **19**, 673–716 (1963).

5. SONDHEIMER, E., and J.B. SIMEONE (eds.): Chemical Ecology. New York: Academic Press. 1970.

6. LAW, J.H., and F.E. REGNIER: Pheromones. Ann. Rev. Biochem. **40**, 533–548 (1971).

7. BIRCH, M.C. (ed.): Pheromones. Amsterdam: North Holland/Elsevier. 1974.

8. SHOREY, H.H.: Animal Communication by Pheromones. New York: Academic Press. 1976.

9. HARBORNE, J.B.: Introduction to Ecological Biochemistry. London: Academic Press. 1977.

10. RITTER, F.J. (ed.): Chemical Ecology: Odour Communication in Animals. Amsterdam: North Holland/Elsevier. 1979.

11. MÜLLER-SCHWARTZE, D., and R. SILVERSTEIN (eds.): Chemical Signals: Vertebrates and Aquatic Invertebrates. New York: Plenum Press. 1980.
12. MUKHERJEA, M.: Pheromones and Pheromone Lipids. Indian Biol. 13, 136–148 (1981).
13. RITTER, F.J., C.J. PERSOONS, and J. GUT: Pheromones: Exocrine Chemical Messengers in the Service of Reproduction and Communication. Organograma 18, 3–9 (1981).
14. SKUHRAVY, V.: The Use of Pheromones in Ecological Studies. Pr. Nauk. Inst. Chem. Org. Fiz. Politech. Wroclaw (22) 1043–1056 (1981).
15. BARBIER, M.: Les Pheromones: Aspects Biochimique et Biologique. Paris: Masson. 1982.
16. BAILLET, J., and J. PAILLARD: Antipheromones? In: M.K. AGARWAL ed. Hormone Antagonists, p. 719–724. Berlin: Walter de Gruyter. 1982.
17. BRAND, J.M., J.C. YOUNG, and R.M. SILVERSTEIN: Insect Pheromones: A Critical Review of Recent Advances in their Chemistry, Biology and Application. Prog. Chem. Org. Nat. Prod. 37, 2–190 (1979).
18. BROWNE, L.E., M.C. BIRCH, and D.L. WOOD: Novel Trapping and Delivery Systems for Airborne Insect Pheromones. J. Insect Physiol. 20, 183–193 (1974).
19. BUSER, H.R., and H. ARN: Analysis of Insect Pheromones by Quadrupole Mass Fragmentography and High-Resolution Gas Chromatography. J. Chromat. 106, 83–95 (1975).
20. BYRNE, K.J., W.E. GORE, G.T. PEARCE, and R. SILVERSTEIN: Porapak Q Collection of Airborne Organic Compounds Serving as Models for Insect Pheromones. J. Chem. Ecol. 1, 1–7 (1975).
21. CANE, J.H., and T. JONSSON: Field Method for Sampling Chemicals Released by Active Insects. J. Chem. Ecol. 8, 15–21 (1982).
22. CLAESSON, A., and R.M. SILVERSTEIN: Chemical Methodology in the Study of Mammalian Communication. In: D. MÜLLER-SCHWARTZE and M.M. MOZELL eds., Chemical Signals in Vertebrates, p. 71–93. New York: Plenum Press. 1977.
23. KATZENELLENBOGEN, J.A.: Insect Pheromone Synthesis: A New Methodology. Science 194, 139–148 (1976).
24. HENRICK, C.A.: The Synthesis of Insect Pheromones. Tetrahedron 33, 1845–1889 (1977).
25. MORGAN, E.T., and R.C. TYLER: Microchemical Methods for the Identification of Volatile Pheromones. J. Chromat. 134, 174–177 (1977).
26. BAKER, T.C., L.K. GASTON, M.M. POPE, L.P.S. KUENEN, and R.S. VETTER: A High-Efficient Collection Device for Quantifying Sex Pheromone Volatilized from Female Glands and Synthetic Sources. J. Chem. Ecol. 7, 961–968 (1981).
27. HENRICK, C.A., R.J. ANDERSSON, and R.L. CARNEY: Aspects of Synthesis of Insect Pheromones. Pr. Nauk. Inst. Chem. Org. Fiz. Politech. Wroclaw (22) 887–918 (1981).
28. MORI, K.: Recent Progess in the Pheromone Synthesis. Study in Organic Chemistry 137–153 (1981).
29. – Synthesis of Chiral Insect Pheromones. Pr. Nauk. Inst. Chem. Org. Fiz. Politech. Wroclaw (22) 921–927 (1981).
30. – The Synthesis of Insect Pheromones. In: J. APSIMON ed., The Total Synthesis of Natural Products, p. 1–183. New York: Academic Press. 1981.
31. PEACOCK, J.W., R.A. CUTHBERTH, W.E. GORE, G.N. LANIER, G.T. PEARCE, and R.M. SILVERSTEIN: Collection on Porapak Q of the Aggregation Pheromone of Scolytus multistriatus (Coleoptera: Scolytidae). J. Chem. Ecol. 1, 149–160 (1975).
32. HOGGE, L.R., and D.J.H. OLSON: New Methodology for GC/MS Clarifies Compounds Identity. Ind. Res. Dev. 24, 144–149 (1982).
33. SONNET, P.E.: Tabulation of Selected Methods of Synthesis that are Frequently Employed for Insect Pheromones, Emphasizing the Literature of 1977–1982. In:

H. HUMMEL and C. MILLER eds., Technology of Pheromone Research, p. 371–403. New York: Springer. 1984.

34. SHANI, A., and M.J. LACEY: Convenient Method Applicable to Single Insects for Collection and Measurements of Blend Ratios of Airborne Pheromones from Artificial Sources. J. Chem. Ecol. **10**, 1677–1692 (1984).

35. WEBER, W., and V. SCHURIG: Complexation Gas-Chromatography: A Valuable Tool for the Stereochemical Analyses of Pheromones. Naturwissenschaften **71**, 408–413 (1984).

36. SLESSOR, K.N., G.G.S. KING, D.R. MILLER, M.L. WINSTON, and T.L. CUTFORTH: Determination of Chirality of Alcohol or Latent Alcohol Semiochemicals in Individual Insects J. Chem. Ecol. **11**, 1659–1667 (1985).

37. SMITH, A.B. III; A.M. BELCHER, G. EPPLE, P.C. JURS, and B. LAVINE: Computerized Pattern Recognition: A New Technique for the Analysis of Chemical Communication. Science **228**, 175–177 (1985).

38. CAIN, W.S. (ed.): Odors: Evaluation, Utilization and Control. Ann. N.Y. Acad. Sci. **237**, 1–439 (1974).

39. DOTY, R.L. (ed.): Mammalian Olfaction, Reproductive Process and Behavior. New York: Academic Press. 1976.

40. MICHAEL, R.P., R.W. BONSALL, and D. ZUMPE: Evidence for Chemical Communication in Primates. Vit. Horm. **24**, 137–186 (1976).

41. THIESSEN, D., and M. RICE: Mammalian Scent Glands. Marking and Social Behavior. Physiol. Bull. **83**, 505–539 (1976).

42. MÜLLER-SCHWARTZE, D., and M.M. MOZELL (eds.): Chemical Signals in Vertebrates. New York: Plenum Press. 1977.

43. MILLIGAN, R.S.: Pheromone and Rodent Reproductive Physiology. Symp. Zool. Soc. London **45**, 251–275 (1980).

44. EPPLE, G., N.F. GOLOB, M.S. CEBUL, and A.B. SMITH III: Communication by Scent in Some Callitrichidae (Primates). An Interdisciplinary Approach. Chem. Senses **6**, 377–390 (1981).

45. GOODRICH, B.S.: Communication in Mammals by Means of Smell. Chem. Aust. **48**, 463–467 (1981).

46. BRONSON, F.H.: Pheromonal Influences of Endocrine Regulation of Reproduction. Endocrine Responses to Primer Pheromones in Mammals. In: W. BREIPOHL ed. Olfaction and Endocrine Regulation, p. 103–113. London IRI Press Limited. 1982.

47. SIGNORET, P.J.: Communication Chimique et Reproduction chez les Mammifères Domestiques. Bull. Soc. Zool. France **107**, 573–586 (1982).

48. BRONSON, F.H.: Chemical Communication in House Mice and Deer Mice: Functional Roles in Reproduction of Wild Populations. In: J.F. EISEMBERG and D.G. KLEIMAN eds. Advances in the Study of Mammalian Behavior. Spec. Publ. Amer. Soc. Mammal. vol. 7, p. 198–238 (1983).

49. MÜLLER-SCHWARTZE, D.: Scent Glands in Mammals and their Functions. In: J.F. EISEMBERG and D.G. KLEIMAN eds. Advances in the Study of Mammalian Behavior. Spec. Publ. Amer. Soc. Mammal. vol. 7, p. 150–197. (1983).

50. – and R.M. SILVERSTEIN (eds.): Chemical Signals in Vertebrates. New York: Plenum Press. 1983.

51. VANDENBERG, J.G. (ed.): Pheromones and Reproduction in Mammals. New York: Academic Press. 1983.

52. ALBONE, S. (ed.): Mammalian Semiochemistry: The Investigation of Chemical Signals between Mammals. New York: John Wiley. 1984.

53. MARCHLEWSKA-KOJ, A.: Pheromones and Mammalian Reproduction. Oxford Rev. Reprod. Biol. **6**, 266–302 (1984).

54. BRONSON, F.H.: Mammalian Reproduction: An Ecological Perspective. Biol. Reprod. **32**, 1–26 (1985).

55. DOTY, R.L.: Odor-Guided Behavior in Mammals. Experientia **42**, 257–270 (1986).
56. EFFLE, G.. Communication by Chemical Signals (in press).
57. JACOBSON, M.: Insect Sex Pheromones. New York: Academic Press. 1972.
58. SHOREY, H.H.: Behavioral Responses to Insect Pheromones. Ann. Rev. Entomol. **18**, 349–380 (1973).
59. NOIROT, CH., P.E. HOWSE, and G. LE MASNE (eds.): Pheromones and Defensive Secretions in Social Insects. Paris: I.U.S.S.I. 1975.
60. SCHNEIDER, D.: Pheromone Communication in Moths and Butterflies. Adv. Behav. Biol. **15**, 173–193 (1975).
61. SHOREY, H.H., and J.J. McKELVEY JR. (eds.): Chemical Control of Insect Behavior. Theory and Application. New York: Wiley Interscience Publication. 1977.
62. HOLLDÖBLER, B.: Ethological Aspects of Chemical Communication in Ants. Adv. Study Behav. **8**, 75–115 (1978).
63. PARRY, K., and E.D. MORGAN: Pheromones of Ants: A Review. Physiol. Entomol. **4**, 161–189 (1979).
64. ALVES, L.F.: Química de Lepidópteros. Quimica Nova 3, 6–29 (1980).
65. RENWICK, J.A.A., and J.P. VITÉ: Biology of Pheromones. In: R. WEGLER ed. Chemie der Pflanzenschutz- und Schädlingsbekämpfungsmittel, p. 1–28. Berlin: Springer, 1980.
66. BAKER, R., and J.W.S. BRADSHAW: Insect Pheromones and Related Behaviour Modifying Chemicals. Aliphat. Relat. Nat. Prod. Chem. **2**, 46–75 (1981).
67. BLUM, M.S.: Sex Pheromones in Social Insects. Syst. Assoc. Spec. **19**, 163–174 (1981).
68. – Chemical Defense of Arthropods. New York: Academic Press. 1981.
69. BESTMANN, H., and O. VOSTROWSKY: Insektenpheromone. Naturwissenschaften **69**, 457–471 (1982).
70. HOWARD, R.W., and G.J. BLOMQUIST: Chemical Ecology and Biochemistry of Insect Hydrocarbons. Ann. Rev. Entomol. **27**, 149–172 (1982).
71. ROELOFS, W.L., and R.L. BROWN: Pheromones and Evolutionary Relationship of Tortricidae. Ann. Rev. Ecol. Syst. **13**, 395–422 (1982).
72. SILVERSTEIN, R.M.: Chemical Communication in Insects. Pure Appl. Chem. **54**, 2479–2488 (1982).
73. BAKER, R., and J.W.S. BRADSHAW: Insect Pheromones and Related Natural Products. Aliphat. Relat. Nat. Prod. Chem. **3**, 66–106 (1983).
74. GOTHE, R.: Pheromones in Ixodid and Argasid Ticks. Part I. Ixodid Ticks. Vet. Med. Rev. **99**, 16–37 (1983).
75. VAN DER MEER, R.K.: Semiochemicals and the Red Imported Fire Ant (*Solenopsis invicta* Buren) (Hymenoptera: Formicidae). Fla. Entomol. **66**, 139–161 (1983).
76. BAKER, R., and H.R. HERBERT: Insect Pheromones and Related Natural Products. Nat. Prod. Re. **1**, 299–318 (1984).
77. BELL, W.J., and R.T. CARDÉ (eds.): Chemical Ecology of Insects. Sunderland: Sinauer Assoc. 1984.
78. LEWIS, T. (ed.): Insect Communication. New York: Academic Press. 1984.
79. ROELOFS, W., and L. BJOSTAD: Biosynthesis of Lepidopteran Pheromones. Bioorgan. Chem. **12**, 279–298 (1984).
80. SCHNEIDER, D.: Pheromone and Insect Reproduction. Adv. Invert. Reprod. **3**, 435–439 (1984).
81. ATTYGALE, A.B., and E.D. MORGAN: Ant Trail Pheromones. Adv. Insect Physiol. **18**, 1–30 (1985).
82. BLUM, M.S.: Exocrine System. In: M.S. BLUM ed. Fundamentals of Insect Physiology, p. 535–579. New York: John Wiley & Sons. 1985.
83. JONES, O.T.: Chemical Mediation of Insect Behavior. Prog. Pest. Biochem. Toxicol. **5**, 311–373 (1985).

84. SONENSHINE, D.E.: Pheromones and other Semiochemicals of the Acari. Ann. Rev. Entomol. **30**, 1–28 (1985).
85. O'CONNEL, R.J.: Chemical Communication in Invertebrates. Experientia **42**, 232–241 (1986).
86. PATRIDGE, B.L., N.R. LILEY, and N.E. STACEY: The Role of Pheromones in the Sexual Behavior of Goldfish. Anim. Behav. **24**, 291–299 (1976).
87. SCHEUER, P. (ed.): Marine Natural Products: Chemical and Biological Perspective. vol. 5. New York: Academic Press. 1983.
88. – Chemical Communication of Marine Invertebrates. Bioscience **27**, 664–668 (1977).
89. UHAZI, L.S., R.D. TANAKA, and A.J. MACINNIS: *Schistosoma mansoni*: Identification of Chemicals that Attract or Trap its Snail Vector, *Biomphalaria glabrata*. Science **201**, 924–926 (1978).
90. COLOMBO, L., P.C. BELVEDERE, A. MARCONATO, and F. BENTIVEGNA: Pheromone in Teleost Fish. In: C.J.J. RICHTER and H.J.T. GOSS eds. Reproductive Physiology of Fish, p. 84–94. Wagenigen: 1982.
91. LILEY, N.R.: Chemical Communication in Fish. Can. J. Fish. Aquat. Sci. **39**, 22–35 (1982).
92. PFEIFFER, W.: Chemical Signals in Communication. In: T.J. HARA ed. Chemoreception in Fish, p. 307–326. Amsterdam: Elsevier. 1982.
93. SCHEUER, P.: Marine Ecology: Some Chemical Aspects. Naturwissenschaften **69**, 528–533 (1982).
94. LILEY, N.R., and N.E. STACEY: Hormones, Pheromones, and Reproductive Behavior in Fish. In: W.S. HOAR; D.J. HANDAL and E.M. DONALDSON eds. Fish Physiology, vol. IXB, p. 1–63. New York: Academic Press. 1983.
95. STACEY, N.: Hormones and Pheromones in Fish Sexual Behavior. Bioscience **33**, 552–556 (1983).
96. PANDEY, A.K.: Chemical Signals in Fish. Theory and Application. Acta Hydrochem. Hydrobiol. **12**, 463–478 (1984).
97. BONE, L.W.: Activation of Male *Nipponstrogylus brasiliensis* by Female Pheromone. Proc. Helminthol. Soc. Wash. **47**, 228–234 (1980).
98. BOILLY-MARER, Y.: Les Phéromones Sexuelles chez les Annélides. Bull. Soc. Zool. France **107**, 619–624 (1982).
99. BONE, L.W.: Reproductive Chemical Communication of Helminths. I. Platyhelminths. Int. J. Invert. Reprod. **5**, 261–268 (1982).
100. – Reproductive Chemical Communication of Helminths. II. Aschelminths. Int. J. Invert. Reprod. **5**, 311–321 (1982).
101. BROSSUT, R.: La Communication Chimique chez les Invertébrés. Bull. Soc. Zool. France **107**, 607–618 (1982).
102. PERKINS, C., and B. FRIED: Intraspecific Pairing of Planaria, *Dugesia tigrina* and *D. dorotocephala* (Platyhelminths: Turbellaria), and Observations on Lipophilic Excretory-Secretory Worm Products. J. Chem. Ecol. **8**, 901–909 (1982).
103. BONE, L.W., and K.P. BOTTJER: Characterization of and Male Adaptation to Pheromone of Female *Trichostrongylus colubriformis* (Nematoda). J. Chem. Ecol. **10**, 1749–1758 (1984).
104. GOLDEN, J.W., and D.L. RIDDLE: A *Caenorhabditis elegans* Dauer-Inducing Pheromone and an Antagonist Component of the Food Supply. J. Chem. Ecol. **10**, 1265–1280 (1984).
105. ZUCKERMAN, B.N., and H.B. JANSSON: Nematode Chemotaxis and Possible Mechanisms of Host/Prey Recognition. Ann. Rev. Phytopatol. **22**, 95–113 (1984).
106. HUETTEL, R.N.: Chemical Communication in Nematodes. J. Nematol. **18**, 3–8 (1986).
107. LE MAGNEN, J.: Les phénomenes olfacto-sexuels chez l'homme. Arch. Sci. Physiol. **6**, 125–160 (1952).

108. WIENER, H.: External Chemical Messengers. I. Emission and Reception in Man. N.Y. State J. Med. **66**, 3153–3170 (1966).
109. – External Chemical Messengers. III. Mind and Body in Schizophrenia. N.Y. State J. Med. **67**, 1287–1310 (1967)).
110. AMOORE, J.E.: Evidence for the Chemical Olfactory Code in Man. Ann. N.Y. Acad. Sci. **237**, 137–143 (1974).
111. COMFORT, A.: The Likelihood of Human Pheromones. In: M.C. BIRCH ed. Pheromones, p. 386–396. Amsterdam: North Holland/Elsevier. 1974.
112. KOELEGA, H.S., and E.P. KOSTER: Some Experiments on Sex Differences in Odor Perception. Ann. N.Y. Acad. Sci. **237**, 234–246 (1974).
113. WALLACE, W.: Individual Discrimination of Human by Odor. Physiol. Behav. **19**, 577–579 (1977).
114. BONSALL, R.W., and R.P. MICHAEL: Volatile Odoriferous Acids in Vaginal Fluid. In: E.S.E. HAFEZ and T.N. EVANS eds. The Human Vagina, p. 167–177. Amsterdam: North Holland/Elsevier. 1978.
115. DOTY, R.L.: Olfactory Communication in Humans. Chem. Senses **6**, 351–376 (1981).
116. RUSSELL, M.J.: Human Olfactory Communication. In: D. MÜLLER-SCHWARTZE and R.M. SILVERSTEIN eds. Chemical Signals in Vertebrates, p. 259–273. New York: Plenum Press. 1983.
117. MYKYTOWYCZ, R.: Olfaction-A Link with the Past. J. Human Evol. **14**, 75–90 (1985).
118. DOTY, R.L.: Gender and Endocrine-Related Influences on Human Olfactory Perception. In: H.L. MEISELMAN and R.S. RIVLIN eds. Clinical Measurements of Taste and Smell, p. 377–413. New York: MacMillan Publishing Co. 1986.
119. ROPARTZ, PH.: Chemical Signals in Agonistic and Social Behavior of Rodents. In: D. MÜLLER-SCHWARTZE and M.M. MOZELL eds. Chemical Signals in Vertebrates, p. 169–184. New York: Plenum Press. 1977.
120. GARRY, N.E.: Pheromones that Affect the Behavior and Physiology of Honeybees. In: M.C. BIRCH ed. Pheromones, p. 200–221. Amsterdam: North Holland/Elsevier. 1974.
121. KARLSON, P., and M. LÜSCHER: "Pheromones" a New Term for a Class of Biologically Active Substances. Nature **183**, 55–56 (1959).
122. NORDLUNG, D.A.: Semiochemicals: A Review of the Terminology. In: D.A. NORDLUNG, R.L. JONES and W.J. LEWIS eds. Semiochemicals: Their Role in Pest Control, p. 13–28. New York: John Wiley & Sons. 1981.
123. – and W.J. LEWIS: Terminology of Chemical Releasing Stimuli in Intraspecific and Interspecific Interactions. J. Chem. Ecol. **2**, 211–220 (1976).
124. MOORE, R.E.: Chemotaxis and the Odor of Seaweeds. Lloydia **39**, 181–191 (1976).
125. – Toxins from Blue-Green Algae. Bioscience **27**, 797–802 (1977).
126. – Volatile Compounds from Marine Algae. Acc. Chem. Res. **10**, 40–47 (1977).
127. DARDEN, W.H.: Some Properties of Male-Inducing Pheromones from *Volvox aureus*. Microbios **28**, 27–39 (1980).
128. JAENICKE, L.: Chemical Signals in the Sexual Life of Tallophytes. J. Sci. Ind. Res. **39**, 819–825 (1980).
129. KAMIYA, Y., and A. SAKURAI: Mating Pheromones of Heterobasidiomycetous Yeasts. Naturwissenschaften **68**, 128–133 (1981).
130. POMMERVILLE, J.: The Role of Sexual Pheromones in Allomyces. In: D.H. O'DAY and P.A. HORGEN eds. Sexual Interactions in Eukaryotic Microbes, p. 53–72. 1981.
131. GILLES, R., C. GILLES, and L. JAENICKE: Pheromone-Binding and Matrix-Mediated Events in Sexual Induction of *Volvox carteri*. Z. Naturforsch. **39 C**, 584–592 (1984).
132. RICE, E.: Allelopathy. New York: Academic Press. 1984.
132 a. ALONZO, T. (ed.): The Chemistry of Allelopathy: Biochemical Interactions Among Plants. American Chemical Society Symposium Series n° 268. Washington, D.C. 1985.

133. WHITTAKER, R.H., and P.P. FEENY: Allelochemics: Chemical Interactions between Species. Science **171**, 757–770 (1971).

134. BROWN, W.L., JR., T. EISNER, and R.H. WHITTAKER: Allomones and Kairomones: Transspecific Chemical Messengers. Bioscience **20**, 21–22 (1970).

135. BEAUCHAMP, G.K., R.L. DOTY, D.G. MOULTON, and R.A. MUGFORD: The Pheromone Concept in Mammalian Comunication: A Critique. In: R.L. DOTY ed. Mammalian Olfaction, Reproductive Processes and Behavior, p. 143–160. New York: Academic Press. 1976.

136. – – – – Response by Beauchamp *et al.* J. Chem. Ecol. **5**, 301–305 (1979).

137. BLUM, M.S.: Behavioral Responses of Hymenoptera to Pheromones, Allomones and Kairomones. In: H.H. SHOREY and J.J. McKELVEY, JR. eds. Chemical Control of Insect Behavior: Theory and Application, p. 149–167. New York: John Wiley & Sons. 1977.

138. KATZ, R.A., and H.H. SHOREY: In Defense of the Term Pheromone. J. Chem. Ecol. **5**, 299–301 (1979).

139. MARTIN, I.G.: Homeochemics: Intraspecific Chemical Signals. J. Chem. Ecol. **6**, 517–519 (1980).

140. PASTEELS, J.M.: Is Kairomone a Valid and Useful Term? J. Chem. Ecol. **8**, 1079–1081 (1982).

141. RUTOWSKI, R.L.: The Function of Pheromones. J. Chem. Ecol. **7**, 481–484 (1981).

142. SMITH, R.L.: Homeochemics? Please Reconsider. J. Chem. Ecol. **7**, 649 (1981).

143. WELDON, P.J.: In Defense of Kairomone as a Class of Chemical Releasing Stimuli. J. Chem. Ecol. **6**, 719–725 (1980).

144. WILSON, E.O.: Sociobiology: The New Synthesis. Cambridge: Harvard University Press. 1975.

145. PASTEELS, J.M.: Écomones: Message Chimiques des Écosystèmes. Ann. Soc. R. Zool. Belg. **103**, 103–117 (1973).

146. HASKINS, C.P., R.E. HEWITT, and E.F. HASKINS: Release of Aggressive Compounds and Capture Behavior in the Ant *Myrmecia gulosa* F. by Exocrine Products of the Ant *Camponotus.* J. Entomol. **47**, 125–139 (1970).

147. HUMMEL, H.E., and R.L. METCALF: Female Cabbage Looper Sex Pheromone Attracts Male *Scirtes orbiculatus.* Ann. Entomol. Soc. Amer. **74**, 339–340 (1981).

148. CORBET, S.A.: Mandibular Gland Secretion of the Larvae of the Flour Moth *Anagasta kuhniella*, Contains an Epideitic Pheromone and Elicits Movement in Hymenopteran Parasite. Nature **232**, 481–484 (1971).

149. STERNLICHT, M.: Parasitic Wasps Attracted by the Sex Pheromone of their Coccid Host. Entomophaga **18**, 339–342 (1973).

150. LEWIS, W.J., D.A. NORDLUNG, R.C. GUELDNER, P.E.A. TEAL, and J.H. TUMLINSON: Kairomones and their Use for Management of Entomophagous Insects. XIII. Kairomone Activity for *Trichogramma* sp. of Abdominal Tips. Excretion and Synthetic Sex Pheromone Blend of *Heliothis zea* (Boddie) Moth. J. Chem. Ecol. **8**, 1323–1331 (1982).

151. EBERHARD, W.G.: Aggressive Chemical Mimicry by a Bolas Spider. Science **198**, 1173–1175 (1977).

152. PAYNE, T.L., J.C. DICKENS, and J.V. RICHERSON: Insect Predator-Prey Coevolution via Enantiomeric Specificity in a Kairomone-Pheromone System. J. Chem. Ecol. **10**, 487–492 (1984).

153. THIESSEN, D.D.: Thermoenergetics and the Evolution of Pheromone Communication. Prog. Psychobiol. Physiol. Psychol. **7**, 91–191 (1977).

154. – The Thermoenergetics of Communication and Social Interactions Among Mongolian Gerbils. In: L. ROSENBLUM and H. MOLTZ eds. Symbiosis in Parent-Young Interactions, p. 113–144. New York: Plenum Press 1983.

155. – Thermal Constraints and Influences on Communication. In: J.S. ROSENBLATT; R.A. HINDE; C. DEER and M.C. BUSNEL eds. Advances in the Study of Behavior, vol. 13, p. 147–189. New York: Academic Press. 1983.

156. – A. HARRIMAN: Thermal and Osmolarity Properties of Pheromonal Communication in the Gerbil, *Meriones unguiculatus*. In: D. MÜLLER-SCHWARTZE and R.M. SILVERSTEIN eds Chemical Signals in Vertebrates, p. 291–308. New York: Plenum Press. 1983.

157. – T.M. BARTH: Ventral Scent Marking in *Meriones unguiculatus* May Contribute to Thermoregulation, J. Comp. Psychol. **99**, 306–310 (1985).

158. CULVENOR, C.C.J., and J.A. EDGAR: Dihydropyrrolizine Secretions Associated with Coremata of *Utetheisa* Moths (Fam. Arctiidae). Experientia **28**, 627–628 (1972).

159. EDGAR, J.A., and C.C.J. CULVENOR: Pyrrolizidine Ester Alkaloid in Danaid Butterflies. Nature **248**, 614–616 (1974).

160. – – Pyrrolizidine Alkaloids in *Parsonsia* Species (Family Apocynaceae) which Attract Danaid Butterflies. Experientia **31**, 393–394 (1975).

161. – Pyrrolizidine Alkaloids Sequestered by Solomon Islands Danaine Butterflies. The Feeding Preferences of the Danaine and Ithomiinae. J. Zool. (London) **196**, 385–399 (1982).

162. – C.C.J. CULVENOR, and G.S. ROBINSON: Hairpencil Dihydropyrrolizines of Danainae from New Hebrides. J. Aust. Entomol. Soc. **12**, 144–150 (1973).

163. HENDRY, L.B.: Insect Pheromones: Diet Related? Science **192**, 143–145 (1976).

164. – J.K. WICHMANN, D.M. HINDENLANG, R.O. MUMMA, and M.E. ANDERSON: Evidence for Origin of Insect Sex Pheromones Presence in Food Plants. Science **188**, 59–63 (1975).

165. – – – K.M. WEAVER, and S.H. KORZENIOWSKI: Plants – The Origin of Kairomones Utilized by Parasitoids of Phytophagous Insects. J. Chem. Ecol. **2**, 271–283 (1976).

166. – J.G. KOSTELC; D.M. HINDENLANG, J.K. WICHMANN, C.J. FIX, and S.H. KORZENIOWSKI: Chemical Messengers in Insects and Plants. In: J.W. WALLACE and R.L. MANSELL eds., Biochemical Interactions Between Plants and Insects (Recent Advances in Phytochemistry vol. 10), p. 351–384. New York: Plenum Press. 1976.

167. MILLER, J.R., T.G. BAKER, R.T. CARDÉ, and W.L. ROELOFS: Reinvestigation of Oak Leaf Roller Sex-Pheromone Components and the Hypothesis that they Vary with Diet. Science **192**, 140–143 (1976).

168. HINDENLANG, D.M., and J.K. WICHMANN: Reexamination of Tetradecenyl Acetates in Oak Leaf Roller Sex Pheromone in Plants. Science **195**, 86–89 (1977).

169. SCHNEIDER, D., M. BOPPRÉ, J. ZWEIG, S.B. HORSLEY, T.W. BELL, J. MEINWALD, K. HANSEN, and E.W. DIEHL: Scent Organ Development in *Creatonotos* Moths: Regulation by Pyrrolizidine Alkaloids. Science **215**, 1264–1265 (1982).

170. FERGUSON, J.E., and R.L. METCALF: Cucurbitacins. Plant Derived Defense Compounds for Diabrotictes (Coleoptera: Chrysomelidae). J. Chem. Ecol. **11**, 311–318 (1985).

171. PEARCE, G.T., W.E. GORE, R.M. SILVERSTEIN, J.W. PEACOCK, R.A. CUTHBERT, G.N. LANIER, and J.W. SIMEONE: Chemical Attractants for the Smaller European Elm Bark Beetle *Scolytus multistriatus* (Coleoptera: Scolytidae). J. Chem. Ecol. **1**, 115–124 (1975).

172. BRAND, J.M., J.W. BRACKE, A. MARKOVETZ, D.L. WOOD, and L.E. BROWNE: Production of Verbenol Pheromone by a Bacterium Isolated from Bark Beetles. Nature **254**, 136–137 (1975).

173. BRAND, J.M., J.W. BRACKE, L.N. BRITTON, A.J. MARKOVETZ, and S.J. BARRAS: Bark Beetle Pheromones: Production of Verbenone by a Mycangial Fungus of *Dendroctonus frontalis*. J. Chem. Ecol. **2**, 195–199 (1976).

174. BRAND, J.M., J. SCHULTZ, S.J. BARRAS, L.J. EDSON, T.L. PAYNE, and R.L. HEDDEN: Bark Beetle Pheromones. Enhancement of *Dendroctonus frontalis* (Coleoptera: Scolytidae). Aggregation Pheromone by Yeast Metabolites in Laboratory Bioassays. J. Chem. Ecol. **3**, 657–666 (1977).

175. GUELDNER, R.C., and G. WIYGUL: Rhythms in Pheromone Production of the Male Boll Weevil. Science **199**, 984–986 (1978).

176. WIYGUL, G., and J.E. WRIGHT: Sex Pheromone Production in Male Boll Weevils, Fed Cotton Squares and Laboratory Diet. Entomol. Exp. Appl. **34**, 333–335 (1983).

177. WIYGUL, G., and P.P. SIKOROWSKI: The Effect of Staphylococcal Enterotoxin B on Pheromone Production in Fat Bodies Isolated from Male Boll Weevils. J. Invertebr. Pathol. **47**, 116–119 (1986).

178. APPEL, A.G., and M.K. RUST: Temperature-Mediated Sex Pheromone Production and Response of the American Cockroach. J. Insect Physiol. **29**, 301–305 (1983).

179. WEBSTER, R.P., and R.T. CARDÉ: Relationship among Pheromone Titre, Calling and Age in the Omnivorous Leafroller Moth (*Platynota stultana*). J. Insect Physiol. **28**, 925–933 (1982).

180. ADAMS, T.S., J.W. DILLWITH, and G.J. BLOMQUIST: The Role of 20-Hydroxyecdysone in House Fly Sex Pheromone Biosynthesis. J. Insect Physiol. **30**, 287–294 (1984).

181. HEDIN, P.A., O.H. LINDIG, and G. WIYGUL: Enhancement of Boll Weevil *Anthonomus grandis* Boh. (Coleoptera: Curculionidae) Pheromone Biosynthesis with JH III. Experientia **38**, 375–376 (1982).

182. BRIDGES, J.R.: Effects of Juvenile Hormone on Pheromone Synthesis in *Dendroctonus frontalis*. Environ. Entomol. **11**, 417–420 (1982).

183. WEBSTER, R.P., and R.T. CARDÉ: The Effects of Mating, Exogenous Juvenile Hormone and a Juvenile Hormone Analogue on Pheromone Titre, Calling and Oviposition in the Omnivorous Leafroller Moth. (*Platynota stultana*). J. Insect Physiol. **30**, 113–118 (1984).

184. PIERCE, A.M., H.D. PIERCE, JR., J.H. BORDEN, and A.C. OEHLSCHLAGER: Enhanced Production of Aggregation Pheromones in Four Stored-Product Coleopterans Feeding on Methoprene-Treated Oats. Experientia **42**, 164–165 (1986).

185. ROBINSON, G.E.: Effects of a Juvenile Hormone Analogue on Honey Bee Foraging Behaviour and Alarm Pheromone Production. J. Insect Physiol. **31**, 277–282 (1985).

186. HUGHES, P.R., and J.A.A. RENWICK: Neural and Hormonal Control of Pheromone Biosynthesis in the Bark Beetle, *Ips paraconfusus*. Physiol. Entomol. **2**, 117–123 (1977).

187. HEDIN, P.A.: A Study of Factors that Control Biosynthesis of the Compounds which Comprise the Boll Weevil Pheromone. J. Chem. Ecol. **3**, 279–289 (1977).

188. WIYGUL, G., M.W. MACGOWN, P.P. SIKOROWSKI, and J.E. WRIGHT: Localization of Pheromone in Male Boll Weevil (*Anthonomus grandis*). Ent. Exp. Appl. **31**, 330–331 (1982).

189. WIYGUL, G., and P.P. SIKOROWSKI: The Effect of Glucose and ATP on Sex Pheromone Production in Fat Bodies from Male Boll Weevils *Anthonomus grandis* Boheman (Coleoptera: Curculionidae). Comp. Biochem. Physiol. **81 B**: 1073–1075 (1985).

190. ALBONE, E.S., and G.C. PERRY: Anal Sac Secretion of the Red Fox *Vulpes vulpes*: Volatile Fatty Acids and Diamines: Implication for Fermentation Hypothesis of Chemical Recognition. J. Chem. Ecol. **2**, 101–111 (1976).

191. – P.E. GOSDEN, and G.C. WARE: Bacteria as a Source of Chemical Signals in Mammals. In: D. MÜLLER-SCHWARTZE and M.M. MOZELL eds., Chemical Signals in Vertebrates, p. 35–43. New York: Plenum Press. 1976.

192. – G. ELINGTON: The Anal Sac Secretion of the Red Fox (*Vulpes vulpes*): its Chemistry and Microbiology. A Comparison with the Anal Sac Secretion of the Lion (*Panthera leo*). Life Sci. **14**, 387–400 (1974).

193. MÜLLER-SCHWARTZE, D.: Pheromone in Black-Tailed Deer (*Odocoileus hemionus columbianus*). Anim. Behav. **19**, 141–152 (1971).

194. RAYMER, J., D. WIELER, M. NOVOTNY, C. ASA, U.S. SEAL, and L.D. MECH: Chemical Investigation of Wolf (*Canis lupus*) Anal Sac Secretion in Relation to Breeding Season. J. Chem. Ecol. **11**, 593–608 (1985).

195. MORSE, D., and E. MEIGHEN: Detection of Pheromone Biosynthetic and Degradative Enzymes *in Vitro*. J. Biol. Chem. **259**, 475–480 (1984).

196. WRIGHT, R.H.: After Pesticides – What? Nature **204**, 121–125 (1964).

197. SILVERSTEIN, R.M. and J.C. YOUNG: Insect Generally Use Multicomponent Pheromones. In: M. BEROZA ed. Pest Management with Insect Sex Attractants. American Chemical Society Symposium Series n° 23, p. 1–29. Washington, D.C. American Chemical Society. 1976.

198. FRANCKE, W., W. REITH, G. BERGSTRÖM, and J. TENGÖ: Pheromone Bouquet of the Mandibular Glands in *Andrena haemorrhoa* F. (Hym., Apoidea). Z. Naturforsch. **36 C**, 928–932 (1981).

199. BERGSTRÖM, G., J. TENGÖ, W. REITH, and W. FRANCKE: Multicomponent Mandibular Gland Secretions in Three Species of *Andrena* Bees. (Hym, Apoidea). Z. Naturforsch. **37 C**, 1124–1129 (1982).

200. DUFFIELD, R.M., W.E. LABERGE, and J.W. WHEELER: Exocrine Secretions of Bees. VII. Aliphatic Esters in the Dufour's Gland Secretion of *Svastra obliqua obliqua* (Hymenoptera: Anthophoridae). Comp. Biochem. Physiol. **78 B**, 47–50 (1984).

201. BRADSHAW, J.W.S., R. BAKER, and P.E. HOWSE: Multicomponent Alarm Pheromones of the Weaver Ant. Nature **258**, 230–231 (1975).

202. – – – Multicomponent Alarm Pheromones in the Mandibular Glands of Major Workers of the African Weaver Ant, *Oecophylla longinoda*. Physiol. Entomol. **4**, 15–25 (1979).

203. FRANCKE, W., W. MACKENROTH, W. SCHRODER, S. SCHULTZ, J. TENGÖ, E. ENGLES, W. ENGELS, R. KITTMANN, and D. SCHNEIDER: Identification of Cyclic Enol Ethers from Insects: Alkyldihydropyranes from Bees and Alkyldihydro-4H-Pyran-4-ones from a Male Moth. Z. Naturforsch. **40 C**, 145–147 (1985).

204. TENGÖ, J., I. GROTH, G. BERGSTRÖM, W. SCHRODER, S. KROHN, and W. FRANCKE: Volatile Secretions in Three Species of *Dufourea* (Hymenoptera: Halictidae) Bees: Chemical Composition and Phylogeny. Z. Naturforsch. **40 C**, 657–660 (1985).

205. KAIB, M., O. BRUISMA, and R.H. LEUTHOLD: Trail Following in Termites: Evidence for a Multicomponent System. J. Chem. Ecol. **8**, 1193–1205 (1982).

206. BARTELT, R.J., R.L. JONES, and H.M. KULMAN: Evidence for a Multicomponent Sex Pheromone in the Yellow Headed Spruce Sawfly. J. Chem. Ecol. **8**, 83–94 (1982).

207. – – – Hydrocarbon Components of the Yellow Headed Spruce Sawfly Sex Pheromone. A Series of (Z,Z)-9,19-Dienes. J. Chem. Ecol. **8**, 95–114 (1982).

208. BERGSTRÖM, G.: Complexity of Volatile Signals in Hymenoptera Insects. Some Central Problems Regarding Analytical Technique and Biological Interpretations in Work with Multicomponent Secretions in Bees, Bumblebees and Ants. In: F.J. RITTER ed. Chemical Ecology: Odour Communication in Animals, p. 187–200. Amsterdam: Elsevier/North Holland, 1979.

209. HALDANE, J.B.S.: Animal Communication and the Origin of Human Language; Sci. Prog. **43**, 385–401 (1955).

210. WILSON, E.O.: Chemical Communication within Animal Species. In: E. SONDHEIMER and J.B. SIMEONE eds. Chemical Ecology, p. 135–155. New York: Academic Press. 1970

211. PEDERSEN, P.E., W.S. STEWART, C.A. GREER, and G.M. SHEPHERD: Evidence for Olfactory Function in Utero. Science **221**, 478–480 (1983).

212. FRIEDMAN, L., and J.G. MILLER: Odor Incongruity and Chirality. Science **172**, 1044–1046 (1971).

213. BRONSON, F.H.: Pheromonal Influences on Reproductive Activities in Rodents. In: M.C. BIRCH ed. Pheromones, p. 344–365. Amsterdam: North Holland/Elsevier. 1974.

66 L.F. ALVES:

214. PAYNE, T.L.: Pheromone Perception. In: M.C. BIRCH ed. Pheromones, p. 35–61. Amsterdam: North Holland/Elsevier. 1974.
215. STODDART, D.M.: The Role of Odor in the Social Biology of Small Rodents. In: M.C. BIRCH ed. Pheromones, p. 297–315. Amsterdam: North Holland/Elsevier. 1974.
216. ROELOFS, W.L., and R.T. CARDÉ: Responses of Lepidoptera to Synthetic Sex Pheromone Chemicals and their Analogues. Ann. Rev. Entomol. 22, 377–405 (1977).
217. SCHOONHOVEN, L.M.: Insect Chemosensory Responses to Plant and Animal Hosts. In: H.H. SHOREY and J.J. McKELVEY, JR. eds. Chemical Control of Insect Behavior: Theory and Application, p. 7–14. New York: Wiley Interscience Publication. 1977.
218. SEABROOK, W.D.: Insect Chemosensory Responses to Other Insects. In: H.H. SHOREY and J.J. McKELVEY JR. eds. Chemical Control of Insect Behavior: Theory and Application, p. 15–43. New York: Wiley Interscience Publication. 1977.
219. CHAPMAN, O.L., J.A. KLUN, K.C. MATTES, R.S. SHERIDAN, and S. MAINI: Chemoreceptors in Lepidoptera: Stereochemical Differentiation by Dual Receptors and Achiral Pheromone. Science 201, 926–928 (1978).
220. ROELOFS, W.: Threshold Hypothesis for Pheromone Perception. J. Chem. Ecol. 4, 685–699 (1978).
221. SEABROOK, W.D.: Neurobiological Contributions to Understanding Insect Pheromone Systems. Ann. Rev. Entomol. 23, 471–485 (1978).
222. CHARTON, M., and B.I. CHARTON: Significance of "Volume" and "Bulk" Parameters in Quantitative Structure-Activity Relationships. J. Org. Chem. 44, 2284–2288 (1979).
223. KOSTELC, J.G., B.J. GARCIA, G.W. GOKEL, and L.B. HENDRY: Macrocyclic Polyethers as Probes into Pheromone Receptor Mechanisms of a Sciarid Fly, Lycoriella mali. Fitch. J. Chem. Ecol. 5, 179–185 (1979).
224. PRIESNER, E.: Progress in the Analysis of Pheromone Receptor Systems Ann. Zool. Ecol. Anim. 11, 533–546 (1979).
225. OHLOFF, G., C. VITAL, H.R. WOLF, K. JOB, E. JÉGOU, J. POLONSKY, and E. LEDERER: Stereochemistry-Odor Relationship in Enantiomeric Ambergris Fragrances. Helv. Chim. Acta 63, 1932–1946 (1980).
226. STODDART, D.M.: Aspects of the Evolutionary Biology of Mammalian Olfaction. Symp. Zool. Soc. London 45, 1–13 (1980).
227. – Vertebrate Olfaction. Endeavour (New Series) 5, 9–14 (1980).
228. VOGT, R.G., and L.M. RIDDIFORD: Pheromone Binding and Inactivation by Moth Antennae. Nature 293, 161–163 (1981).
229. – – Pheromone Deactivation by Antennal Protein of Lepidoptera. Pr. Nauk. Inst. Chem. Org. Fiz. Politech. Wroclaw (22) 955–967 (1981).
230. NISHINO, C., S. MANABE, and H. TAKAYNAGI: Chiral Influences of Sex Pheromonal Substances on Responses of the Male American Cockroach. Experientia 40, 1137–1140 (1984).
231. BARTELL, R.J.: Pheromone-Mediated Behavior of Male Lightbrown Apple Moth, Epiphyas postvittana, Correlated with Adaptation of Pheromone Receptors, Physiol. Entomol. 10, 121–126 (1985).
232. KANAUJIA, S., and K.E. KAISSLING: Interactions of Pheromone with Moth Antennae: Adsorption, Desorption and Transport. J. Insect Physiol. 31, 71–81 (1985).
233. PACE, U., E. HANSKI, Y. SALOMON, and D. LANCET: Odorant-Sensitivity Adenylate Cyclase May Mediate Olfactory Reception. Nature 316, 255–258 (1985).
234. VOGT, R.G., L.M. RIDDIFORD, and G.D. PRESTWICH: Kinetic Properties of a Sex Pheromone-Degrading Enzyme: The Sensillar Esterase of Antheraea polyphemus. Proc. Nat. Acad. Sci. USA 82, 8827–8831 (1985).
235. MORA, O.A., J.E. SANCHEZ-CRIADO, and S. GUISADO: Role of Vomeronasal Organ on the Estrus Cycle Reduction by Pheromones in the Rat. Rev. Esp. Fisiol. 41, 305–310 (1985).

236. O'CONNELL, R.J.: Responses to Pheromone Blends in Insect Olfactory Receptor Neurons. J. Comp. Physiol. 156 A, 747–761 (1985).
237. MANABE, S., and C. NISHINO: A Theoretical Interpretation for M/F Ratio Index in Electroantennogram of the American Cockroach. Comp. Biochem. Physiol. 83 A, 341–346 (1986).
238. OHLOFF, G.: Chemistry of Odor Stimuli. Experientia 42, 271–279 (1986).
239. BLANEY, W.M., L.M. SCHOONHOVEN, and M.S.J. SIMMONDS: Sensitivity Variations in Insect Chemoreceptors. Experientia 42, 13–19 (1986).
240. TUMLINSON, J.H., R.M. SILVERSTEIN, J.C. MOSER, R.G. BROWNLEE, and J.M. RUTH: Identification of the Trail Pheromone of Leaf-Cutting Ant *Atta texana*. Nature 234, 348–349 (1971).
241. MATSUMURA, F., H.C. COPPEL, and A. TAI: Isolation and Identification of Termite Trail-Following Pheromone. Nature 219, 963–964 (1968).
242. PRIESNER, E.: Receptor of Di-Unsaturated Pheromone Analogues in the Male Summer Fruit Tortrix Moth. Z. Naturforsch. 38 C, 874–877 (1983).
243. – The Pheromone Receptor System of Male *Eulia ministrata* L. with Notes on Other Cnephasiid Moths. Z. Naturforsch. 39 C, 849–852 (1984).
244. GRANT, A.J., and R.J. O'CONNELL: Neurophysiological and Morphological Investigation of Pheromone-Sensitive Sensilla on the Antenna of Male *Trichoplusia ni*. J. Insect Physiol. 32, 503–515 (1986).
245. BRAHMACHARY, R.L., and J. DUTTA: Phenylethylamine as a Biochemical Marker in the Tiger. Z. Naturforsch. 34 C, 632–633 (1979).
246. – – On the Pheromone of Tigers. Experiments and Theory. Amer. Natu. 118, 561–567 (1981).
247. MACDONALD, D., K. KRANTZ, and R.T. APLIN: Behavioral, Anatomical and Chemical Aspects of Scent Marking Amongst Capybaras. (*Hydrochoerus hydrochaeris*) (Rodentia: Carviomorpha). J. Zool. 202, 341–360 (1984).
248. AYORINDE, F., J.W. WHEELER, C. WEMMER, and J. MURTAUGH: Volatile Components of the Occipital Gland Secretion of the Bactrian Camel (*Camelus bactrianus*). J. Chem. Ecol. 8, 177–183 (1982).
249. WHEELER, J.W., L.E. RASMUSSEN, F. AYORINDE, I.O. BUSS, and G.L. SMITH: Constituents of Temporal Gland Secretion of the African Elephant *Loxodonta africana*. J. Chem. Ecol. 8, 821–835 (1982).
250. CLARK, A.B.: Scent Markers as Social Signals in *Galago crassicaudatus*. I. Sex and Reproductive Status as Factors in Signals and Responses J. Chem. Ecol. 8, 1133–1151 (1982).
251. – Scent Markers as Social Signals in *Galago crassicaudatus*. II. Discrimination between Individuals by Scent. J. Chem. Ecol. 8, 1153–1165 (1982).
252. BURGER, B.V., M. LE ROUX, H.S.C. SPIES, V. TRUTER, R.C. BIGALKE, and P.A. NOVELLIER: Mammalian Pheromone Studies, V. Compounds from the Preorbital Gland of the Grysbok, *Raphicerus melanotis*. Z. Naturforsch. 36, 344–346 (1981).
253. GARSTKA, W.R. and D. CREWS: Female Sex Pheromone in the Skin and Circulation of a Greater Snake. Science 214, 681–683 (1981).
254. MÜLLER-SCHWARTZE, D., and C. MÜLLER-SCHWARTZE: Subspecies Specificity of Response to a Mammalian Social Odor. J. Chem. Ecol. 1, 125–131 (1975).
255. MÜLLER-SCHWARTZE, D., L. KÄLLQUIST, T. MOSSING, A. BRUNDIN, and G. ANDERSSON: Responses of Reindeer to Interdigital Secretions of Conspecifics. J. Chem. Ecol. 4, 325–335 (1978).
256. WELLINGTON, J., G.K. BEAUCHAMP, and C. WOJCIECHOWSKI-METZLER: Stability of Chemical Communication in Urine. Individual Identification and Age of Samples. J. Chem. Ecol. 9, 235–245 (1983).
257. HESTERMAN, E.R., and MYKYTOWYCZ, R.: Misidentification by Wild Rabbits *Oryctolagus cuniculus*, of Group Members Carrying the Odor of Foreign Inguinal Gland Secretion. I. Experiments with All-Male Groups. J. Chem. Ecol. 8, 419–427 (1982).

258. COOPER, W., and L.J. VITT: Interspecific Odour Discrimination by a Lizard (*Eumeces laticeps*). Anim. Behav. **34**, 367–376 (1986).
259. HARRINGTON, J.E.: Discrimination between Individuals by Scent in *Lemur fulvus*. Anim. Behav. **24**, 207–212 (1976).
260. – Discrimination between Males and Females by Scent in *Lemur fulvus*. Anim. Behav. **25**, 147–151 (1977).
261. MYRBERG, A.A.: The Role of Chemical and Visual Stimuli in the Preferential Discrimination of Young by the Cichlid Fish *Cichlasoma nigrofasciatum* (Gunther). Z. Tierpsychol. **37**, 274–297 (1975).
262. BARNETT, C.: Chemical Recognition of the Mother by the Young of the Cichlid Fish *Cichlasoma citrinellum*. J. Chem. Ecol. **3**, 461–466 (1977).
263. ITAGAKI, H., and J.H. THORP: Laboratory Experiments to Determine if Crayfish Can Communicate Chemically in a Flow-Through System. J. Chem. Ecol. **7**, 115–126 (1981).
264. ROSE, R.D.: On the Nature of Chemical Communication by Crayfish in Laboratory Controlled Flow-Through System. J. Chem. Ecol. **8**, 1065–1071 (1982).
265. THORP, J.H., and H. ITAGAKI: Verification versus Falsification of Existing Theory: Analysis of Possible Chemical Communication in Crayfish. J. Chem. Ecol. **8**, 1073–1077 (1982).
266. ROSE, R.D.: Experimental Design and Ecological Realism. J. Chem. Ecol. **10**, 1281–1292 (1984).
267. ROSE, R.D.: Chemical Communication in Crayfish: Physiological, Ecology Realism and Experimental Design. J. Chem. Ecol. **10**, 1289–1292 (1984).
268. THORP, J.H.: Theory and Practice in Crayfish Communication Studies. J. Chem. Ecol. **10**, 1283–1287 (1984).
268 a. HAZLETT, B.A.: Chemical Detection of Sex and Condition in the Crayfish *Orconectis virilis*. J. Chem. Ecol. **11**, 181–189 (1985).
269. FOX, G.: Potentials for Pheromones in Chimpanzee Vaginal Fatty Acids. Folia Primatol. **37**, 255–266 (1982).
270. BELCHER, A.M., A.B. SMITH III, P.C. JURS, B. LAVINE, and G. EPPLE: Analysis of Chemical Signals in a Primate Species (*Saguinus fuscicolis*): Use of Behavioral, Chemical and Pattern Recognition Methods. J. Chem. Ecol. **12**, 513–531 (1986).
271. GOODWIN, M., K.M. GOODWIN, and F. REGNIER: Sex Pheromone in the Dog. Science **203**, 559–561 (1979).
272. BONSALL, R.W., and R.P. MICHAEL: The Externalization of Vaginal Fatty Acids by the Female Rhesus Monkeys. J. Chem. Ecol. **6**, 499–509 (1980).
273. MICHAEL, R.P., and D. ZUMPE: Influence of Olfactory Signals on the Reproductive Behavior of Social Groups of Rhesus Monkeys (*Macaca mulata*). J. Endocr. **95**, 189–205 (1982).
274. STEVENS, K., G.C. PERRY, and S. LONG: Effect of Ewe Urine and Vaginal Secretions on Ram Investigative Behavior. J. Chem. Ecol. **8**, 23–29 (1982).
275. NISHIMURA, K., T. OKANO, K. UTSUMI, and M. YUHARA: Partial Separation of Mounting Inducing Pheromones from Vaginal Mucus of Holstein Heifer. Congr. Proc. Int. Congr. Anim. Reprod. Artif. Insemnin. 10th Paper n° 291, 1984 (Chemical Abstracts **103**: 154.772k).
276. STEEL, E.: Odour Recognition by Male Hamster: Discrimination of the Hormonal State of Females by Odour from Vaginal Secretions. J. Endocr. **105**, 255–262 (1985).
277. CRUMP, D., A.A. SWIGAR, J.R. WEST, R.M. SILVERSTEIN, D. MÜLLER-SCHWARTZE, and R. ALTIERI: Urine Fractions that Release Flehmen in Black-Tailed Deer, *Odocoileus hemionus columbianus*. J. Chem. Ecol. **10**, 203–215 (1984).
278. JOHNSTON, R.E., and F. BRONSON: Endocrine Control of Female Mouse Odors that Elicit Luteinizing Hormone Surge and Attraction in Males. Biol. Reprod. **27**, 1174–1180 (1982).

279. ONODA, N., T. ARIKI, K. IMAMURA, and M. IINO: Neocortical Response to Odors of Sex Steroid Hormones in the Dog. Proc. Japan Acad. **58 B**, 222–225 (1982).

280. TAYLOR, G.T., D. REGAN, and J. HALLER: Sexual Experience, Androgens and Female Choice of a Mate in Laboratory Rats: J. Endocr. **96**, 43–52 (1983).

281. STRALENDORFF, F.V.: Urinary Signaling Pheromone and Specific Behavioral Response in Tree Shrews (*Tupaia belangeri*). J. Chem. Ecol. **12**, 99–106 (1986).

282. JORGENSON, J.W., M. NOVOTNY, M. CARMACK, G.B. COPLAND, S.R. WILSON, S. KATONA and W. WHITTEN: Chemical Scent Constituients in the Urine of the Red Fox (*Vulpes vulpes* L.) During the Winter Season. Science **199**, 796–798 (1978).

283. TAKEYOSHI, S.: Sex Odor Components in Male Goat. Estrus Goats Are Interersted in 4-Ethyl Fatty Acids Secreted by Mature Male Goats. Kagaku to Seibutsu **21**, 428–430 (1983) (in Japanese Chemical Abstracts **100**:18.366s).

284. SASADA, H., T. SIGYAMA, K. YAMASHITA, and J. MASAKI: Identification of Specific Odor Components in Mature Male Goats During the Breeding Season. J. Nippon Chikusan Gakkai Ho **54**, 401–408 (1983). (in Japanese Chemical Abstracts **99**:188.519x).

285. RAYMER, J., D. WIELER, M. NOVOTNY, C. ASA, U.S. SEAL, and L.D. MECH: Volatile Constituents of Wolf (*Canis lupus*) Urine as Related to Gender and Season. Experientia **40**, 707–709 (1984).

286. EISENBERG, J.F., and D.G. KLEIMAN: Olfactory Communication in Mammals. Ann. Rev. Ecol. Syst. **3**, 1–32 (1972).

287. MÜLLER-SCHWARTZE, D., L. MOREHOUSE, R. CORRADI, C. ZHAO, and R.M. SILVERSTEIN: Odor Images: Responses of Beaver to Castoreum Fractions. In: D. DUVALL, D. MÜLLER-SCHWARTZE and R.M. SILVERSTEIN eds. Chemical Signals in Vertebrates vol IV: Ecology, Evolution and Comparative Biology. New York: Plenum Press (in press).

288. SINGER, A.G., W. AGOSTA, R.J. O'CONNELL, C. PFAFFMAN, D.V. BOVEN, and P.H. FIELD: Dimethyldisulfide: An Attractant Pheromone in Hamster Vaginal Secretion. Science **191**, 948–950 (1976).

289. O'CONNELL, R.J., A.G. SINGER, C. PFAFFMAN, and W.C. AGOSTA: Pheromones of Hamster Vaginal Discharges. Attraction to Fentogram Amounts of Dimethyldisulfide and to Mixture of Volatile Components. J. Chem. Ecol. **5**, 575–585 (1979).

290. SINGER, A.G., R.J. O'CONNELL, F. MACRIDES, A.F. BENCSATH, and W. AGOSTA: Methylthiolbutyrate: A Reliable Correlate of Estrus in the Golden Hamster. Physiol. Behav. **30**, 139–143 (1983).

291. MICHAEL, R.P., R.W. BONSALL, and M. KUTNER: Volatile Fatty Acids, "Copulins", in Human Vaginal Secretions. Psychoendocrinol. **1**, 153 163 (1975).

292. – – Chemical Signals and Primate Behavior. In: D. MÜLLER-SCHWARTZE and M.M. MOZELL eds. Chemical Signals in Vertebrates, p. 251–271. New York: Plenum Press. 1977.

293. NOVOTNY, M., F.J. SCHWENDE, D. WIESLER, J.W. JORGENSON, and M. CARMACK: Identification of a Testosterone-Dependent Unique Volatile Constituent of Male Mouse Urine: 7-*exo*-Ethyl-5-Methyl-6,8-Dioxabicyclo-[3.2.1.]-3-Octene. Experientia **40**, 217–219 (1984).

294. NOVOTNY, M., S. HARVEY, B. JEMIOLO, and J. ALBERTS: Synthetic Pheromones that Promote Inter-Male Aggression in Mice. Proc. Nat. Acad. Sci. *USA* **82**, 2059–2061 (1985).

295. SCHULTZ, T.H., S.M. KRUSE, and R.A. FLATH: Some Volatile Constituents of Dog Urine. J. Chem. Ecol. **11**, 169–175 (1985).

296. BROWNLEE, R.G., R.M. SILVERSTEIN, D. MÜLLER-SCHWARTZE, and A. SINGER: Isolation, Identification and Function of the Chief Component of the Male Tarsal Scent in Black-Tailed Deer. Nature **221**, 284–285 (1969).

297. MÜLLER-SCHWARTZE, D., R.M. SILVERSTEIN, C. MÜLLER-SCHWARTZE, A.G. SINGER, and N.J. VOLKMAN: Response to a Mammalian Pheromone and its Geometrical Isomers. J. Chem. Ecol. **2**, 389–398 (1976).

298. – U. RAVID, A. CLAESSON, A.G. SINGER, R.M. SILVERSTEIN, C. MÜLLER-SCHWARTZE, N.J. VOLKMAN, K.F. ZEMANEK, and R.G. BUTLER: The "Deer-Lactone". Source, Chiral Properties, and Responses by Black-Tailed Deer. J. Chem. Ecol. **4**, 247–256 (1978).

299. – R. ALTIERI, and N. PORTER: Alert Odor from Skin Glands in Deer. J. Chem. Ecol. **10**, 1707–1729 (1984).

300. – L. KÄLLQUIST, and T. MOSSING: Social Behavior and Chemical Communication in Reindeer (*Rangifer t. tarandus*). J. Chem. Ecol. **5**, 483–517 (1979).

301. BRUNDIN, A., and G. ANDERSSON: Seasonal Variation of Three Ketones in the Interdigital Gland Secretion of Reindeer (*Rangifer tarandus*). J. Chem. Ecol. **5**, 881–889 (1979).

302. MÜLLER-SCHWARTZE, D., W.B. QUAY, and A. BRUNDIN: The Caudal Gland in Reindeer (*Rangifer tarandus* L.): Its Behavioral Role, Histology and Chemistry. J. Chem. Ecol. **3**, 591–601 (1977).

303. SKEEN, J.T., and D. THIESSEN: Scent of Gerbil Cuisine. Physiol. Behav. **19**, 11–14 (1977).

304. HARRIMAN, A.E., and D.D. THIESSEN: Removal of Harderian Exudates by Sandbathing Contributes to Osmotic Balance in Mongolian Gerbils. Physiol. Behav. **31**, 317–323 (1983).

305. – – Harderian Letdown in Male Mongolian Gerbils (*Meriones unguiculatus*) Contributes to Proceptive Behavior, Horm. Behav. **19**, 213–219 (1983).

306. KRUSE, S.M., and W.E. HOWARD: Canid Sex Attractant Studies. J. Chem. Ecol. **9**, 1503–1510 (1983).

307. MYKYTOWYCZ, R.: Olfaction in Relation to Reproduction in Domestic Animals. In: D. MÜLLER-SCHWARTZE and M.M. MOZELL eds. Chemical Signals in Vertebrates, p. 207–224. New York: Plenum Press. 1977.

308. BOOTH, W.D.: Sexual Dimorphism Involving Steroidal Pheromones and their Binding Protein in the Submaxillary Gland of the Gottingen Miniature Pig. J. Endocr. **100**, 195–202 (1984).

309. PERRY, G.C., R.L.S. PATTERSON, H.J.H. MACFIE, and C.G. STINSON: Pig Courtship Behavior: Pheromonal Property of Androstene Steroids in Male Submaxillary Secretion. Anim. Prod. **31**, 191–199 (1980).

310. MÜLLER-SCHWARTZE and S. HECKMAN: The Social Role of Scent Marking in Beaver (*Castor canadensis*). J. Chem. Ecol. **6**, 81–95 (1980).

311. – – B. STAGGE: Behavior of Free-Ranging Beaver (*Castor canadensis*) at scent marks. Acta Zool. Fennica **174**, 111–113 (1983).

312. THIESSEN, D.D., H.C. FRIEND, and G. LINDZEY: Androgen Control of Territorial Marking in the Mongolian Gerbil. Science **160**, 432–434 (1968).

313. – G. LINDZEY, and J. NYBY: The Effect of Olfactory Deprivation and Hormones on Territorial Marking in the Male Mongolian Gerbil (*Meriones unguiculatus*). Horm. Behav. **1**, 315–325 (1970).

314. – F.E. REGNIER, M. RICE, M. GOODWIN, N. ISAACKS, and N. LAWSON: Identification of a Ventral Scent Marking Pheromone in the Male Mongolian Gerbil (*Meriones unguiculatus*). Science **184**, 83–85 (1974).

315. JAEGER, R.G., J.M. GOY, M. TARVER, and C.E. MARQUEZ: Salamander Territoriality: Pheromonal Markers as Advertisement by Males. Anim. Behav. **34**, 860–864 (1986).

316. ANDERSEN, K.K., and D.T. BERNSTEIN: Some Chemical Constituents of the Scent of the Striped Skunk (*Mephites mephites*). J. Chem. Ecol. **1**, 493–499 (1975).

317. SCHILDKNECHT, H., I. WILZ, F. ENZMANN, N. GRUND, and M. ZIEGLER: Mustelans,

the Malodorous Substance from the Anal Gland of the Mink (*Mustela vison*) and the Polecat (*Mustela putorius*). Angew. Chem. Int. Ed. Eng. 15, 242–243 (1976).

318. – C. BIRKNER: Struktur und Wirkung der Musteliden-Ökomone III. Analyse der Analbeutelsekrete Mitteleuropäischer Musteliden. Chem. Ztg. 107, 267–270 (1983)).

319. CRUMP, D.R., and P.J. MOORS: Anal Gland Secretion of the Stoat (*Mustela erminea*) and the Ferret (*Mustela putorius* forma *furo*): Some Additional Thietane Components. J. Chem. Ecol. 11, 1037–1043 (1985).

320. – Thietanes and Dithiolanes from the Anal Glands of the Stoat (*Mustela erminea*). J. Chem. Ecol. 6, 341–347 (1980).

321. SCHILDKNECHT, H., E. BIRKNER, and D. KRAUSS: Struktur und Wirkung der Musteliden-Ökonome II. Erweiterte Analyse des Analbeutelsekretes des Nerzes *Mustela vison* L. Chem. Ztg. 105, 273–286 (1981).

322. CRUMP, D.R.: Anal Gland Secretion of the Ferret (*Mustela putorius* forma *furo*). J. Chem. Ecol. 6, 837–844 (1980)).

323. SCHILDKNECHT, H., and H. HILLER: Struktur und Wirkung der Musteliden-Ökomone. IV. Analyse Verhaltensaktiver Drüsensekrete des Dachses (*Meles meles*). Chem. Ztg. 108, 1–5 (1984).

324. – –, J. UBL: Struktur und Wirkung der Säugetier-Ökomone V.: Analyse Verhaltensaktiver Drüsensekrete des Nordamerikanischen Waschbären (*Procyon lotar* L.). Chem. Ztg. 109, 135–138 (1985).

325. ANDERSEN, K.K., D.T. BERSTEIN, R.L. CARET, and L.J. ROMANCZYK, JR.: Chemical Constituents of the Defensive Secretion of the Striped Skunk (*Mephites mephites*). Tetrahedron 38, 1965–1970 (1982).

326. EISNER, T., W.E. CONNER, K. HICKS, K.R. DODGE, H.I. ROSENBERG, T.H. JONES, M. COHEN, and J. MEINWALD: Stink of Stinkpot Turtle Identified: ω-Phenylalkanoic Acids. Science 196, 1347–1349 (1977).

327. STEEL, E.: Effect of the Odour Vaginal Secretion on Non-Copulatory Behaviour of Male Hamster (*Mesocricetus auratus*). Anim. Behav. 32, 597–608 (1984).

328. MARCHLEWSKA-KOJ, A.: Male Pheromone Effect on the Enzyme Activity of the Uterus and on the Efficiency of Pregnancy in Mice. Symp. Zool. Soc. London 45, 277–288 (1980).

329. DESSI-FULGHERI, F., C. LUPO, G. CIAMP, M. CANONACO, and K. LARSSON: Exposure to Odor During Development and Hypothalamic Metabolism of Testosterone. In: J. BALTHAZAR, E. PREEVE, and R. GILES eds., Hormonal Behavior in Higher Vertebrates, p. 305–312. Berlin: Springer 1982.

330. MARCHLEWSKA-KOJ, A.: Pregnancy Block Elicited by Male Urinary Peptides in Mice: J. Reprod. Fertil. 61, 221–224 (1981).

331. BRONSON, F.H., and C. DESJARDINS: Endocrine Responses to Sexual Arousal in Male Mice. Endocrinol. 111, 1286–1291 (1982).

332. KEVERNE, E.B., and C. DE LA RIVA: Pheromones in Mice: Reciprocal Interaction Between the Nose and Brain. Nature 296, 148–150 (1982).

333. BELTRAMINO, C., and S. TALEISNIK: Release of LH in the Female Rat by Olfactory Stimuli. Effect of the Removal of the Vomeronasal Organs or Lesioning of the Accessory Olfactory Bulbs. Neuroendocrinology 36, 53–58 (1983).

334. FURUDATE, S., and K. NAKANO: Mechanism of Pregnancy Inhibition in Mice by Pheromone. Kitasato Igaku 13, 16–24 (1983). (in Japanese Chem. Abstr. 99:99.995 g).

335. KEVERNE, E.B.: Pheromonal Influences on the Endocrine Regulation of Reproduction. Trends Neurosci. 6, 381–384 (1983).

336. EPPLE, G., and Y. KATZ: Social Influences on Estrogen Excretion and Ovarian Ciclycity in Saddle Back Tamarins. (*Saguinus fuscicolis*). Amer. J. Primatol. 6, 215–227 (1984).

337. KEVERNE, E.B., J.A. EBERHARD, U. YODYINGUUAD, and D.H. ABBOTT: Social Influences on the Sex Differences on the Behavior and Endocrine State of Talapoin Monkeys. Prog. Brain Res. **61**, 331–347 (1984).

338. SAHU, S.C., and C.J. DOMINIC: Effect of *alpha*-Methyl Dopa Administration on the Pheromonal Block to Implantation in Mice. Curr. Sci. **52**, 179–181 (1983).

339. DRICKAMER, L.C., and B.C. SHIRO: Effects of Adrenalectomy with Hormone Replacement Therapy on the Presence of a Sexual Maturation-Delaying Chemosignal in the Urine of Grouped Females. Endocrinology **115**, 255–260 (1984).

340. GANGRADE, B.K., and C.J. DOMINIC: Evaluation of the Involvement of the Adrenals in the Pheromonal Influences on the Estrus Cycle of Laboratory Mice. Exp. Clin. Endocrinol. **84**, 13–19 (1984).

341. GANGRADE, B.K., and C.J. DOMINIC: Studies on the Male-Originating Pheromones Involved in the Whitten Effect and Bruce Effect in Mice. Biol. Reprod. **31**, 89–96 (1984).

342. MARCHLEWSKA-KOJ, A., and J. DROZDOWSKA: Testosterone Control of the Pregnancy Block Pheromone in Mice. Folia Biol. (Krakow) **32**, 301–306 (1984).

343. PANDEY, S.C., and S.D. PANDEY: Role of Clitorial Glands in Production of the Estrus-Suppressing Pheromone in Wild Mice. J. Reprod. Biol. Comp. Endocrinol. **4**, 19–23 (1984).

344. SLOB, A.K., G. VAN ES, and J.J. VAN DER WERFFTEN BOSCH: Social Factors and Puberty in Female Rats. J. Endocr. **104**, 309–313 (1985).

345. BELTRAMINO, C., and S. TALEISNIK: Ventral Premammillary Nuclei Mediate Pheromonal-Induced LH Release Stimuli in the Rat. Neuroendocrinology **41**, 119–124 (1985).

346. PANDEY, S.D., and S.C. PANDEY: Effect of an Antiandrogen on Attraction Function of Preputial Glands in the Wild Mouse, *Mus musculus* L. Physiol. Behav. **35**, 851–854 (1985).

347. PANDEY, S.D., and S.C. PANDEY: Regulation of Estrus-Suppressing Pheromone in Wild Mice by Ovarian Hormones. Indian J. Exp. Biol. **23**, 188–190 (1985).

348. ROSSER, A.E., and E.B. KEVERNE: The Importance of Central Noradrenergic Neurones in the Formation of an Olfactory Memory in the Prevention of Pregnancy Block. Neuroscience (Oxford) **15**, 1141–1147 (1985).

349. SAHU, S.C., and C.J. DOMINIC: Effect of Nembutal Anesthesia on the Pheromonal Block to Implantation (the Bruce Effect) in the Laboratory Mice. J. Adv. Zool. **6**, 62–67 (1985).

350. COHEN-TANNOUDJI, J., A. LOCATELLI, and J.P. SIGNORET: Non-Pheromonal Stimulation by the Male of LH Release in the Anestrous Ewe. Physiol. Behav. **36**, 921–924 (1986).

351. JEMIOLO, B., S. HARVEY, and M. NOVOTNY: Promotion of the Whitten Effect in Female Mice by Synthetic Analogs of Male Urinary Constituents. Proc. Nat. Acad. Sci. *USA* **83**, 4576–4579 (1986).

352. NOVOTNY, M., B. JEMIOLO, S. HARVEY, D. WIESLER, and A. MARCHLEWSKA-KOJ: Adrenal-Mediated Endogenous Metabolites Inhibit Puberty in Female Mice. Science **231**, 722–724 (1986).

353. RAYMER, J., D. WIESLER, M. NOVOTNY, C. ASA, U.S. SEAL, and L.D. MECH: Chemical Scent Constituents in Urine of Wolf (*Canis lupus*) and their Dependence on Reproductive Hormones. J. Chem. Ecol. **12**, 297–314 (1986).

354. SCHWENDE, F., D. WIESLER, and M. NOVOTNY: Volatile Compounds Associated with Estrus in Mouse Urine: Potential Pheromone. Experientia **40**, 213–215 (1984).

355. MUGFORD, R.A., and N.W. NOWELL: The Dose Response to Testosterone Propionate of Preputial Gland Pheromones and Aggression in Mice. Horm. Behav. **3**, 39–46 (1972).

356. FAGO, B., and D.A. STEVEN: Pheromonal Influences on Rodent Agonist Behavior. In: D. MÜLLER-SCHWARTZE and M.M. MOZELL eds., Chemical Signals in Vertebrates, p. 185–206. New York: Plenum Press. 1977.
357. EPPLE, G.: Relationship between Aggression, Scent Marking and Gonadal State in a Primate Tamarin *Saguinus fuscicollis*. In: D. MÜLLER-SCHWARTZE and R.M. SILVERSTEIN eds., Chemical Signals in Vertebrates and Aquatic Invertebrates, p. 87–105. New York: Plenum Press 1980.
358. EBLING, F.J.G.: The Role of Odours in Mammalian Aggression. In: F.F. BRAIN and D. BENTON eds., The Biology of Aggression, p. 301–321. Amsterdam: Styhoff and Noorhoff Int. Pub. 1981.
359. NOWELL, N.W.; A.J. THODHY and R. WOODLEY: The Source of an aggression-promoting Olfactory Cue by α-Melanocite Stimulating Hormone, in the Male Mouse. Peptides 1. 69–72 (1980).
360. INGERSOLL, D.W., G. BOBOTAS, C.T. LEE, and A. LUKTON: *Beta*-Glucuronidase of Latent Aggression-Promoting Cues in Mouse Bladder Urine. Physiol. Behav. 29, 789–793 (1982).
361. TAYLOR, G.T., J. HALLER, R. RUPICH, and J. WEISS: Testicular Hormones and Inter-Male Aggressive Behaviour in the Presence of a Female Rat. J. Endocr. 100, 315–321 (1984).
362. McGLONE, J.J.: Olfactory Cues and Pig Agonistic Behavior: Evidence for a Submissive Pheromone. Physiol. Behav. 34, 195–198 (1985).
363. PARROTT, R.F., W.D. BOOTH, and B.A. BALDWIN: Aggression During Sexual Encounters Between Hormone-Treated Gonadectomized Pigs in the Presence or Absence of Boar Pheromones. Aggressive Behav. 11, 245–252 (1985).
364. NOVIKOV, S.N., and V.V. BABALYAN: Recipient Genotype and Efficiency of the Action of Pheromone Controlling the Aggressive Behavior of the House Mouse *Mus musculus*. Dokl. Acad. Nauk. SSSR 278, 1479–1481 (1984) (in Russian Chemical Abstracts 102, 73.325u).
365. GAWIENOWSKI, A.M., I.J. BERRY, and J.J. KENNELLY: Aversion Substances of the Rat Coagulating Glands. J. Chem. Ecol. 8, 379–382 (1982).
366. SCHILLING, A., M. PERRET, and J. PREDINE: Sexual Inhibition in a Prosimian Primate: A Pheromone-Like Effect. J. Endocr. 102, 143–151 (1984).
367. JOHNSTON, R.E.: Testosterone Dependence of Scent Marking by Male Hamster (*Mesocricetus auratus*). Behav. Neurol. Biol. 31, 96–99 (1981).
368. KEVERNE, E.B., J.A. EBERHART, and R.E. MELLER: Plasma Testosterone, Sexual and Aggressive Behavior in Social Groups of Talapoin Monkeys. In: H.S. STEKLIS and A.S. KLUG eds. Hormones and Drugs in Social behavior of Primates, p. 33–59. New York: Spectrum. 1983.
369. GOLDFOOT, D.A.: Olfaction, Sexual Behavior, and the Pheromone Hypothesis in Rhesus Monkeys: A Critique. Amer. Zool. 21, 153–164 (1981).
370. LOMBARDI, J., and J.G. VANDENBERG: Pheromonally Induced Sexual Maturation in Females: Regulation by the Social Environment of the Male. Science 196, 545–546 (1977).
371. FRENCH, J.A., D.A. ABBOTT, and C.T. SNOWDON: The Effect of Social Environment on Estrogen Excretion, Scent Marking, and Sociosexual Behavior in Tamarins (*Saguinus fuscicolis*). Amer. J. Primatol. 6, 155–167 (1984).
372. GLICK, B.B.: Male Endocrine Responses to Females: Effects of Social Cues in Cynomolgus Macaques. Amer. J. Primatol. 6, 229–239 (1984).
373. NICHOLS, D.J., and P.F.D. CHEVIN: Adrenocortical Responses and Changes During the Estrous Cycle in Mice: Effect of Male Presence, Male Urine and Housing Conditions. J. Endocr. 91, 263–269 (1981).

374. NISHIMURA, K., K. UTSUMI, and Y. MASATAKA: Isolation of Puberty Accelerating Pheromone from Male Mouse Urine. Kachiku Hanshokugaku Zasshi **29**, 24–31 (1983) (in Japanese Chem. Abstr. **99**:188.498q).

375. McCLINTOCK, M.K.: Estrous Synchrony: Modulation of Ovarian Cycle Length by Female Pheromones. Physiol. Behav. **32**, 701–705 (1984).

376. LAWTON, A.D., and J.M. WHITSETT: Inhibition of Sexual Maturation by a Urinary Pheromone in Male Prairie Deer Mice. Horm. Behav. **13**, 128–138 (1979).

377. MOLTZ, H., and L.C. LEIDAHL: Bile, Prolactin and the Maternal Pheromone. Science **196**, 81–83 (1977).

378. LEE, T.M., B. HALPERN, C. LEE, and H. MOLTZ: Reduced Prolactin Binding to Liver Membranes During Pheromonal Emission in the Rat. Pharmacol. Biochem. Behav. **17**, 1149–1154 (1982).

379. KILPATRICK, S.J., T.M. LEE, and H. MOLTZ: The Maternal Pheromone of the Rat: Testing Some Assumptions Underlying a Hypothesis. Physiol. Behav. **30**, 539–543 (1983).

380. LEE, T.M., and H. MOLTZ: The Maternal Pheromone and Brain Development in the Prewealing Rat. Physiol. Behav. **33**, 385–390 (1984).

381. – – The Maternal Pheromone and Deoxycholic Acid in Relation to Brain Myelin in the Prewealing Rat. Physiol. Behav. **33**, 391–395 (1984).

382. – – The Maternal Pheromone and Deoxycholic Acid in the Survival of Prewealing Rats. Physiol. Behav. **33**, 931–935 (1984).

383. SCHUMACKER, S.K., and H. MOLTZ: Prolonged Responsiveness to the Maternal Pheromone in the Postwealing Rat. Physiol. Behav. **34**, 471–473 (1985).

384. RYAN, E.P.: Pheromone: Evidence in a Decapod Crustacean. Science **151**, 340–341 (1966).

385. DUNHAM, P.: Sex Pheromone in Crustacea. Biol. Rev. **53**, 555–583 (1978).

386. CHRISTOFFERSON, J.P.: Evidence for the Controled Released of a Crustacean Sex Pheromone. J. Chem. Ecol. **4**, 633–639 (1978).

387. GLEASON, R.A., M.A. ADAMS, and A.B. SMITH: Characterization of a Sex Pheromone in the Blue Crab *Callinectis sapidus*. J. Chem. Ecol. **10**, 913–921 (1984).

388. VAN DEN HURK, R., and J.G.D. LAMBERT: Ovarian Steroid Glucuronides Function as Sex Pheromones for Male Zebrafish, *Brachdanio rerio*. Can. Zool. **61**, 2381–2387 (1983).

389. MEYER, J.H., and N.R. LILEY: The Control of Production of a Sexual Pheromone in the Female Guppy *Poecilia reticulata*. Can. Zool. **60**, 1505–1510 (1982).

390. GOLUBEV, A.V.: Role of Chemical Stimuli in Group and Spawning Behavior of *Brachdanio rerio* (Hamilton-Bluchman) (Cyprinidae). Vopr. Ikhtiol. **24**, 1020–1027 (1984) (in Russian Chem. Abstr. **102**:182.814y).

391. KINNEL, R., A.J. DUGGAN, T. EISNER, J. MEINWALD, and I. MIURA: Panacene: An Aromatic Bromoallene from a Sea Hare (*Aplysia brasiliana*). Tetrahedron Lett. 3913–3916 (1977).

392. IRELAND, C.I., and D.J. FAULKNER: The Defensive Secretion of the Opisthobranch Mollusk *Onchidella binneyi*. Bioorg. Chem. **7**, 125–131 (1978).

393. SLEEPER, H.L., and W. FENICAL: Navenones A–C: Trail-Breaking Alarm Pheromones from the Marine Opisthobranch *Navanax inermis*. J. Amer. Chem. Soc. **99**, 2367–2368 (1977).

394. – V.L. PAUL, and W. FENICAL: Alarm Pheromones from the Marine Opisthobranch *Navanax inermis*. J. Chem. Ecol. **6**, 57–70 (1980).

395. BURRESON, B.J., P.J. SCHEUER, J. FINER, and J. CLARDY: 9-Isocyanopupukeanane, a Marine Invertebrate Allomone with a New Sesquiterpene Skeleton. J. Amer. Chem. Soc. **97**, 4763–4764 (1975).

396. SIUDA, J.F.: Chemical Defense Mechanism of Marine Organism. Identification of

8-Hydroxy-4-Quinolone from the Ink of the Giant Octopus *Octopus dofleini* (Martin). Lloydia **37**, 501–503 (1974).

397. SCHULTE, G., P.J. SCHEUER, and O.J. McCONNELL: Two Furanosesquiterpene Marine Metabolites with Antifeedant Properties. Helv. Chim. Acta **63**, 2159–2167 (1980).

398. – – Defense Allomones of Some Marine Mollusks. Tetrahedron **38**, 1857–1863 (1982).

399. THOMPSON, J.E., R.P. WALKER, S.J. WRATTEN, and D.J. FAULKNER: A Chemical Defense Mechanism for the Nudibranch *Cadlina luteomarginata*. Tetrahedron **38**, 1865–1873 (1982).

400. CARTE, B., and D.J. FAULKNER: Role of Secondary Metabolites in Feeding Associations between a Predatory Nudibranch, Two Grazing Nudibranchs and a Bryzoan. J. Chem. Ecol. **12**, 795–804 (1986).

401. HELLOU, J., R.J. ANDERSEN, and J.E. THOMPSON: Terpenoids from the Dorid Nudibranch *Cadlina luteomarginata*. Tetrahedron **38**, 1875–1879 (1982).

402. TURSCH, B.: Some Recent Developments in the Chemistry of Alcyonaceans. Pure App. Chem. **48**, 1–6 (1976).

403. ALBERICCI, M., J.C. BRAEKMAN, D. DALOZE, and B. TURSCH: Chemical Studies of Marine Invertebrates. XLV. The Chemistry of Three Norsesterterpene Peroxides from the Sponge *Sigmosceptrella laevis*. Tetrahedron **38**, 1881–1890 (1982).

404. HOWE, N.R., and L.G. HARRIS: Transfer of the Sea Anemone Pheromone, Anthopleurine by the Nudibranch *Aeolodia papillosa* J. Chem. Ecol. **4**, 551–561 (1978).

405. TURSCH, B.: Chemical Protection of a Fish (*Abudefduf leucogaster* Bleeker) by a Soft Coral (*Litophyton viridis* May). J. Chem. Ecol. **8**, 1421–1428 (1982).

406. SANGSTER, A.W., S.E. THOMAS, and N.L. TINGLING: Fish Attractants from Marine Invertebrates. Arcamine from *Arca zebra* and Strombine from *Strombus gigas*. Tetrahedron **31**, 1135–1137 (1975).

406a. HARDEN JONES, F.R.: Fish Migration: Strategy and Tactics. In: D.J. ARDLEY ed., Animal Migration, p. 139–165. Cambridge: Cambridge University Press 1981.

407. NORDENG, H.: Is the Local Orientation of Anadromous Fishes Determined by Pheromones? Nature **233**, 411–413 (1971).

408. – A Pheromone Hypothesis for Homeward Migration in Anadromous Salmonids. Oikos **28**, 155–159 (1977).

409. SELSET, R., and K.B. DOEVING: Behaviour of Mature Anadromous Char (*Salmo alpinus* L.) Towards Odorants Produced by Smolts of their Own Population. Acta Physiol. Scand. **108**, 113–122 (1980).

410. FISKENER, B., and K. DOEVING: Olfactory Sensitivity to Group Specific Substances in Atlantic Salmon (*Salmo salar* L.). J. Chem. Ecol. **8**, 1083–1091 (1982).

411. STABELL, B., R. SELSET, and K. SLETTEN: A Comparative Chemical Study of Population-Specific Odorants from Atlantic Salmon. J. Chem. Ecol. **8**, 201–217 (1982).

412. MICHAEL, R.P., and E.B. KEVERNE: Pheromones in the Communication of Sexual Status in Primates. Nature **218**, 746–749 (1968).

413. GOLDFOOT, D.A., S. ESSOCK-VITALE, C.S. ASA, J. THORNTON, and A.I. LESHNER: Anosmia in Male Rhesus Monkeys Does Not Alter Copulatory Activity with Cycling Females. Science **199**, 1095–1096 (1978).

414. GANGRADE, B.K., and C.J. DOMINIC: Effect of Zinc Sulphate Induced Anosmia on Estrus Cycle of Laboratory Mice. Indian J. Exp. Biol. **21**, 425–427 (1983).

415. BOOTH, W.D., and D.A. BALDWIN: Lack of Effect on Sexual Behavior on the Development of Testicular Function after Removal of Olfactory Bulbs in Prepubertal Boar. J. Reprod. Fert. **58**, 173–182 (1980).

416. MURPHY, M.R., and G.E. SCHNEIDER: Olfactory Bulb Removal Eliminates Mating Behavior in the Male Golden Hamster. Science **167**, 302–304 (1970).

417. WILSON, E.O.: The Insect Societies. Cambridge: Harvard University Press. 1971.

418. BRIAN, M.V.: Caste Differentiation and Division of Labor. In: H.R. HERMANN ed. Social Insects, vol. 1, p. 122–222. New York: Academic Press. 1979.

419. BLUM, M.S.: Pheromonal Basis of Insect Sociality. In: Les Mediateurs Chimiques. Les Colloques d'INRA vol. 7, p. 16–20. Versailles: INRA. 1981.

420. JAFFÉ, K.: Evolucion de los Sistemas de Communicacion Quimica en Hormigas (Hymenoptera: Formicidae). Folia Entomol. Mexicana 61, 189–203 (1984).

421. BRAND, J.M., R.M. DUFFIELD, J.G. MACCONNEL, M.S. BLUM, and H.M. FALES: Caste Specificity Compounds in Male Carpenter Ants. Science 179, 388–389 (1973).

422. PASTEELS, J.M., J.C. VERHAEGHE, J.G. BRAEKMAN, D. DALOZE, and B. TURSCH: Caste-Dependent Pheromones in the Heads of Ant Tetramorium caespitum. J. Chem. Ecol. 6, 467–472 (1980).

423. LÖFQVIST, J., and G. BERGSTRÖM: Volatile Communication Substances in Dufour's Gland of Virgin Females and Old Queens of the Ant Formica polyctena. J. Chem. Ecol. 6, 309–320 (1980).

424. HOWARD, R.W., C.A. MACDANIEL, D.R. NELSON, G.J. BLOMQUIST, L.T. GELBAUM, and L.H. ZALKOW: Cuticular Hydrocarbons of Reticulitermes virginicus and their Role as Potential Species and Caste Recognition Cues. J. Chem. Ecol. 8, 1227–1239 (1982).

425. CARLIN, N.F., and B. HÖLLDOBLER: Nestmate and Kin Recognition in Interspecific Mixed Colonies of Ants. Science 222, 1027–1029 (1983).

426. KLAHN, J.E., and G.J. GAMBOA: Social Wasps: Discrimination between Kin and Non-Kin Brood. Science 221, 482–484 (1983).

427. BELL, W.J.: Recognition of Resident and Non-Resident Individuals in Intraspecific Nest Defense of a Primitively Eusocial Halictine Bees. J. Comp. Physiol. 93, 195–202 (1974).

428. BUCKLE, G.R., and L. GREENBERG: Nestmate Recognition in Sweat Bees (Lasioglossum zephyrum): Does an Individual Recognize its Own Odour or Only Odours of its Nestmates? Anim. Behav. 29, 802–809 (1981).

429. GREENBERG, L.: Genetic Component of Bee Odor in Kin Recognition. Science 206, 1095–1097 (1979).

430. HEFETZ, A., G. BERGSTRÖM, and J. TENGÖ: Species, Individual and Kin Specific Blends in Dufour's Gland Secretions of Halictine Bees. Chemical Evidence. J. Chem. Ecol. 12, 197–208 (1986).

431. SHIMRON, O., A. HEFETZ, and J. TENGÖ: Structural and Communicative Functions of Dufour's Gland Secretion in Eucera palestinae (Hymenoptera; Anthophoridae). Insect Biochem. 15, 635–638 (1985).

432. BRAND, J.M., and V. PRETORIUS: Individual Variation in the Major Alarm Pheromone Components of Two Crematogaster species. Biochem. Syst. Ecol. 14, 341–343 (1986).

433. GAMBOA, G.J., H.K. REEVE, I.D. FERGUSON, and T.L. WACKER: Nestmate Recognition in Social Wasps: The Origin and Acquisition of Recognition Odours. Anim. Behav. 34, 685–695 (1986).

434. NORDLUNG, D.A., R.L. JONES, and W.J. LEWIS (eds.): Semiochemicals: Their Role in Pest Control. New York: John Wiley 7 Sons 1981.

435. SILVERSTEIN, R.M.: Pheromones: Background and Potential for Use in Pest Control. Science 213, 1326–1332 (1981).

436. LEONHARDT, B.A., and M. BEROZA (eds.): Insect Pheromone Technology: Chemistry and Application. American Chemical Society Symposium Series n° 190. Washington, D.C. 1982.

437. HIRAI, K., H.H. SHOREY, and L.K. GASTON: Competition among Courting Male Moths: Male-to-Male Inhibitory Pheromone. Science 202, 644–645 (1978).

438. PESCHKE, K.: Immature Males of Aleochara curtula Avoid Intrasexual Aggressions by Producing the Female Sex Pheromone. Naturwissenschaften 72, 274–275 (1985).

439. BURNET, B., K. CONNOLLY, M. KEARNEY, and R. COOK: Effects of Male Paragonial Gland Secretion on Sexual Receptivity and Courtship Behavior of Female *Drosophila melanogaster*. J. Insect Physiol. **19**, 2421–2431 (1973).

440. GILBERT, L.E.: Postmating Female Odor in *Heliconius* Butterflies: A Male-Contributed Antiaphrodisiac? Science **193**, 419–420 (1976).

441. SCHLEIN, Y., R. GALUM, and M.N. BEN-ELIAHU: Abstinons. Male-Produced Deterrents of Mating in Flies. J. Chem. Ecol. **7**, 285–290 (1981).

442. JALLON, J.M., C. ANTHONY, and O. BEN MAR: Un Antiaphrodisiacque Produit par les Males des *Drosophila melanogaster* et Transferé aux Femelles lors de la Copulation Comp. Rend. Hebd. Séanc. Acad. Sci. Paris **292**, 1147–1149 (1981).

443. TOMPKINS, L., and J.C. HALL: *Drosophila* Males Produce a Pheromone which Inhibits Courtship. Z. Naturforsch. **36 C**, 694–696 (1981).

444. – – The Different Effects on Courtship of Volatile Compounds from Mated and Virgin *Drosophila* Females. J. Insect Physiol. **27**, 17–21 (1981).

445. JALLON, J.M.: A Few Chemical Words Exchanged During Courtship and Mating of *Drosophila melanogaster*. Behav. Genet. **14**, 441–478 (1984).

446. ZAWISTOWSKI, S., and R.C. RICHMOND: Inhibition of Courtship and Mating of *Drosophila melanogaster* by the Male-Produced Lipid, Cis-Vaccenyl Acetate. J. Insect Physiol. **32**, 189–192 (1986).

447. VAN DER MEER, R.K., M.S. OBIN, S. ZAWISTOWSKI, K.B. SHEEHAN, and R.C. RICHMOND: A Reevaluation of the Role of Cis-Vaccenyl Acetate, Cis-Vaccenol and Esterase 6 in the Regulation of Mated Female Sexual Attractiveness in *Drosophila melanogaster*. J. Insect Physiol. **32**, 681–686 (1986).

448. BELLES, X., A. GALOFRE, and A. GINEBREDA: Pheromones: Relations between Chemical Structures and Taxonomy. An Example for Some of the Lepidopteran Families. Afinidad **42**, 147–147 (1985) (in Spanish Chem. Abstr. **103**:85489j).

449. EDGAR, J.A.: Danaine (Lep.) and 1,2 Dehydropyrrolizidine Alkaloid – Containing Plants – with Reference to Observations Made in the New Hebrides. Phil. Trans. R. Soc. London **272 B**, 467–476 (1973).

450. PLISKE, T.: Attraction of Lepidoptera to Plants Containing Pyrrolizidine Alkaloids. Environ. Entomol. **4**, 455–473 (1975).

451. – Courtship Behavior and Use of Chemical Communication by Males of Certain Species of Ithomiinae Butterflies (Nymphalidae: Lepidoptera) Ann. Entomol. Soc. Amer. **68**, 935–942 (1975).

452. – Courtship Behavior of the Monarch Butterfly, *Danaus plexippus* L. Ann. Entomol Soc. Amer. **68**, 143–151 (1975).

453. SCHNEIDER, D., M. BOPPRÉ, H. SCHNEIDER, W.R. THOMPSON, C.J. BORIACK, R.L. PETTY, and J. MEINWALD: A Pheromone Precursor and its Uptake in Male *Danaus* Butterflies. J. Comp. Physiol. **97**, 245–256 (1975).

454. EDGAR, J.A., C.C.J. CULVENOR, and T.E. PLISKE: Isolation of a Lactone, Structurally Related to the Esterifying Acids of Pyrrolizidine Alkaloids, from the Costal Fringes of Male Ithomiinae. J. Chem. Ecol. **2**, 263–270 (1976).

455. PLISKE, T.E., J.A. EDGAR, and C.C.J. CULVENOR: The Chemical Basis of Attraction of Ithomiinae Butterflies to Plants Containing Pyrrolizidine Alkaloids. J. Chem. Ecol. **2**, 255–262 (1976).

456. BOPPRÉ, M., R.L. PETTY, D. SCHNEIDER, and J. MEINWALD: Behaviorally Mediated Contacts between Scent Organs: Another Prerequisite for Pheromone Production in *Danaus chrysippus* (Lepidoptera). J. Comp. Physiol. **126**, 97–103 (1978).

457. ROELOFS, W.L., and R.T. CARDÉ: Sex Pheromones in the Reproductive Isolation of Lepidopterous Species. In: M.C. BIRCH ed., Pheromone, p. 96–114. Amsterdam: North Holland/Elsevier. 1974.

458. LOEFSTEDT, C., and J.N.C. VAN DER PERS: Sex Pheromones and Reproductive Isolation in Four European Small Ermine Moths. J. Chem. Ecol. **11**, 649–666 (1985).

459. KLUN, J.A., J.R. PLIMMER, B.A. BIERL-LEONHARDT, A.N. SPARKS, and O.L. CHAP-
     MAN: Trace Chemicals: The Essence of Sexual Communication System in *Heliothis*
     Species. Science **204**, 1328–1330 (1979).
460. – – – – M. PRIMIANI, O.L. CHAPMAN, G.H. LEE, and G. LEPONE: Sex Pheromone
     Chemistry of Female Corn Earworm Moth, *Heliothis zea*. J. Chem. Ecol. **6**, 165–175
     (1980).
461. – B.A. BIERL-LEONHARDT, J.R. PLIMMER, A.N. SPARKS, N. PRIMIANI, O.L. CHAPMAN,
     G. LEPONE, and G.H. LEE: Sex Pheromone Chemistry of the Female Tobacco Bud-
     worm Moth *Heliothis virescens*. J. Chem. Ecol. **6**, 177–183 (1980).
462. STECK, W., E.W. UNDERHILL, and M.D. CHISHOLM: Attraction and Inhibition in
     Moth Species Responding to Sex-Attractant Lures Containing Z-11-Hexadecen-1-yl-
     acetate. J. Chem. Ecol. **3**, 603–612 (1977).
463. LÖFQVIST, J., and G. BERGSTRÖM: Nerol-Derived Volatile Signals as a Biochemical
     Basis for Reproductive Isolation between Sympatric Populations of Three Species
     of Ant-Lions (Neuroptera: Myrmeleontidae). Insect Biochem. **10**, 1–10 (1980).
464. BIRCH, M.C., and D.L. WOOD: Mutual Inhibition of the Attractant Pheromone
     Response by Two Species of *Ips* (Coleoptera: Scolytidae). J. Chem. Ecol. **1**, 101–113
     (1975).
465. BORDEN, J.H., L. CHONG, J.A. MCLEAN, K.N. SLESSOR, and K. MORI: *Gnathotrichus
     sulcatus*: Synergistic Response to Enantiomers of the Aggregation Pheromone Sulca-
     tol. Science **192**, 894–896 (1976).
466. BIRCH, M.C., and D.M. LIGHT: Inhibition of the Attractant Pheromone Response
     in *Ips pini* and *I. paraconfusus* (Coleoptera: Scolytidae). Field Evaluation of Ipsenol
     and Linalool. J. Chem. Ecol. **3**, 257–267 (1977).
467. SHAPAS, T.J., and W.E. BURKHOLDER: Diel and Age-Dependent Behavioral Patterns
     of Exposure-Concealment in Three Species of *Trogoderma*: Simple Mechanisms for
     Enhancing Reproductive Isolation in Chemically Mediated Mating Systems. J.
     Chem. Ecol. **4**, 409–423 (1978).
468. BORDEN, J.H., J.R. HANDLEY, J.A. MCLEAN, R.M. SILVERSTEIN, L. CHONG, K.N.
     SLESSOR, B.D. JOHNSTON, and H.R. SCHULER: Enantiomer-Based Specificity in Phero-
     mone Communication by Two Sympatric *Gnathotrichus* Species (Coleoptera: Scolyti-
     dae). J. Chem. Ecol. **6**, 445–456 (1980).
469. FRANCKE, W., P. SAUERWEIN, J.P. VITÉ, and D. KLIMETZEK: The Pheromone Bouquet
     of *Ips amitinus*. Naturwissenschaften **67**, 147–148 (1980).
470. BIRCH, M.C., P. SVIHRA, T.D. PAYNE, and J.C. MILLER: Influence of Chemically
     Mediated Behavior on Host Tree Colonization by Four Cohabiting Species of Bark
     Beetle. J. Chem. Ecol. **6**, 395–414 (1980).
471. BORDEN, J.H., and J.A. MCLEAN: Secondary Attraction in *Gnathotrichus retusus*
     and Cross Attraction of *G. sulcatus* (Coleoptera: Scolytidae). J. Chem. Ecol. **5**, 79–88
     (1979).
472. EISNER, T.: Chemical Defense Against Predation in Arthropods. In: E. SONDHEIMER
     and J.B. SIMEONE eds., Chemical Ecology, p. 157–217. New York: Academic Press
     1970.
473. WEATHERSTON, J., and J. PERCY: Arthropod Defensive Secretions. In: M. BEROZA
     ed., Chemicals Controlling Insect Behavior, p. 95–144. New York: Academic Press
     1970.
474. ROTHSCHILD, M.: Some Observations on the Relationship between Plants, Toxic
     Insects and Birds. In: J.B. HARBORNE ed., Phytochemical Ecology, p. 1–12. London:
     Academic Press 1972.
475. – T. REICHSTEIN: Some Problems Associated with the Storage of Cardiac Glycosides
     by Insects. In: M. LUCKNER; K. MOTHES and L. NOVER eds., Secondary Metabolism
     and Coevolution. Nova Acta Leopoldina (Supplementum) p. 507–550. Halle (Saale):
     Deutsche Akademie der Naturforscher Leopoldina 1976.

476. ROESKE, C.N., J.N. SEIBER, L.P. BROWER, and C.M. MOFFITT: Milkweed Cardenolides and their Comparative Processing by Monarch Butterflies. (Danaus plexippus
L.). In: J.W. WALLACE and R.L. MANSELL eds., Biochemical Interactions Between
Plants and Insects, p. 93–167. New York: Plenum Press 1976.
477. TURSCH, B., J.C. BRAEKMAN, and D. DALOZE: Arthropod Alkaloids. Experientia
32, 401–407 (1976).
478. EISNER, T., D. ALSOP, K. HICKS, and J. MEINWALD: Defensive Secretions of Milli-
pedes. In: S. BETTINI ed., Handbook of Experimental Pharmacology vol. 48, "Arthropod Venoms", p. 41–72. Berlin: Springer 1978.
479. – – J. MEINWALD: Secretions of Opilionids, Whip Scorpions and Pseudoscorpions.
In: S. BETTINI ed., Handbook of Experimental Pharmacology, vol. 48, "Arthropod
Venoms", p. 87–99. Berlin: Springer 1978.
480. MEINWALD, J. (ed.): The Organic Chemistry of Animal Defense Mechanisms. Tetrahedron 38, 1853–1970 (1982).
481. PASTEELS, J.M., J.C. GRÉGOIRE, and M. ROWELL-RAHIER: The Chemical Ecology
of Defense in Arthropods. Ann. Rev. Entomol. 28, 263–289 (1983).
482. HOFFMANN, D.: Les Méchanismes de Défense chez les Insects. Bull. Inst. Pasteur
81, 259–264 (1983).
483. JONES, T.H., W.E. CONNER, J. MEINWALD, H.E. EISNER, and T. EISNER: Benzoyl
Cyanide and Mandelonitrile in Cyanogenic Secretion of a Centipede. J. Chem. Ecol.
2, 421–429 (1976).
484. CARREL, J.E., and T. EISNER: Spider Sedation Induced by Defensive Chemicals of
Millipede Prey. Proc. Nat. Acad. Sci. USA 81, 806–810 (1984).
485. SCHILDKNECHT, H., D. BERGER, D. KRAUSS, J. CONNERT, J. GEHLHAUS, and H. ESSEN-
BREIS: Defensive Chemistry of Stenus comma (Coleoptera: Staphylinidae). LXI. J.
Chem. Ecol. 2, 1–11 (1976).
486. PASTEELS, J.M., M. ROWELL-RAHIER, J.C. BRAEKMAN, and D. DALOZE: Chemical
Defences in Leaf Beetles and their Larvae: The Ecological, Evolutionary and Taxonomic Significance. Biochem. Syst. Ecol. 12, 395–306 (1984).
487. PESCHKE, K.: Defensive and Pheromonal Secretions of the Tergal Gland of Aleochara
curtula. II. Release and Inhibition of Male Copulatory Behavior. J. Chem. Ecol.
9, 13–31 (1983).
488. PASTEELS, J.M., and D. DALOZE: Cardiac Glycosides in the Defensive Secretion
of Chrysomelid Beetles: Evidence for their Production by the Insects. Science 197,
70–72 (1977).
489. DALOZE, D., and J.M. PASTEELS: Production of Cardiac Glycosides by Chrysomelid
Beetles and Larvae. J. Chem. Ecol. 5, 63–77 (1979).
490. BROWER, L.P., J.N. SEIBER, C.J. NELSON, S.P. LYNCH, and N.N. HOLLAND: Plant
Determined Variation in the Cardenolide Content, Thin-Layer Chromatography
Profiles and Emetic Potency of Monarch Butterflies, Danaus plexippus L. Reared
on Milkweed Plants in California. 2. Asclepias speciosa. J. Chem. Ecol. 10, 601–639
(1984).
491. – – – – N.P. HOGGARD, and J.A. COHEN: Plant-Determined Variation in Cardenolide Content and Thin-Layer Chromatography Profiles of Monarch Butterflies, Danaus plexippus Reared on Milkweeds Plants in California. 3. Asclepias californica.
J. Chem.Ecol. 10, 1823–1857 (1984).
492. CALVERT, W., L.E. HEDRICK, and L.P. BROWER: Mortality of the Monarch Butterfly
(Danaus plexippus L.): Avian Predator at Five Overwintering Sites in Mexico. Science
204, 847–851 (1979).
493. BROWN, K.S., JR., and J. VASCONCELLOS NETO: Predation on Aposematic Ithomiine
Butterflies by Tanagers (Pipraeidea melanonota). Biotropica 8, 136–141 (1976).
494. BROWER, L.P., N.J. SEIBER, C.J. NELSON, S.P. LYNCH, and P.N. TUSKER: Plant-
Determined Variation in the Cardenolide Content, Thin-Layer Chromatography

Profiles and Emetic Potency of Monarch Butterflies *Danaus plexippus* Reared on the Milkweed *Asclepias eriocarpa* in California. J. Chem. Ecol. **8**, 579–633 (1982).

495. COHEN, J.A., and L.P. BROWER: Cardenolide Sesquetration by the Dogbane Tiger Moth (*Cycnia tenera*: Arctiidae). J. Chem. Ecol. **9**, 521–532 (1983).

496. BROWN, K.S., JR.: Chemical Ecology of Dehydropyrrolizidine Alkaloids in Adult Ithomiinae (Lepidoptera: Nymphalidae). Rev. Bras. Biol. **44**, 435–460 (1985).

497. EMERSON, A.E.: Vestigial Characters of Termites and Processes of Regressive Evolution. Evolution **15**, 115–131 (1961).

498. KRISHNA, K., and F.M. WEESNER (eds.): Biology of Termites. New York: Academic Press 1970.

499. MCMAHAN, E.: Non-Aggressive Behavior in the Large Soldier of *Nasutitermes exitiosus* (Hill) (Isoptera: Termitidae). Insect Soc. **21**, 95–106 (1974).

500. WADHAN, L.J., R. BAKER, and P.E. HOWSE: 4,11-Epoxy-cis-eudesmane, a Novel Oxygenated Sesquiterpene in the Frontal Gland Secretion of the Termite *Amitermes evuncifer* (Silvestri). Tetrahedron Lett.1697–1700 (1974).

501. PRESTWICH, G.D.: Chemical analysis of Soldier Defense Secretions of Several Species of East African Termites. In: CH. NOIROT; P.E. HOWSE and G. LEMASNE eds., Pheromones and Defensive Secretions in Social Insects, p. 149–152. Dijon: I.U.S.S.I. 1975.

502. - M. KAIB, W.F. WOOD, and J. MEINWALD: 1,13-Tetradecadien-3-one and Homologs: New Natural Products Isolated from *Schedorhinotermes* Soldiers. Tetrahedron Lett. 4701–4704 (1975).

503. QUENNEDEY, A.: La Guerre Chimique ches les Termites. La Recherche **6**, 274–276 (1975).

504. WOOD, W.F., W. TRUCKENBRODT and J. MEINWALD: Chemistry of the Defensive Secretion from the African Termite *Odontotermes badius*. Ann. Entomol. Soc. Amer. **68**, 359–360 (1975).

505. EISNER, T., I. KRISTON, and D. ANESHANSLEY: Defensive Behavior of a Termite (*Nasutitermes exitiosus*). Behav. Ecol. Sociob. **1**, 83–125 (1976).

506. PRESTWICH, G.D., S.P. TANIS, J.P. SPRINGER, and J. CLARDY: Nasute Termite Soldier Frontal Gland Secretions. 1. Structure of Trinervi-2-β-3-α-9α-triol-9-O-acetate, a Novel Diterpene from *Trinervitermes* Soldiers. J. Amer. Chem. Soc. **98**, 6061–6062 (1976).

507. - - F.G. PILKIEWICZ, I. MIURA, and K. NAKANISHI: Nasute Termite Soldier Frontal Gland Secretions 2. Structure of Trinervitene Congeners from *Trinervitermes* Soldiers. J. Amer. Chem. Soc. **98**, 6062–6064 (1976).

508. KRISTON, I., J.A. WATSON, and T. EISNER: Non-Combative Behaviour of Large Soldiers of *Nasutitermes exitiosus* (Hill): An Analytical Study. Insect Soc. **24**, 103–111 (1977).

509. PRESTWICH, G.D.: Chemical Composition of the Soldier Secretions of the Termite *Trinervitermes gratiosus*. Insect Biochem. **7**, 91–94 (1977).

510. - B.A. BIERL, E.D. DEVILBISS, and M.P.B. CHAUDHURY: Soldier Frontal Glands of the Termite *Macrotermes subhyalinus*: Morphology, Chemical Composition and Use in Defense. J. Chem. Ecol. **3**, 579–590 (1977).

511. - B.A. SOLHEIM, J. CLARDY, F.G. PILKIEWICZ, I. MIURA, S.P. TANIS, and K. NAKANISHI: Kempene-1 and -2, Unusual Tetracyclic Diterpenes from *Nasutitermes* Termite Soldiers. J. Amer. Chem. Soc. **99**, 8082–8083 (1977).

512. BAKER, R., P.H. BRINER, and D.A. EVANS: Chemical Defense in the Termite *Ancistrotermes cavithorax*: Ancistrodial and Ancistrofuran. Chem. Comm. 410–411 (1978).

513. PRESTWICH, G.D.: Isotrinervi-2-β-ol. Structural Isomers in the Defense Secretions of the Allopatric Populations of the Termite *Trinervitermes gratiosus*. Experientia **34**, 682–684 (1978).

514. - D.F. WIEMER, J. MEINWALD, and J. CLARDY: Cubitene: An Irregular Twelve-

Membered-Ring Diterpene from a Termite Soldier. J. Amer. Chem. Soc. **100**, 2560–2561 (1978).

515. VRKOČ, J., M. BUDĚŠINSKY, and P. SEDMERA: Structure of Trinervitane Diterpenoids from *Nasutitermes rippetii* (Rambur). Coll. Czech. Chem. Comm. **43**, 1125–1133 (1978).

516. – J. KŘEČEK, and I. HRDÝ: Monoterpenic Alarm Pheromones in Two *Nasutitermes* Species. Acta Entomol. Bohemoslov. **75**, 1–8 (1978).

517. PRESTWICH, G.D.: Termite Chemical Defense: New Natural Products and Chemosystematics. Sociobiology **4**, 127–140 (1979).

518. – Chemical Defense by Termite Soldiers. J. Chem. Ecol **5**, 459–480 (1979).

519. – Interspecific Variation in the Defense Secretions of *Nasutitermes* Soldiers. Biochem. Syst. Ecol. **7**, 211–221 (1979).

520. – J.W. LAUHER, and M.S. COLLINS: Two New Tetracyclic Diterpenes from the Defense Secretion of the Neotropical Termite *Nasutitermes octopilis*. Tetrahedron Lett. 3827–3830 (1979).

521. WIEMER, D.F., J. MEINWALD, G.D. PRESTWICH, and I. MIURA: Cembrane A and (3 Z)-Cembrane A: Diterpenes from a Termite Soldier. (Isoptera: Termitidae: Termitinae). J. Org. Chem. **44**, 3950–3952 (1979).

522. WIEMER, D.F., J. MEINWALD, G. PRESTWICH, B.A. SOLHEIM, and J. CLARDY: Biflora-4, 10(19), 15-triene: A New Diterpene from a Termite Soldier (Isoptera: Termitidae: Termitinae). J. Org. Chem. **45**, 191–192 (1980).

523. BAKER, R., M. EDWARDS, D.A. EVANS, and S. WALMSLEY: Soldier-Specific Chemicals of the Termite *Curvitermes strictinasus* (Mathews) (Isoptera. Nasutitermitinae). J. Chem. Ecol. **7**, 127–133 (1981).

524. BAKER, R., H.R. COLES, M. EDWARDS, D.A. EVANS, P.E. HOWSE, and S. WALMSLEY: Chemical Composition of the Frontal Gland Secretion of *Syntermes* Soldiers. (Isoptera, Termitidae). J. Chem. Ecol. **7**, 135–145 (1981).

525. PARTON, A.H., P.E. HOWSE, R. BAKER, and J.L. CLEMENT: Variation in the Chemistry of the Frontal Gland Secretion of European *Reticulitermes* Species. In: P.E. HOWSE and J.L. CLEMENT eds., Biosystematic of Social Insects, p. 193–209. New York: Academic Press 1981.

526. PRESTWICH, G.D., and D. CHEN: Soldier Defense Secretions of *Trinervitermes bettonianus*. (Isoptera, Nasutitermitinae): Chemical Variation in Allopatric Populations. J. Chem. Ecol. **7**, 147–157 (1981).

527. – M.S. COLLINS: Chemotaxonomy of *Subulitermes* and *Nasutitermes* Termite Soldiers Defense Secretions. Evidence Against the Hypothesis of Diphyletic Evolution of the Nasutitermitinae. Biochem. Syst. Ecol. **9**, 83–88 (1981).

528. – R.W. JONES, and M.S. COLLINS: Terpene Biosynthesis by Nasute Termite Soldiers (Isoptera: Nasutitermitinae). Insect Biochem. **11**, 331–336 (1981).

529. SPANTON, S.G., and G. PRESTWICH: Chemical Self-Defense by Termite Workers: Prevention of Autotoxication in Two Rhinotermitids. Science **214**, 1363–1365 (1981).

530. ZALKOW, L.H., R.W. HOWARD, L.T. GELBAUM, M.M. GORDON, H.M. DEUTSCH, and M.S. BLUM: Chemical Ecology of *Reticulitermes flavipes* (Kollar) and *R. virginicus* (Banks) (Rhinotermitidae): Chemistry of the Soldier Cephalic Secretions. J. Chem. Ecol. **7**, 717–731 (1981).

531. PRESTWICH, G.D.: From Tetracycles to Macrocycles. Chemical Diversity in the Defense Secretions of Nasute Termites. Tetrahedron **38**, 1911–1919 (1982).

532. – M.S. COLLINS: Chemical Defense Secretions of the Termite Soldiers of *Acorhinotermes* and *Rhinotermes* (Isoptera, Rhinotermitinae): Ketones, Vinyl Ketones and β-Ketoaldehydes Derived from Fatty Acids. J. Chem. Ecol. **8**, 147–161 (1982).

533. – Chemical Systematics of Termite Exocrine Secretions. Ann. Rev. Ecol. Syst. **14**, 287–311 (1983).

534. GOH, S.H., C.C. CHAN, Y.P. THO, and G.D. PRESTWICH: Extreme Intraspecific Chemical Variability in Soldier Defense Secretions of Sympatric and Allopatric Colonies of *Longipeditermes longipes*. J. Chem. Ecol. **10**, 929–944 (1984).

535. PRESTWICH, G.D.: Interspecific Variation of Diterpene Composition of *Cubitermes* Soldier Defense Secretion. J. Chem. Ecol. **10**, 1219–1231 (1984).

536. – Defense Mechanisms of Termites. Ann. Rev. Entomol. **29**, 201–232 (1984).

537. – M. TEMPESTA, and C. TURNER: Longipenol. A Novel Tetracyclic Diterpene from the Termite Soldier *Longipeditermes longipes*. Tetrahedron Lett. **25**, 1531–1532 (1984).

538. – W.S. ENG, E. DEATON, and D. WICHERN: Structure-Activity among Aromatic Analogs of the Trail Following Pheromone of Subterranean Termites. J. Chem. Ecol. **10**, 1201–1217 (1984).

539. TEMPESTA, M.S., J.K. PAWLAK, T. IWASHITA, Y. NAYA, K. NAKANISHI, and G.D. PRESTWICH: Cubegene, A Diterpenoid with a Novel Carbon Skeleton from a Termite Soldier (Isoptera: Termitidae: Termitinae). J. Org. Chem. **49**, 2077–2079 (1984).

540. TRANIELLO, J.F.A., B.L. THORNE, and G.D. PRESTWICH: Chemical Composition and Efficacy of Cephalic Gland Secretion of *Armitermes chagresi* (Isoptera: Termitidae). J. Chem. Ecol. **10**, 531–543 (1984).

541. VALTEROVA, I., M. BUDĚŠINSKÝ, F. TURECEK, and J. VRKOČ: Minor Diterpene Components of the Defense Secretion from the Frontal Gland of Soldiers of the Species *Nasutitermes costalis* (Holmgren). Collect. Czech. Chem. Comm. **49**, 2024–2039 (1984).

542. – J. KŘEČEK, and J. VROČ: Frontal Gland Secretion and Ecology of the Greater Antillean Termite *Nasutitermes hubbardii* (Isoptera: Termitidae). Acta Entomol. Bohemoslov **81**, 416–425 (1984).

543. GUSH, T.J., B.L. BENTLEY, G.D. PRESTWICH, and B. THRONE: Chemical Variation in Defensive Secretions of Four Species of *Nasutitermes*. Biochem. Syst. Ecol. **13**, 329–336 (1985).

544. PRESTWICH, G.D.: Communication in Insects. II. Molecular Communication in Insects. Q. Rev. Biol. **60**, 437–456 (1985).

545. – Isolation and Identification of Diterpenes from Termite Soldiers. Methods in Enyzmol. **110**, 417–425 (1985).

546. SCHEFFRAHN, R.H., L.K. GASTON, W.L. NUTTING, and M.K. RUST: Chemical heterogeneity of Soldier Defensive Secretions in the Desert Subterranean Termite, *Armitermes wheeleri*. Biochem Syst. Ecol. **14**, 661–664 (1986).

547. CHAU, C.H., S.H. COH, and Y.P. THO: Soldier Defense Secretions of the Genus *Hospitalitermes* in Peninsular Malasia. J. Chem. Ecol. **12**, 701–712 (1986).

548. MEINWALD, J., G. PRESTWICH, K. NAKANISHI, and I. KUBO: Chemical Ecology: Studies from East Africa. Science **199**, 1167–1173 (1978).

549. MOORE, B.P.: Studies on the Chemical Composition and Function of the Cephalic Gland Secretion in Australian Termites. J. Insect Physiol. **14**, 33–39 (1968).

550. LONGHURST, C., R. BAKER, and P.E. HOWSE: Chemical Crypsis in Predatory Ants. Experientia **35**, 870–872 (1979).

551. WESELOH, R.M.: Host Location by Parasitoids. In: D.A. NORDLUNG, R.L. JONES, and W.J. LEWIS eds., Semiochemicals: Their Role in Pest Control, p. 79–95. New York: John Wiley & Sons 1981.

552. JONES, R.L., W.J. LEWIS, M.C. BOWMAN, M. BEROZA, and B.A. BIERL: Host-Seeking Stimulant for Parasite of Corn Earworm: Isolation, Identification and Synthesis. Science **173**, 842–843 (1971).

553. – – M. BEROZA, B.A. BIERL, and A.N. SPARKS: Host-Seeking Stimulants (Kairomones) for the Egg Parasite, *Trichogramma evanescens*. Environ. Entomol. **2**, 593–596 (1973).

554. HENDRY, L.B., P.D. GREANY, and R.J. GILL: Kairomone Mediated Host-Finding Behavior in the Parasitic Wasps, *Orgilus lepidus*. Entomol. Exp. Appl. **16**, 471–477 (1973).

555. GROSS, H.R., JR., W.J. LEWIS, R.L. JONES, and D.A. NORDLUNG: Kairomones and their Use for Management of Entomophagous Insects. III. Stimulation of *Trichogramma achaeae*, *T. pretiosum*, and *Microplitis croceipes* with Host-Seeking Stimuli at Time of Release to Improve their Efficiency. J. Chem. Ecol. **1**, 431–438 (1975).

556. LEWIS, W.J., R.L. JONES, D.A. NORDLUNG, and A.N. SPARKS: Kairomones and their Use for Management of Entomophagous Insects. I. Evaluation for Increasing Rates of Parasitization by *Trichogramma* spp. in the Field. J. Chem. Ecol. **1**, 343–347 (1975).

557. – – – H.R. GROSS, JR.: Kairomones and their Use for Management of Entomophagous Insects. II. Mechanisms Causing Increase in Rate of Parasitization by *Trichogramma* spp. J. Chem. Ecol. **1**, 349–360 (1975).

558. VINSON, S.B., R.L. JONES, P.E. SONNET, B.A. BIERL, and M. BEROZA: Isolation, Identification and Synthesis of Host-Seeking Stimulants for *Cardiochiles nigriceps*, a Parasitoid of Tobacco Budworm. Entomol. Exp. Appl. **18**, 443–450 (1975).

559. NORDLUNG, D.A., W.J. LEWIS, R.L. JONES, and H.R. GROSS, JR.: Kairomones and their Use for Management of Entomophagous Insects. IV. Effects of Kairomones on Productivity and Longevity of *Trichogramma pretiosum* Riley (Hymenoptera: Trichogrammatidae). J. Chem. Ecol. **2**, 67–62 (1976).

560. LEWIS, W.J., R.L. JONES, H.R. GROSS, JR., and D.A. NORDLUNG: The Role of Kairomones and other Behavioral Chemicals in Host Finding by Parasitic Insects. Behav. Biol. **16**, 267–289 (1976).

561. – – D.A. NORDLUNG, and H.R. GROSS, JR.: Kairomones and their Use for Management of Entomophagous Insects. In: Editions du CNRS, Comportment des Insects et Milieu Trophique, p. 454–469. Paris: 1976.

562. JONES, R.L., W.J. LEWIS, H.R. GROSS, JR., and D.A. NORDLUNG: Use of Kairomones to Promote Action by Beneficial Insect Parasites. In: M. BEROZA ed. Pest Management with Insect Sex Attractants, American Chemical Society Symposium Series n° 23, p. 119–134. Washington, D.C. 1976.

563. LEWIS, W.J., D.A. NORDLUNG, H.R. GROSS, JR., and R.L. JONES: Kairomones and their Use for Management of Entomophagous Insects. V. Moth Scales as a Stimulus for Predation of *Heliothis zea* (Boddie) Eggs by *Chrysopa carnea* (Stephens) larvae. J. Chem. Ecol. **3**, 483–487 (1977).

564. NORDLUNG, D.A., W.J. LEWIS, R.L. JONES, H.R. GROSS, JR., and K.S. HAGEN: Kairomones and their Use for Management of Entomophagous Insects. IV. An Examination of the Kairomones for the Predator *Chrysopa carnea* Stephens at the Oviposition Sites of *Heliothis zea* (Boddie). J. Chem. Ecol. **3**, 507–511 (1977).

565. – – J.W. TODD, and R.B. CHALFANT: Kairomones and their Use for Management of Entomophagous Insects. VII. The Involvement of Various Stimuli in the Differential Response of *Trichogramma pretiosum* Riley to Two Suitable Hosts. J. Chem. Ecol. **3**, 513–518 (1977).

566. LEWIS, W.J., M. BEEVERS, D.A. NORDLUNG, H.R. GROSS, JR., and K.S. HAGEN: Kairomones and their Use for Management of Entomophagous Insects. IX. Investigations of Various Kairomone Treatment Patterns for *Trichogramma* spp. J. Chem. Ecol. **5**, 673–680 (1979).

567. STRAND, M.R., and S.B. VINSON: Source and Characterization of an Egg Recognition Kairomone of *Telenomus heliothidis*, a Parasitoid of *Heliothis virescens*. Physiol. Entomol. **7**, 83–90 (1982).

568. NORDLUNG, D.A., W.J. LEWIS, and R.C. GUELDNER: Kairomones and their Use for Management of Entomophagous Insects. XIV. Response of *Telenomus remus*

to abdominal Tips of *Spodoptera frugiperda*, (Z)-9-tetradecen-1-yl acetate and (Z)-9-dodecen-1-yl acetate. J. Chem. Ecol. **9**, 695–701 (1983).

569. GUELDNER, R.C., D.A. NORDLUNG, W.J. LEWIS, J.E. THEAN, and D.M. WILSON: Kairomones and their use for Management of Entomophagous Insects. XV. Identification of Several Acids in Scales of *Heliothis zea* Moths and Comments on their Possible Role as Kairomones for *Trichogramma pretiosum*. J. Chem. Ecol. **10**, 245–251 (1984).

570. BLUM, M.S., T.H. JONES, B. HOLLDÖBLER, H.M. FALES, and T. JAOUNI: Alkaloidal Venom Mace: Offensive Use by a Thief Ant. Naturwissenschaften **67**, 144–145 (1980).

571. TENGÖ, J.: Territorial Behavior of the Kleptoparasite Reduces Parasitic Pressure in Communally Nesting Bees. Short Lecture Presented at the XVII International Congress of Entomology, p. 20–26. Hamburg (1984).

572. KULLEMBERG, B., and B. BERGSTRÖM: Chemical Communication between Living Organisms. Endeavour **34**, 59–65 (1976).

573. TENGÖ, J., and B. BERGSTRÖM: Comparative Analysis of Lemon-Smelling Secretions from Heads of *Andrena* F. (Hymenoptera: Apoidea) Bess. Comp. Biochem. Physiol. **55 B**, 179–188 (1976).

574. – – All-*trans*-farnesyl Hexanoate and Geranyl Octanoate in the Dufour Gland Secretion of *Andrena* (Hymenoptera: Apoidea). J. Chem. Ecol. **1**, 253–268 (1975).

575. – – Odor Correspondence between *Melitta* Females and Males of their Nest Parasites *Nomada flavopicta* K. (Hymenoptra: Apoidea). J. Chem. Ecol. **2**, 57–65 (1976).

576. – – Cleptoparasitism and Odor Mimetism in Bees: Do *Nomada* Males Imitate the Odor of *Andrena* Females? Science **196**, 1117–1119 (1977).

577. – – Comparative Analyses of Complex Secretions from Heads of *Andrena* Bees. Comp. Biochem. Physiol. **57 B**, 197–200 (1977).

578. – – A.K. BORG-KARLSON, I. GROTH, and W. FRANCKE: Volatile Compounds from Cephalic Secretions of Females in two Cleptoparasite Bees Genera, *Epeolus* (Hym. Anthophoridae) and *Coelioxys* (Hym., Megachilidae). Z. Naturforsch **37 C**, 376–380 (1982).

579. HEFETZ, A., G.C. EICKWORT, M.S. BLUM, J. CANE, and G.E. BOHART: A Comparative Study of the Exocrine Products of Cleoptoparasitic Bees (*Holcopasites*) and their Hosts (*Calliopsis*) (Hymenoptera: Anthophoridae, Andrenidae). J. Chem. Ecol. **8**, 1389–1397 (1982).

580. REGNIER, F.E., and E.O. WILSON: Chemical Communication and "Propaganda" in Slave-Maker Ants. Science **172**, 267–269 (1971).

581. WILSON, E.O.: *Leptothorax duloticus* and the Beginnings of Slavery in Ants. Evolution **29**, 108–119 (1975).

582. HOWARD, R.W., C.A. McDANIEL, and G.J. BLOMQUIST: Cuticular Hydrocarbons of the Eastern Subterranean Termite *Reticulitermes flavipes* (Kollar) (Isoptera: Rhinotermitidae). J. Chem. Ecol. **4**, 233–245 (1978).

583. – – – Chemical Mimicry as an Integrating Mechanism: Cuticular Hydrocarbons of Termotophile and its Host. Science **210**, 431–433 (1980).

584. HADLEY, N.F.: Surface Waxes and Integumentary Permeability. Amer. Sci. **68**, 546–553 (1980).

585. HOWARD, R.W., C.A. McDANIEL, and G.J. BLOMQUIST: Chemical Mimicry as an Integrating Mechanism for Three Termitophiles Associated with *Reticulitermes virginicus* (Banks). Psyche **89**, 157–167 (1982).

586. BLOMQUIST, G.J., and L.L. JACKSON: Chemistry and Biochemistry of Insect Waxes. Prog. Lipid Rese. **17**, 319–345 (1979).

587. COMFORT, A.: Likelihood of Human Pheromones. Nature **230**, 432–433 (1971).

588. McCLINTOCK, M.: Menstrual Synchrony and Suppression. Nature **229**, 244–245 (1971).

589. BROOKSBANK, R W I , R. BROWN, and J.A. GUSTAFSSON: The Detection of 5α-Androst-16-en-3α-ol in Human Males Axillary Sweat. Experientia **30**, 864–865 (1974).

590. MICHAEL, R.P., R.W. BONSALL, and P. WARNER: Human Vaginal Secretions: Volatile Fatty Acid Content. Science **186**, 1217–1218 (1974).

591. DOTY, R.L., M. FORD, G. PRETI, and G.R. HUGGINS: Changes in the Intensity and Pleasantness of Human Vaginal Odors During the Menstrual Cycle. Science **190**, 1316–1318 (1975).

592. AMOORE, J.E., and L.J. FORRETIER: Specific Anosmia to Trimethyl Amine: The Fish Primary Odor. J. Chem. Ecol. **2**, 49–56 (1976).

593. CLAUS, R., and W. ALSING: Occurrence of 5α-Androst-16-en-3-one, a Boar Pheromone, in Man and its Relationship to Testosterone. J. Endocr. **68**, 483–484 (1976).

594. RUSSELL, M.J.: Human Olfactory Communication. Nature **260**, 520–522 (1976).

595. GRAHAN, C.A., and W.C. McGREW: Menstrual Synchrony in Female Undergraduates Living on a Coeducational Campus. Psychoendocrinol. **5**, 245–252 (1980).

596. DOTY, R.L., M.M. ORNDORFF, J. LEYDEN, and A. KLIGMAN: Communication of Gender from Human Axillary Odors: Relationship to Perceived Intensity and Hedonicity. Behav. Biol. **23**, 373–380 (1978).

597. KIRK-SMITH, M.D., and D.A. BOOTH: Effect of Androstenone on Choice of Location in Other's Presence. Proc. 7th Int. Symp. Olfaction and Taste, p. 397–400 (1980).

598. PRETI, G., J.G. KOLSTEC, J. TONZETICH, and G.R. HUGGINS: Detecting Ovulation by Monitoring Dodecanol Concentration in Saliva. Chemical Abstracts **99**, 35.687a (1980).

599. RUSSELL, M.J., G.N. SCHWITZ, and K. THOMPSON: Olfactory Influences on the Human Menstrual Cycle. Pharmacol. Biochem. Behav. **13**, 737–738 (1980).

600. LEYDEN, J.J., McGINLEY, K.J., E. HOLZLE, J.N. LABOWS, and A.M. KLIGMAN: The Microbiology of the Human Axilla and its Relationship to Axillary Odor. J. Invest. Dermatol. **77**, 413–416 (1981).

601. SCHLEIDT, M., B. HOLD, and G. ATTILI: A Cross-Cultural Study on the Attitude Towards Personal Odors. J. Chem. Ecol. **7**, 19–31 (1981).

602. LABOWS, J.N., K.J. McGINLEY, and A.M. KLIGMAN: Perspective on Axillary Odors. J. Soc. Cosmet. Chem. **34**, 193–202 (1982).

603. BIRD, S., and D.B. GOWER: Estimation of the Odours Steroid, 5α-Androst-16-en-3-one, in Human Saliva. Experientia **39**, 790–792 (1983).

604. FILSINGER, E.E., J.J. BRAUN, W.C. MONTE, and D.E. LINDER: Human (*Homo sapiens*) Responses to the Pig (*Sus scrofa*) Sex Pheromone 5α-Androst-16-en-3-one. J. Comp. Psychol. **98**, 219–222 (1984).

605. NIXON, A., P. JACKMAN, A.I. MALLET, and D.B. GOWER: Steroid Metabolism by Human Axillary Bacteria. Biochem. Soc. Trans. **12**, 1114–1115 (1984).

606. WYSOCKI, C.J., and G.K. BEAUCHAMP: Ability to Smell Androstenone Is Genetically Determined. Proc. Natl. Acad. Sci. USA **81**, 4899–4902 (1984).

607. STODDART, D.M.: Is Incense a Pheromone? Interdiscip. Sci. Rev. **10**, 237–247 (1985).

(Received April 20, 1987)

Phenolic Compounds of the Mulberry Tree and Related Plants

By T. NOMURA, Faculty of Pharmaceutical Sciences, Toho University, Miyama, Funabashi, Chiba 274, Japan

With 70 Figures

Contents

I. Introduction

Moraceae comprise a large family of sixty genera and nearly 1400 species (*1*), including important groups such as *Artocarpus*, *Morus*, and *Ficus*. In particular, *Morus* (mulberry) is a small genus of trees and shrubs found in temperate and subtropical regions of the Northern Hemisphere and has been widely cultivated in China and Japan for its leaves which serve as indispensable food for silkworms. Many varieties of *Morus* are cultivated in Japan; these varieties are described as belonging to three species: *Morus alba* L. ("Karayamaguwa" in Japanese), *M. bombycis* Koidz. ("Yamaguwa" in Japanese), and *M. lhou* (ser.) Koidz. ("Roguwa" in Japanese) (*2*). In addition, the root bark of the mulberry tree (Mori Cortex, *Morus alba* L. and other plants of the genus *Morus*) has been used as an antiphlogistic, diuretic, and expectorant in Chinese herbal medicine (*3, 4*), and the crude drug is known as "Sohakuhi" in Japanese. In the pharmaceutical field, a few papers have been published reporting the hypotensive effect of this extract (*5–11*). The first of these reports was presented by Fukutome in 1938 who asserted that oral administration of the hot water extract of the mulberry tree showed a remarkable hypotensive effect in rabbits (*5*). Ohishi reported the hypotensive effect of the ethanol extract of mulberry root bark (*6*), while Suzuki and Sakuma (*7*) reported that the hypotensive activity seemed to be due to phenolic substances and that the effect disappeared on acetylation. Later, Katayanagi *et al.* reported that the ether extract of the root bark given to rabbits (6 mg/ Kg, iv) showed a marked hypotensive effect and that the active constituents seemed to be a mixture of unstable phenolic compounds (*8*). Tanemura ascribed the activity of mulberry tree root bark to acetylcholine and its analogues presumably contained in the alcohol soluble fraction,

and that the hypotensive constituents produced a yellowish-brown pre-
cipitate on treatment with DRAGENDORFF reagent (9).

As for the components of *Morus* root bark, occurrence of triterpen-
oids (12), diglycelide (13), and piperidine alkaloids (14) had been reported prior to the beginning of our work, whereas the hypotensive
constituents had not been identified. In view of the reports cited in
the previous paragraph we assumed that the hypotensive components
were a mixture of unstable phenolic compounds and therefore under-
took a study of the phenolic constituents of the root bark of the cultivated mulberry tree. This article reviews the chemistry and biological
activities of phenolic constituents obtained from the mulberry tree and
related plants.

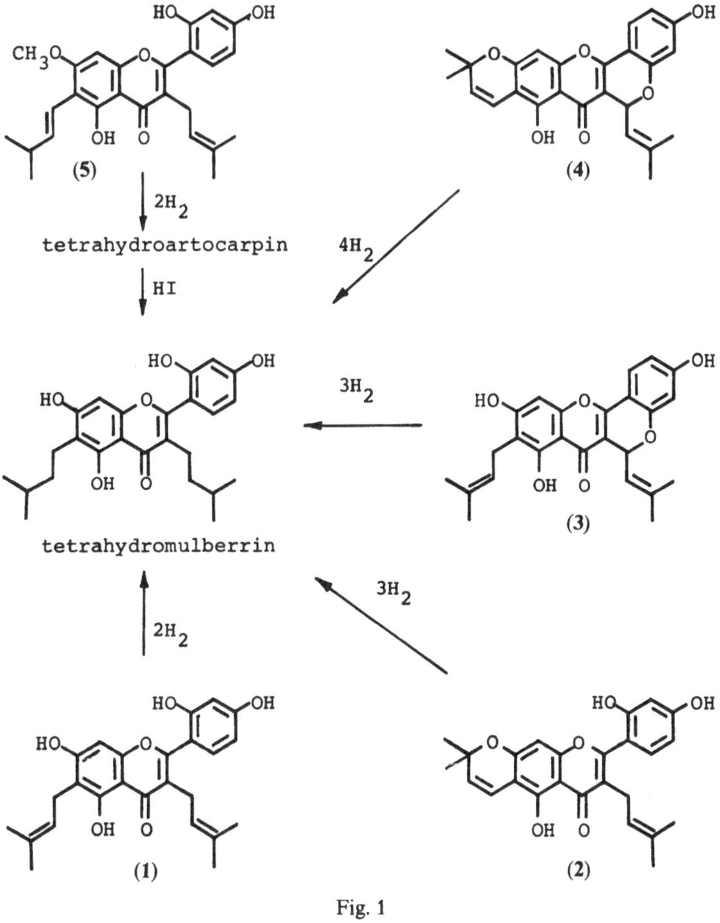

Fig. 1

In earlier work on the phenolic constituents of *Morus* root bark, Uno isolated coumarins and related compounds (*15, 16*) such as umbelliferone and 5,7-dihydroxychromone, while Shibata reported isolation of a large quantity of β-tocopherol (more than 15 mg% for dried material) (*17*). Venkataraman *et al.* isolated a series of flavonoids with isoprenoid substituents from Moraceae which they described in review articles (*18, 19*). Details were given by Deshpande *et al.* who described four flavones with isoprenoid substituents which the named mulberrin, mulberrochromene, cyclomulberrin, and cyclomulberrochromene, from the root bark of *Morus alba* L. Structures (**1**), (**2**), (**3**), and (**4**), were assigned to the four compounds, the position of C-alkylation at C-6 of the flavone moiety presumably having been deduced by chemical correlation with artocarpin (**5**) (*20*) (Fig. 1).

Mulberranol (**6**) was also isolated from the same source (*21*). The bark of *M. rubra* (red mulberry), an American species, gave four flavones designated as rubraflavones A, B, C, and D whose structures were demonstrated to be (**7**), (**8**), (**9**), and (**10**), respectively, mainly on the basis of ^{1}H-NMR and mass spectral data. These rubraflavones are the first naturally occurring flavones with a C_{10} substituent (*22*). The heartwood of *Artocarpus heterophyllus* Lamarck (Jack tree, *A. integrifolia*), yielded artocarpin (**5**) (*23, 24*), artocarpesin (**11**) (*25*), cycloartocarpesin (**12**) (*26*), oxydihydroartocarpesin (**13**) (*26*), cycloheterophyllin (**14**) (*27*), heterophyllin (**15**) (*27*), cycloartocarpin (**16**) (*28*), and isocycloheterophyllin (**17**) (*29*) (Fig. 2). The heartwood of *A. chaplasha* Roxb. gave chaplasin (**18**) containing an oxepine ring whose structure was derived by UV, ^{1}H-NMR, and mass spectra (MS) and by synthesis of racemic dihydrochaprashin from compound (**19**) obtained by the action of dichlorodicyanobenzoquinone (DDQ) on dihydroartocarpin (**20**) (*30*) (Fig. 3). From the heartwood of *A. integer* Thunb., an Indonesian species, cyclointegrin (**21**), integrin (**22**), oxyisocyclointegrin (**23**), were isolated; cyclointegrin was the first naturally occuring flavone containing an oxocin ring system (*31*) (Fig. 2). All of these flavonoids are characterized by β-resorcylic acid orientation of the hydroxyl groups in the B ring. From the chemotaxonomic point of view the flavonoids containing an isoprenoid at the C-3 position characterize some species of the genera *Artocarpus* and *Morus* of the family Moraceae.

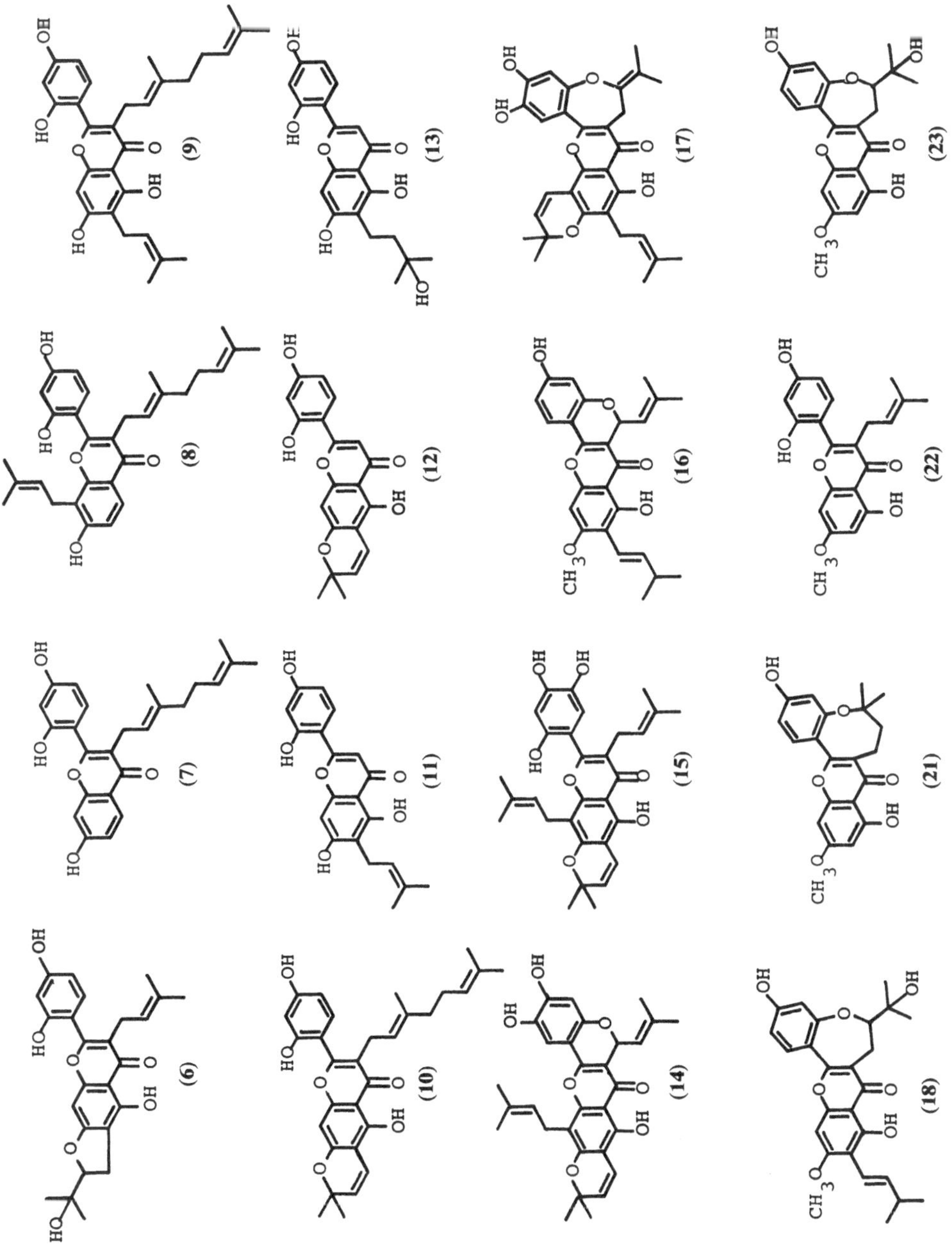

Fig. 2

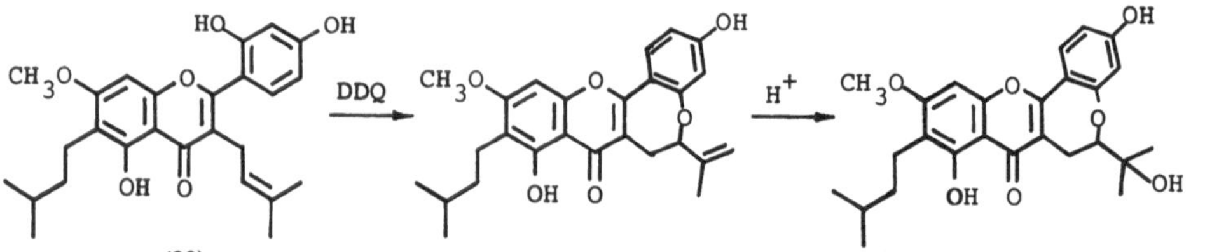

Fig. 3

II. Flavonoids Carrying Isoprenoid Substituents Isolated from the Japanese Cultivated Mulberry Tree

1. Isolation of Flavonoids Carrying Isoprenoid Substituents

As shown in Fig. 4 thirteen new flavonoids with isoprenoid substituents were obtained from the extract of the root bark of the Japanese cultivated mulberry tree. Morusin (24, 0.04% yield from dried *Morus* root bark) (*32*), kuwanons G (25, 0.2% yield) (*33*), H (26, 0.13% yield) (*34*), and mulberrofuran G (27, 0.08% yield) (*35*) are the main phenolic constituents, the others are minor.

2. Structures of Flavonoids Carrying Isoprenoid Substituents

Morusin (24) (*32*), $C_{25}H_{24}O_6$, was obtained in two crystalline forms, one being pale yellow prisms, mp 214–216 °C, and the other, being pale yellow needles, 161–163 °C. Substance (24) gave an intense green color with methanolic $FeCl_3$, and was positive in the color reactions characteristic of flavones. The UV spectrum showed maxima at 270 nm (39 800), 320 (shoulder, 7900), and 350 (shoulder, 6500). UV spectra of flavones usually exhibit two absorption maxima known as Band I (usually 300–380 nm) and Band II (240–280 nm). Band I is associated with absorption due to the B ring cinnamoyl system and Band II with absorption involving the A ring benzoyl system (*36*). In the case of 24, Band I is shifted to shorter wave length and is much less intense than that of 5,7,2′,4′-tetrahydroxyflavone (*24*). This result suggests that the B and C rings are twisted relative to each other due to steric hindrance (*20, 24*). The ^{1}H-NMR spectrum of 24 had no signal characteristic of H-3 while the signal of H-6′ in the B ring was at higher field

References, pp. 191–201

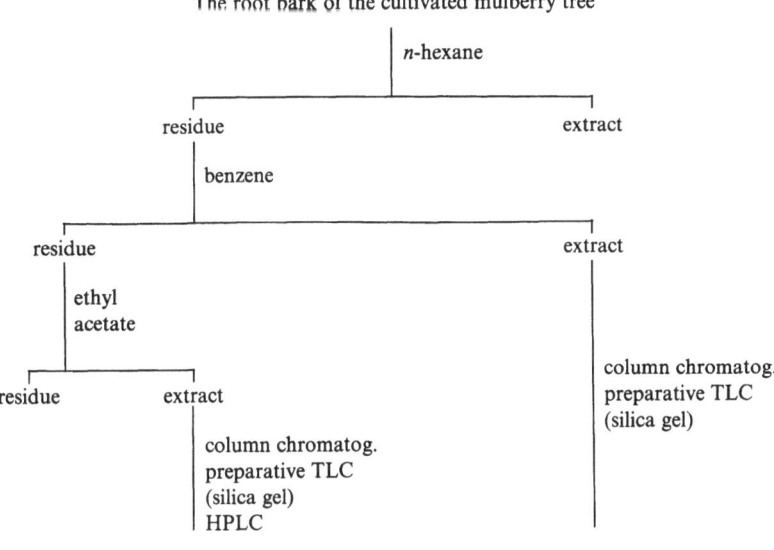

The root bark of the cultivated mulberry tree

n-hexane

residue   extract

benzene

residue   extract

ethyl
acetate
                                           column chromatog.
                                           preparative TLC
residue   extract   (silica gel)

column chromatog.
preparative TLC
(silica gel)
HPLC

kuwanon G (= albanin F = moracenin B)   morusin (= mulberrochromene)
kuwanon H (= albanin G = moracenin A)   cyclomorusin (= cyclomulberrochromene)
kuwanon I → P, S → U, W   compound A
moracenin D   oxydihydromorusin (= morusinol)
mulberrofuran C   kuwanon A → F
mulberrofuran F   mulberrofuran A, B, L
mulberrofuran G (= albanol A)
mulberrofuran H → J

Fig. 4

than H-6' of flavones which have no prenyl (γ,γ-dimethylallyl) group
at C-3. Hence **24** was a 3-isoprenylated flavone (*20*). The MS of **24**
showed the following fragment ions: *m/z* 405 (**28**), 203 (**29**, formed
from the ion at *m/z* 405 by retro Diels-Alder reaction) (Fig. 5). The
above results and analysis of the ^{1}H-NMR spectrum of **24** (Fig. 6)
led to formulas **2** or **24** for the structure of morusin. Angular structure (**24**) was supported by comparing of the chemical shifts of the
olefinic protons in the chromene portion of the diacetate (**24a**) with
those in triacetate (**24b**) (Table 1). These changes are of the same sign
and the same order of magnitude as those observed by many investiga-
tors for similar compounds (*37*). Hence, morusin is represented by formula (**24**). The ^{13}C-NMR spectrum of **24** was analyzed by the off-
resonance decoupling technique and by comparison with model compounds (Table 2) (*38, 39*).

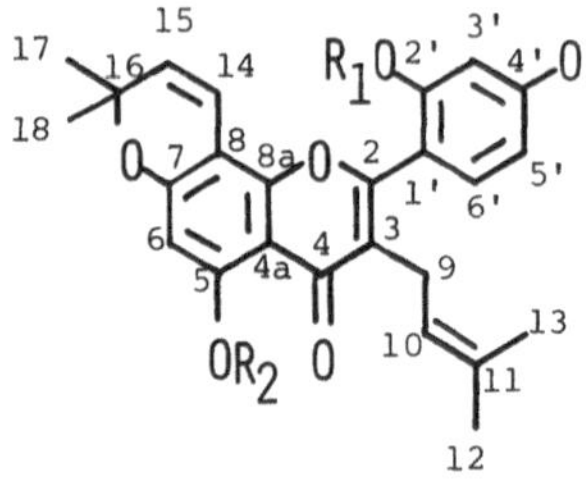

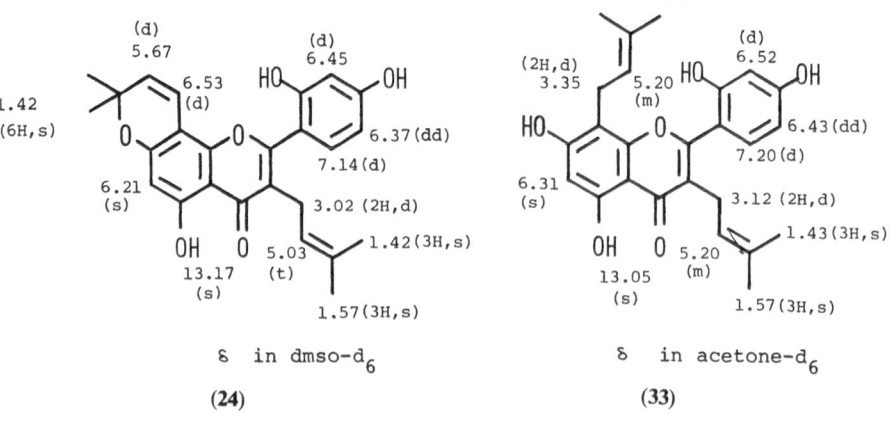

24: $R_1 = R_2 = H$
24a: $R_1 = Ac$, $R_2 = H$
24b: $R_1 = R_2 = Ac$
24c: $R_1 = R_2 = CH_3$

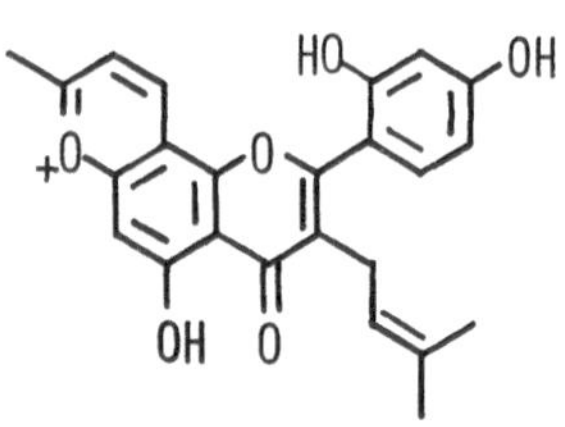

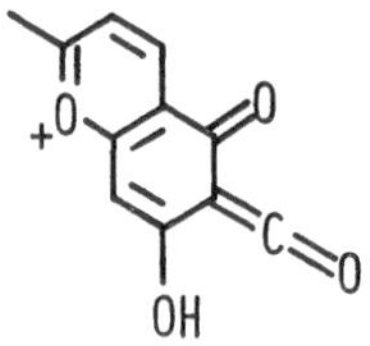

(28)

(29)

Fig. 5

δ in dmso-d_6

(24)

δ in acetone-d_6

(33)

Fig. 6. ^{1}H-NMR chemical shifts values (ppm) of morusin (**24**) and kuwanon C (**33**)

Cyclomorusin (**30**) (*32*), $C_{25}H_{22}O_6$, mp 246–248 °C, was negative in the Gibbs test and gave a characteristic MS fragment ion at m/z 363 (*32*). The structure was confirmed by oxidation of **24** with manganese dioxide to **30′**.

Compound A (**31**) (*32*), $C_{25}H_{24}O_7$, mp 258–260 °C, afforded a diacetate (**31a**, negative to $FeCl_3$), a triacetate (**31b**), and a monomethyl ether (**31c**, positive to $FeCl_3$). Treatment of **31** with thionyl chloride in pyridine gave an anhydro product (**32**), whose ^{1}H-NMR spectrum

Table 1. *Acetylation Shifts for H-14 and H-15 of* **24**, **35**, *and* **36**

	C_{14}-H	C_{15}-H
24a*	6.52	5.49
24b*	6.60	5.59
Δ	−0.08	−0.10
35**	6.75	5.69
35a**	6.53	5.88
Δ	+0.22	−0.19
36**	6.77	5.72
36a**	6.46	5.90
Δ	+0.31	−0.18

* Measured in $CDCl_3$.
** Measured in acetone-d_6.

in pyridine-d_5 showed the following signals: δ1.88 (3H, s, C-11-CH$_3$), 2.88–3.15 (2H, m, C-9-Hx2), 4.83 (1H, m, C-10-H), 4.93, 5.06 (each 1H, s, C-11-CH$_2$). Compound (**32**) afforded a diacetate (**32a**) which had no IR absorption in the hydroxyl region. Further evidence supporting structure **31** was its formation by oxidation of **24** with DDQ, in light of similar reactions, reported earlier by VENKATARAMAN *et al.* (*30*).

Kuwanon C (**33**) (*40*), $C_{25}H_{26}O_6$, mp 148–150 °C, showed the ^{1}H-NMR spectrum shown in Fig. 6. Its structure was confirmed by its oxidation with DDQ which afforded morusin (**24**) and compound (**32**).

Oxydihydromorusin (morusinol, **34**) (*38, 40*), $C_{25}H_{26}O_7$, mp 215–216 °C, was isolated from *Morus* root bark by our group and by KONNO *et al.* Its structure was confirmed by hydration of **24** with methanol-hydrochloric acid. The correlations of the five flavones with isoprenoid substituents (**24**, **30**, **31**, **33**, and **34**) are shown in Fig. 7.

Kuwanons A (**35**) (*40*), $C_{25}H_{24}O_6$, and B (**36**), (*40*), $C_{25}H_{24}O_6$, mp 250–254 °C, are flavonoids with a 2,2-dimethylchromene ring as ring B and a γ,γ-dimethylallyl group attached to C-3 based on the spectral data (UV, ^{1}H-NMR, MS). That H-14 of the pyran ring is *peri* to the hydroxyl group in ring B of in **35** and **36** is based on the chemical shift changes of, H-14 and H-15 when **35** and **36** are acetylated (Table 1). The changes are of the same sign and of the same order of magnitude as those observed for similar compounds in which the hydroxyl group is *peri* to H-14 (*37*). These results indicate that both kuwanons A and B are represented by formula **35** or **36**. Final proof for the structures was obtained by a photo-oxidation process described in Section III. Thus irradiation of **36** in chloroform using a high-pres-

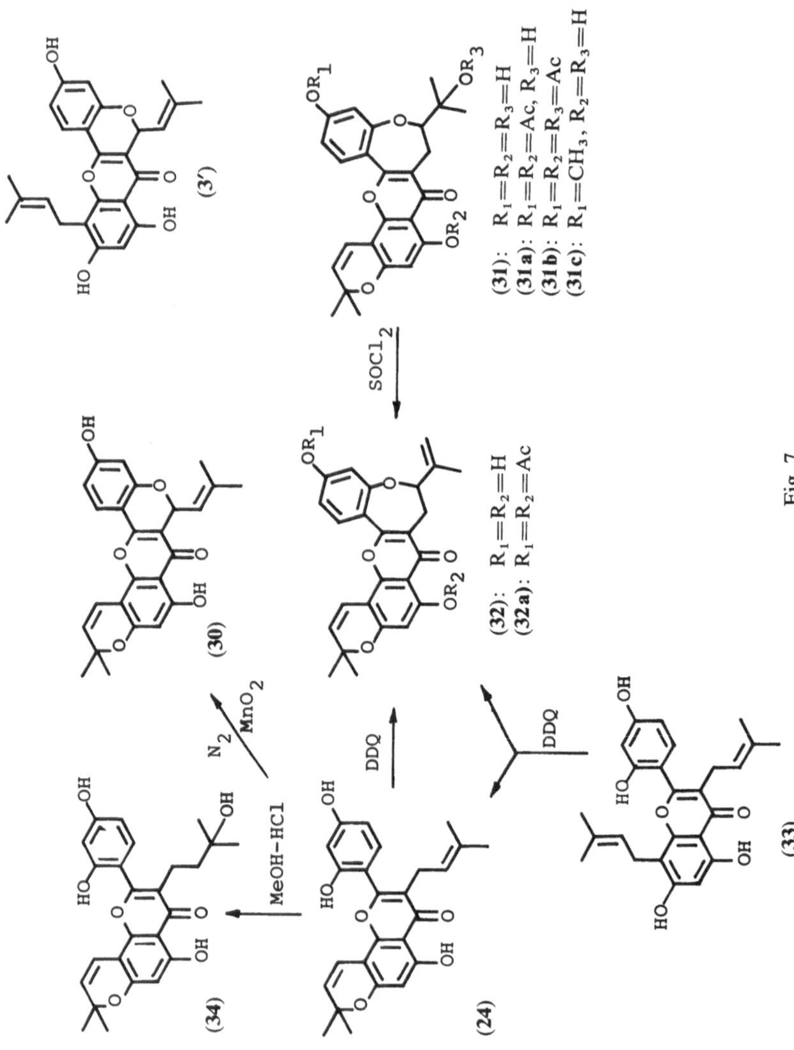

Fig. 7

sure mercury lamp gave kuwanon B hydroperoxide (**36b**), while this photo-oxidation did not occur in kuwanon A (**35**). In the light of the result of photo-oxidation of morusin (**24**) (*41*), it was concluded that kuwanon B (**36**) has a hydroxyl group at C-2′ and an isolated double bond in the γ,γ-dimethylallyl group attached to C-3, thus leading to formula **35** for kuwanon A and formula **36** for kuwanon B (Fig. 8).

References, pp. 191–201

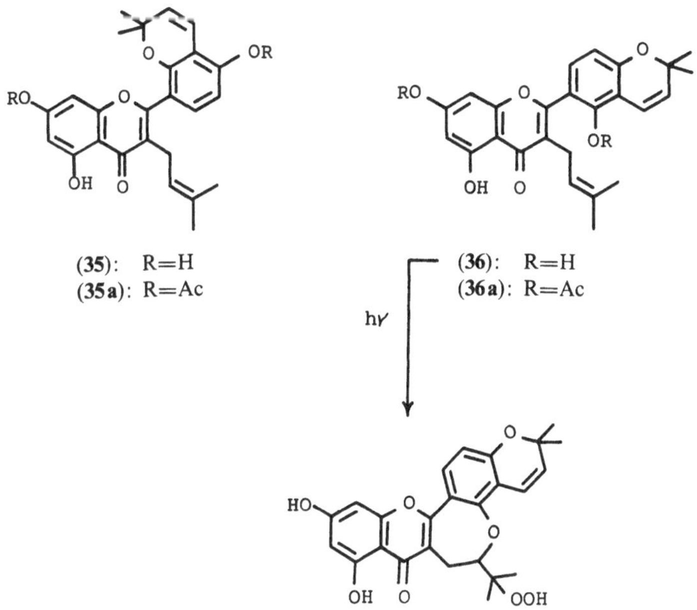

(35): R=H
(35a): R=Ac

(36): R=H
(36a): R=Ac

hν

(36b)

Fig. 8

Kuwanon D (37) (42), $C_{25}H_{26}O_6$, mp 230–232 °C, gave a positive Mg-HCl test and $NaBH_4$ test (43). The spectral data suggested that 37 was a 5,7,2′4′-tetraoxygenated flavanone. The MS had the base peak at m/z 339 ($C_{19}H_{15}O_6$) corresponding to oxonium ion (38) which is expected for flavanone derivatives which contain a pyran ring fused to the aromatic nucleus (44). The ^{1}H-NMR spectrum of 37 showed the absence of signals for olefinic protons. Kuwanon D (37) was recovered unchanged on catalytic hydrogenation using Adams catalyst. The absence of double bonds indicated that the kuwanon D was hexacyclic structure, thus leading to formula 37.

The "cyclol unit" (39) (45) in formula 37 was supported by the ^{1}H-NMR spectrum which displayed the following signals in pyridine-d_5 δ0.78 (3H, s, C-8″-CH$_3$), 1.27 (3H, s, C-3″-CH$_3$ or C-8″-CH$_3$), 1.33 (3H, s, C-3″-CH$_3$ or C-8″-CH$_3$), 1.40–2.10 (4H, m, H5″a, b and H6″a, b-Hx2), 2.28 (1H, m, H-7″), 2.52 (1H, dd, $J=7.5$ and 9.6 Hz, H-2″), 3.02 (1H, d, $J=9.6$ Hz, H-1″). The chemical shifts and the splitting patterns of these signals are similar to those of the corresponding protons in compound, such as cannabicyclol (40) (46) and hydroxyeriobrucinol (41) (44).

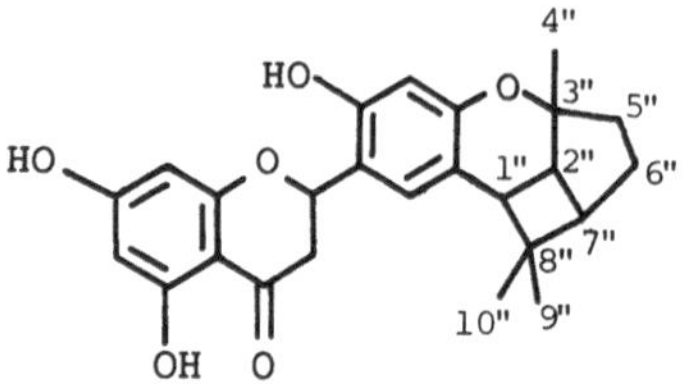

Fig. 9

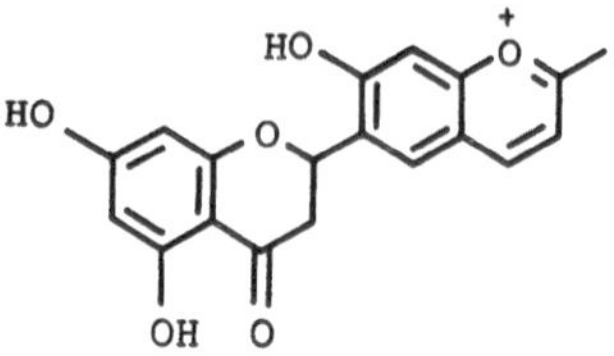

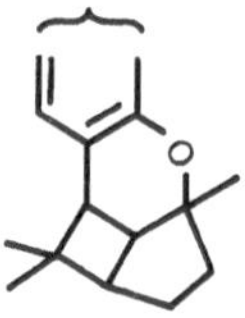

Kuwanons E (42) (42), $C_{25}H_{28}O_6$, mp 132–136 °C, F (43) (42), $C_{25}H_{26}O_6$, S (44) (47), $C_{25}H_{26}O_5$, mp 78–80 °C, T (45) (47), $C_{25}H_{26}O_6$, mp 191–193 °C, and U (46) (48) have been isolated from the root bark of cultivated mulberry tree. Structures were deduced on the basis of spectral data. Kuwanon F (43) was derived from 42

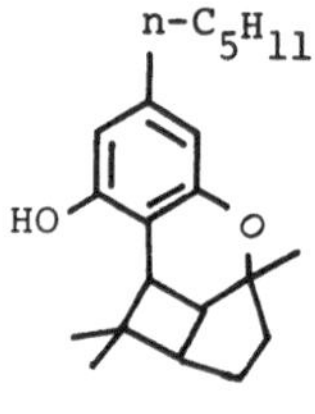

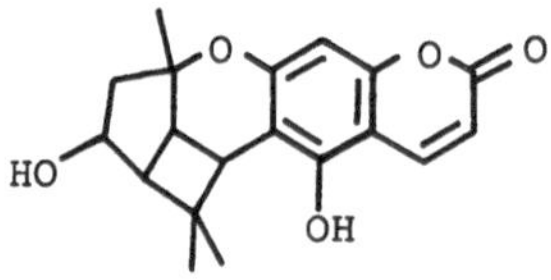

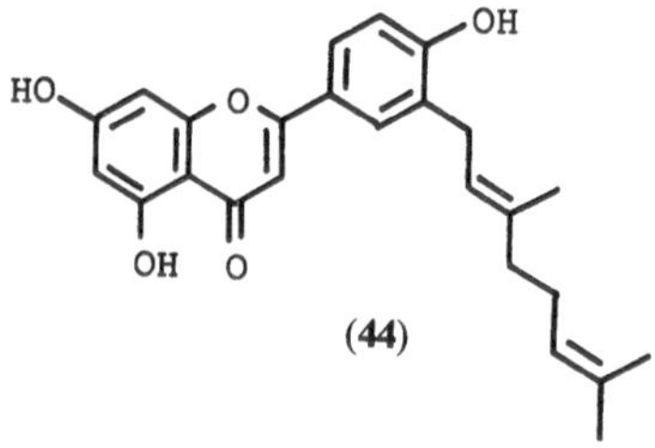

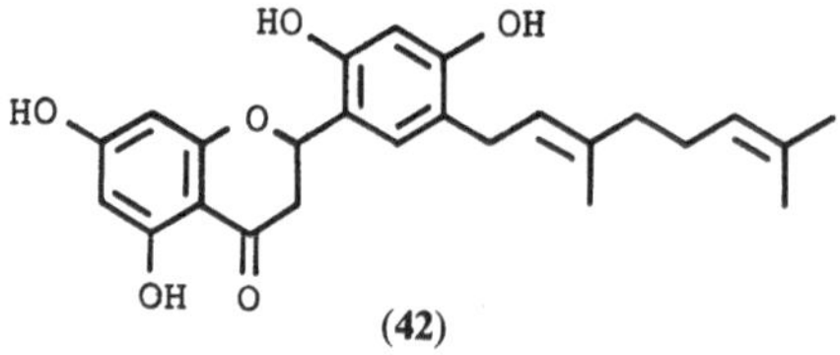

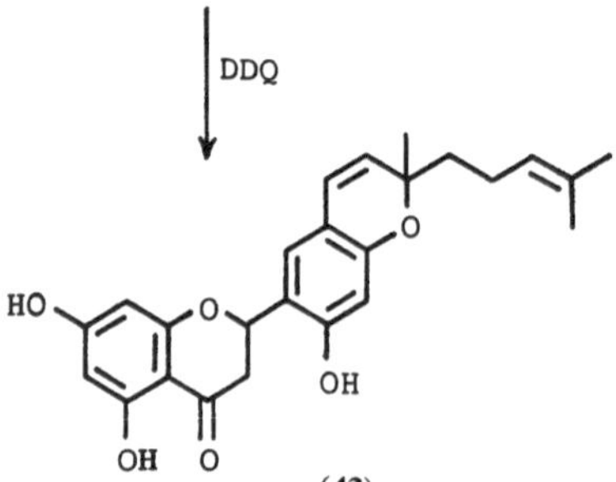

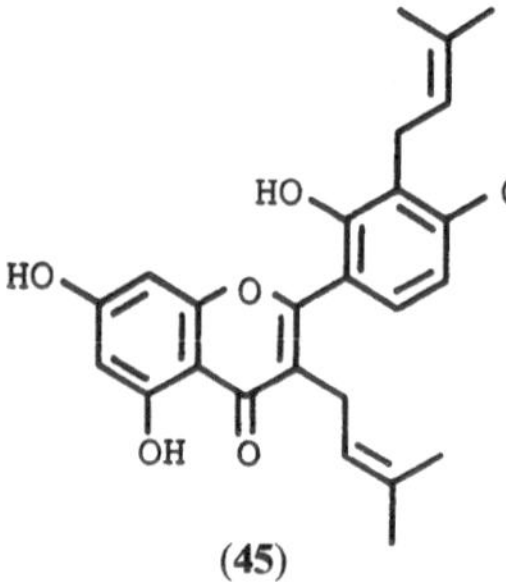

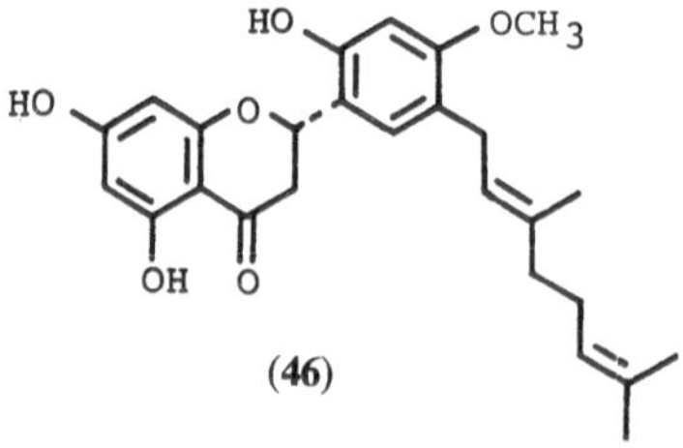

Table 2. ^{13}C-NMR Data of Flavonoids Carrying Isoprenoid Substituents

C	MO (24)	ODM (34)§	CA (31)	KC (33)	KT (45)	(C)	KD (37)	(C)	KE (42)	KS (44)
2	158.6[a]	158.8[a]	159.7[a]	158.9[a]	161.6[a]		75.3		76.0	165.4
3	120.2	121.6	117.2	119.4	120.2		42.9		42.6	104.0
4	181.8	182.2	181.7	181.8	181.8		197.3		198.5	183.0
4a	100.5[b]	100.8[b]	101.4[b]	103.4[b]	103.7		103.1		103.3	105.3
5	151.9[a]	152.1[a]	152.2[a]	155.0	161.5[a]		165.4[a]		167.6	159.6
6	98.9	99.1	99.9	97.9	98.3		97.2		97.0	99.7
7	162.0[c]	160.6[c]	163.6[c]	161.7[c]	164.0		168.6		167.8	164.9
8	104.4[b]	104.6[b]	104.5[b]	105.5[b]	93.4		96.2		96.4	94.7
8a	160.7[c]	160.6[c]	161.4[c]	160.3[a]	157.9[b]		164.8[a]		165.0	158.0
1′	110.9	111.2	114.2	111.3	111.8		119.4		120.8	123.2
2′	156.7[a]	156.7[a]	151.1[a]	156.5[a]	153.3		155.2[b]		154.1	128.8
3′	103.0	103.1	112.3	102.7	115.8		105.5		103.7	137.2
4′	161.1[c]	162.0[c]	162.4[c]	161.2[c]	157.8[b]		155.4[b]		156.6	163.4
5′	107.0	107.1	109.0	106.7	106.9		115.7		117.3	116.3
6′	131.3	131.4	130.7	131.2	127.5		128.9		128.8	125.0
9	23.7	28.0	25.5	23.5	23.6	(1″)	46.8	(1″)	25.8	28.9
10	121.6	42.1	90.9	121.7[d]	121.4	(3″)	84.1	(2″)	123.9	122.9
11	131.3	69.3	71.9	131.2[e]	131.1	(2″)	34.6	(3″)	136.8	131.8
12	25.4	29.0	25.2	25.4	25.4[e]	(5″)	38.5	(4″)	16.2	16.3
13	17.3	29.0	27.6	17.3	17.2[d]	(6″)	38.9	(5″)	40.5	40.5
14	114.3	114.5	115.4	21.1	22.2	(7″)	39.4	(6″)	28.3	27.5
15	127.6	127.9	127.6	122.1[d]	123.0	(8″)	39.8	(7″)	125.3	126.6
16	78.0	78.4	78.3	130.7[e]	130.0	(4″)	19.4	(8″)	132.1	129.9
17	27.7	28.0	28.1	25.4	25.5[e]	(9″)	25.7	(9″)	27.5	25.8
18	27.7	28.0	18.1	17.3	17.7[d]	(10″)	27.2	(10″)	17.8	17.8
Solvent	A	A	B	A	A		B		C	D

MO morusin, ODM oxydihydromorusin, CA compound A, KC kuwanon C, KT kuwanon T, KD kuwanon D, KE kuwanon E, KS kuwanon S.
§ Data from KONNO et al. (Ref. 38).
A dmso-d_6, B pyridine-d_5, C CD$_3$OD, D acetone-d_6.
[a–d] Assignments may be interchangeable in vertical column.

by oxidation with DDQ (Fig. 9). ^{13}C-NMR spectra of these flavonoids with isoprenoid substituents are shown in Table 2 (49).

3. Synthesis of Tetrahydrokuwanon C Tetramethyl Ether

As mentioned in the preceding section, morusin (24), cyclomorusin (30), compound A (31), kuwanon C (33), and oxydihydromorusin (34) were correlated as shown in Fig. 7. Structures were confirmed by synthesis of tetrahydrokuwanon C tetramethyl ether (33a) from phloroglucinol by the route shown in Fig. 10 (50). Condensation of 47 (51) with 5-methylhexanoic acid in the presence of boron trifluoride etherate gave 2-hydroxy-3-isopentyl-4,6-dimethoxy-isoheptophenone (48) in about 65% yield. The structure assigned to 48 was based the following results. Compound (48) gave a positive Gibbs test and the IR spectrum had a band characteristic of a conjungted carbonyl at 1630 cm^{-1}. In the low-field region of the ^{1}H-NMR spectrum, a sharp signal was observed at δ 13.97; the spectrum also indicated the presence of an isopentyl and a 5-methylhexanoyl group. These results excluded a 2,6-dimethoxy-3-isopentyl-4-hydroxy-isoheptophenone structure for this condensation product. The flavone derivative (33a) derived from 48 and 2,4-dimethoxybenzoyl chloride by the Baker-Venkataraman method (52) was identical with tetrahydrokuwanon C tetramethyl ether (33a) obtained from kuwanon C (33). These results confirmed the structures of morusin, cyclomorusin, compound A, kuwanon C, and oxydihydromorusin as 24, 30, 31, 33, and 34, respectively (Fig. 10).

As mentioned earlier, Deshpande et al. had reported the isolation and structure determination of four flavones with isoprenoid substituents from Morus alba root bark, mulberrin (1), mulberrochromene (2), cyclomulberrin (3), and cyclomulberrochromene (4) (20). The difference between the flavones with isoprenoid substituents isolated by Deshpande et al. and those, such as morusin (24), cyclomorusin (30), and kuwanon C (33), isolated by our group is lay in the mode of attachment of the C_5 unit to ring A.

Chari et al. (53) on reappraisal of the ^{13}C-NMR data of Wenkert and Gottlieb (54) for mulberrin (1) and mulberrochromene (2) revised structures of these two compounds as well as those of cyclomulberrin (3) and cyclomulberrochromene (4). Signals at 98.0 ppm and 98.9 ppm in the ^{13}C-NMR spectra had been assigned to C-8 of the formulas derived to mulberrin and mulberrochromene respectively. However Chari et al. pointed out that these chemical shift values were ca. 4.0 ppm downfield of the C-8 signals in 5,7-dihydroxyflavones and that they were in the range one might expect for compounds with unsubstituted C-6 when compared with the C-8 signal of linear chromenoflavones and the C-4 signal of chromenoxanthone derivatives. Hence, Chari et al. concluded that the four mutually interrelated Morus flavones with isoprenoid substituents isolated by Deshpande et al. were C-8- rather than C-6-substituted compounds, and proposed that mulberrin was identical with kuwanon C (33), mulberrochromene with morusin (24), and cyclomulberrochromene with cyclomorusin (30).

In order to clarify these points, we studied the ^{13}C-NMR spectra of morusin (24) and related flavonoids with isoprenoid substituents

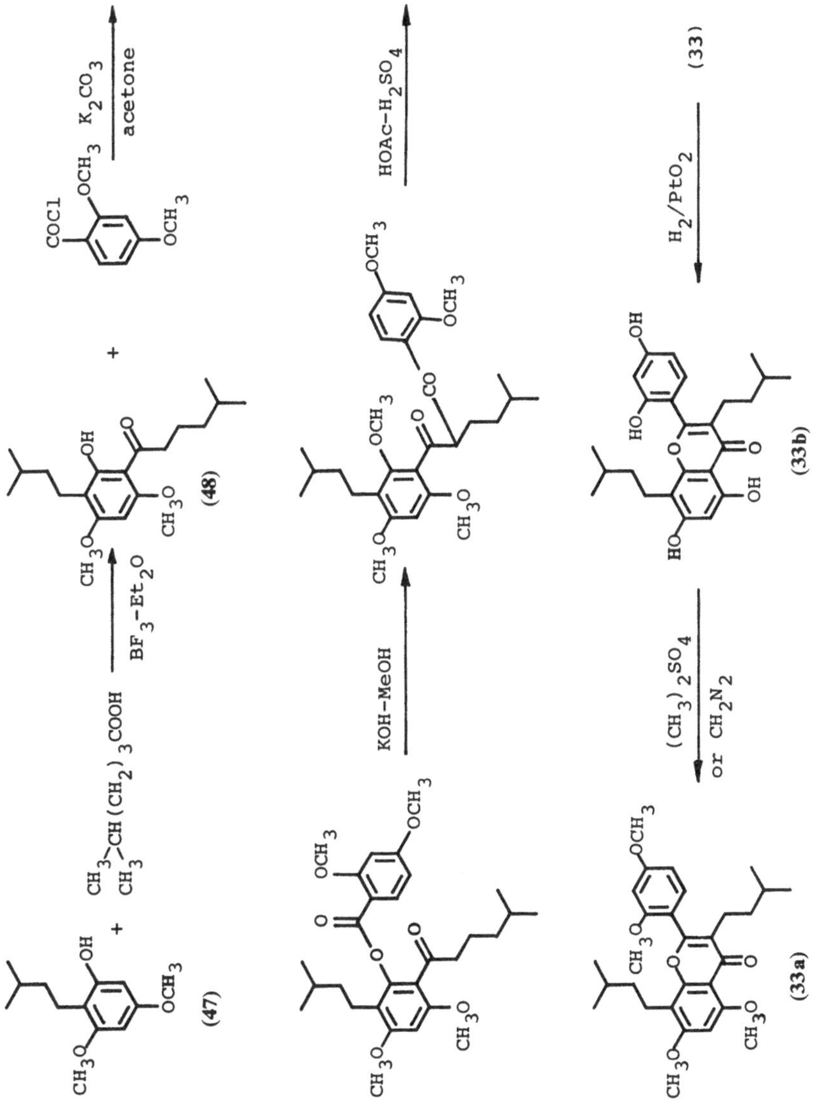

Fig. 10 Synthesis of tetrahydrokuwanon C tetramethyl ether (33a)

(Table 2) (55). In the spectra of these flavonoids, the C-6 signals appeared in the range of 97.9–100.2 ppm. Direct comparisons were carried out between mulberrin and kuwanon C (33), and between mulberrochromene and morusin (24). Mulberrin and mulberrochromene were identical with kuwanon C and morusin, respectively, so that the linear

structures (1–4) of the four *Morus* flavones with isoprenoid substituents, mulberrin, mulberrochromene, cyclomulberrin, and cyclomulberrochromene were reversed to the angular structures, 33, 24, 3′, and
30, respectively (*55*). To explain the correlation with artocarpin (5),
Chari *et al.* invoked an isomerization during the conversion of artocarpin (5) to tetrahydromulberrin (Fig. 1), the only conceivable change
being a Wessely-Moser rearrangement in the demethylation step (*53*).
In order to examine this possibility tetrahydrokuwanon C tetramethyl
ether (33a) was demethylated with hydroiodic acid. The only product
was tetrahydrokuwanon C (33b) (*55*), and not 3,6-diisopentyl-5,7,2′,4′-
tetrahydroxyflavone.

III. Photo-Oxidative Cyclization of Prenylflavones

1. Photo-Oxidative Cyclization of Morusin

In the course of our examination the constituents of the *Morus*
root bark, we met with a novel photo-oxidative cyclization (*41*) to
which reference has already been made in Section II.2, when a solution
of morusin (24) in chloroform was irradiated using a high-pressure
mercury lamp or a tungsten lamp, morusin hydroperoxide (49),
mp 204–206 °C, $C_{25}H_{24}O_8$, was obtained in *ca.* 80% yield. The reaction did not occur in the dark and was dependent on the solvent.
It proceeded in chloroform or benzene solution, but not in methanol,
ethanol or *tert*-butyl alcohol solution. When a solution containing morusin diacetate (24a) as well as 10,11,14,15-tetrahydromorusin was irradiated in chloroform, the photoreaction did not occur and the starting
materials were recovered unchanged. However, 14,15-dihydromorusin
hydroperoxide (50) was obtained when a solution of 14,15-dihydromor-
usin (24d) was irradiated. These findings indicated that the photoreaction requires the presence of the isolated double bond in the side chain
attached to C-3 and the hydroxyl group in the B ring.

Two isomers of morusin monomethyl ether, 2′-*O*-methylmorusin
(24e) and 4′-*O*-methylmorusin (24f), are formed on methylation of
24 with ethereal diazomethane in isopropyl alcohol and could be distinguished from each other by the Gibbs test. When a solution of 24e
was irradiated, starting material was recovered unchanged, but irradiation of 24f, gave morusinhydroperoxide monomethyl ether (49a). The
structure of 49 was established as follows. The UV spectrum (λ_{max}
280 and 335 nm) resembled that of compound A (31) rather than that

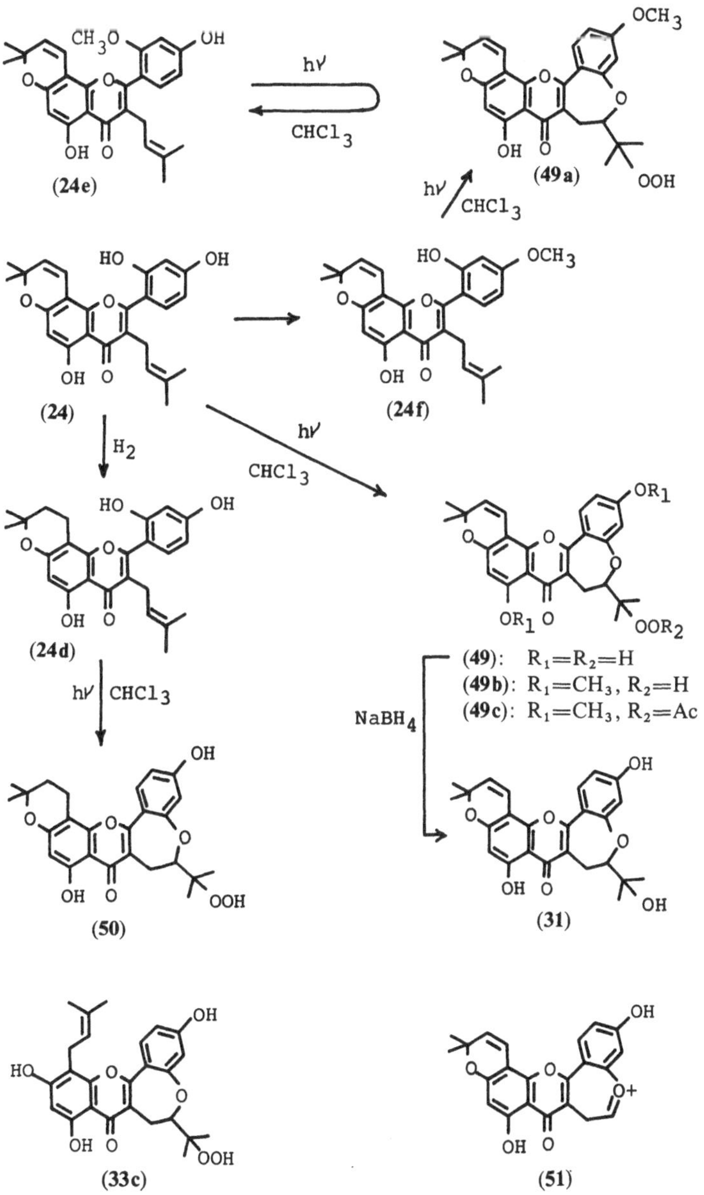

Fig. 11

of **24** or **30**. Compound (**49**) gave a dimethyl ether (**49b**) which gave a negative test with $FeCl_3$ and had hydroxyl absorption in the IR spectrum. Acetylation of **49b** gave dimethyl ether monoacetate (**49c**) whose IR spectrum showed absorption of the acetylperoxyl group at 1780 cm^{-1} (*56*). The ^{1}H-NMR spectrum of **49** exhibited a characteristic AMX pattern, i.e. $\delta 2.59$ (1 H, dd, $J=10$ and 18 Hz, H-9a), 3.46 (1 H, dd, $J=2$ and 18 Hz, H-9b), and 4.38 (1 H, dd, $J=2$ and 10 Hz, H-10). The MS of **49** gave fragments at m/z 436 (M$^+$ $-$O) (*57*), 421 (M$^+$ $-$O$-$ CH$_3$), 377 (M$^+$-C$_3$H$_7$O$_2$, **51**). Heating **49** in dimethylsulfoxide solution or reduction with sodium borohydride, diphenylsulfide, triphenyl-phosphine, or trimethylamine in methanol gave compound A (**31**). Kuwanon B hydroperoxide (**36b**) and kuwanon C hydroperoxide (**33c**) were derived from **36** and **33**, by analogous photoreactions (Figs. 8 and 11).

2. Mechanism of the Photo-Oxidative Cyclization of Morusin

Matsuura and his co-workers (*58, 59*) found that 5-hydroxyflavone derivatives resist photoreaction and ascribed this stability to hydrogen bonding between the C-5 hydroxyl and the C-4 carbonyl group, which results in intramolecular hydrogen abstraction in the excited state to yield a tautomer. An experimental result (*60*) which seem to support this theory is the photoreaction of morusin trimethyl ether (**24c**), but although morusin (**24**) itself is a flavonoid with intramolecular hydrogen bond between the C-5 hydroxyl and the C-4 carbonyl group (*32*), photoreaction occurred in chloroform or benzene solution (*41*). In this respect, the photo-oxidative cyclization of **24** was unprecedented, and its mechanism was investigated by our group (*61*).

The following three possible mechanisms (*62*) can be proposed for the primary step of the photo-oxidative cyclization of **24** as outlined in Fig. 12.

1) a reaction mechanism involving "singlet oxygen",
2) a reaction *via* a phenoxy radical,
3) a reaction *via* a contact charge transfer complex.

On considering the experimental results reaction mechanisms involving singlet oxygen or a phenoxy radical cannot explain the photo-oxidative cyclization of morusin (**24**). The third possibility is more reasonable and can be sketched as follows. Ground state morusin (**24**) interacts with an oxygen molecule to form a contact charge transfer complex (*63*). Irradiation of the complex produces an excited charge transfer state which leads to reactive species such as those drawn in Fig. 13. While conclusive proof supporting this mechanism has not

1 A reaction mechanism involving "singlet oxygen"

$$AH \xrightarrow{h\nu} {}^{1}AH* \longrightarrow {}^{3}AH*$$

$${}^{3}AH* + {}^{3}O_2 \longrightarrow AH + {}^{1}O_2$$

$${}^{1}O_2 + AH \longrightarrow AOOH$$

2. A reaction via a phenoxy radical

$$AH \xrightarrow{h\nu} A\cdot \xrightarrow{{}^{3}O_2} AOO\cdot \longrightarrow AOOH$$

3. A reaction via a contact charge transfer complex

$$AH + {}^{3}O_2 \longrightarrow (AH ---- O_2) \xrightarrow{h\nu} (AH^{\dot{+}} ---- O_2^{\dot{-}})$$

$$\longrightarrow A\cdot + \cdot OOH$$

Fig. 12

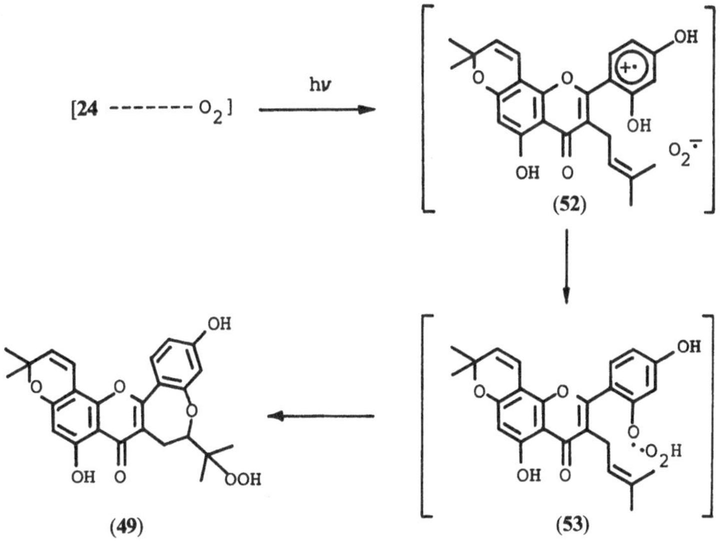

Fig. 13

yet been obtained, the following experimental results will be explained using this hypothesis. The photo-oxidative cyclization of 24 is dependent on the solvent and proceeds in chloroform, dichloromethane, or benzene solution whereas starting material is recovered unchanged in methanol, ethanol, or *tert*-butyl alcohol solution (41). It is tempting to speculate that the contact charge transfer complex cannot be formed in a solvent in which the photoreaction does not occur. A similar oxidative cyclization was observed by Shani and Mechoulam in the case of the synthesis of cannabielsoic acid A from cannabidiolic acid (64).

3. Oxidative Cyclization of Morusin

In connection with the photo-oxidative cyclization of morusin (24), oxidative cyclizations of 24 using one-electron transfer oxidizing agents (manganese dioxide, silver oxide) were also studied (65, 66, 67). A solution of 24 in dry benzene containing manganese dioxide was allowed to stand at room temperature in the dark. The reaction afforded morusin hydroperoxide (49) and compound A (31). The yield of 49 in this reaction was considerably lower than in the photo-oxidative cyclization. To elucidate the mechanism, the reaction was carried out in the presence of a radical quencher such as 2,4,6-tri-tert-butylphenol and gave compounds 54, 55, 56, and 57. Considering these results, a possible mechanism for the formation of 31 and 49 using one-electron transfer oxidizing agents in the dark is outlined in Fig. 14 (65, 67).

The photo-sensitized oxidation of morusin (24) and hematoporphyrin in benzene containing 25% methanol by irradiation with a tungsten lamp produced morusin hydroperoxide (49), 61 and 62 (Fig. 15) (66, 67). Blank runs without any dye gave no reaction, the starting material being recovered completely unchanged. When Rose Bengal was used as a sensitizer, 31, 49, and 61 were obtained. To elucidate the mechanism, the reactions were carried out in the presence of a radical quencher such as 2,4,6-tri-tert-butylphenol or a singlet oxygen quencher such as triethylene diamine (Dabco) (68). When a solution of 24, radical quencher, and Rose Bengal in benzene containing 25% methanol was irradiated, compounds 55 and 61 were obtained, but 31 and 49 were not formed. On the other hand, when a solution of morusin (24), Dabco, and Rose Bengal was irradiated, 61 and 62 were not formed, but instead 31 and 49 were isolated. In the light of these results, it is probable that 61 and 62 are formed via hydroperoxide 63 obtained by an "ene" reaction (62). As mentioned above, the oxidative cyclization of morusin (24) occurred in the presence of one-electron transfer oxidizing agents in the dark and also by photo-sensitized oxidation. In these two cases, the reaction proceeded via the same phenoxy radical (58), but in the latter case, the radical was generated by the abstraction of the phenolic hydrogen using an excited dye triplet to produce morusin hydroperoxide (49) (62) (Fig. 15).

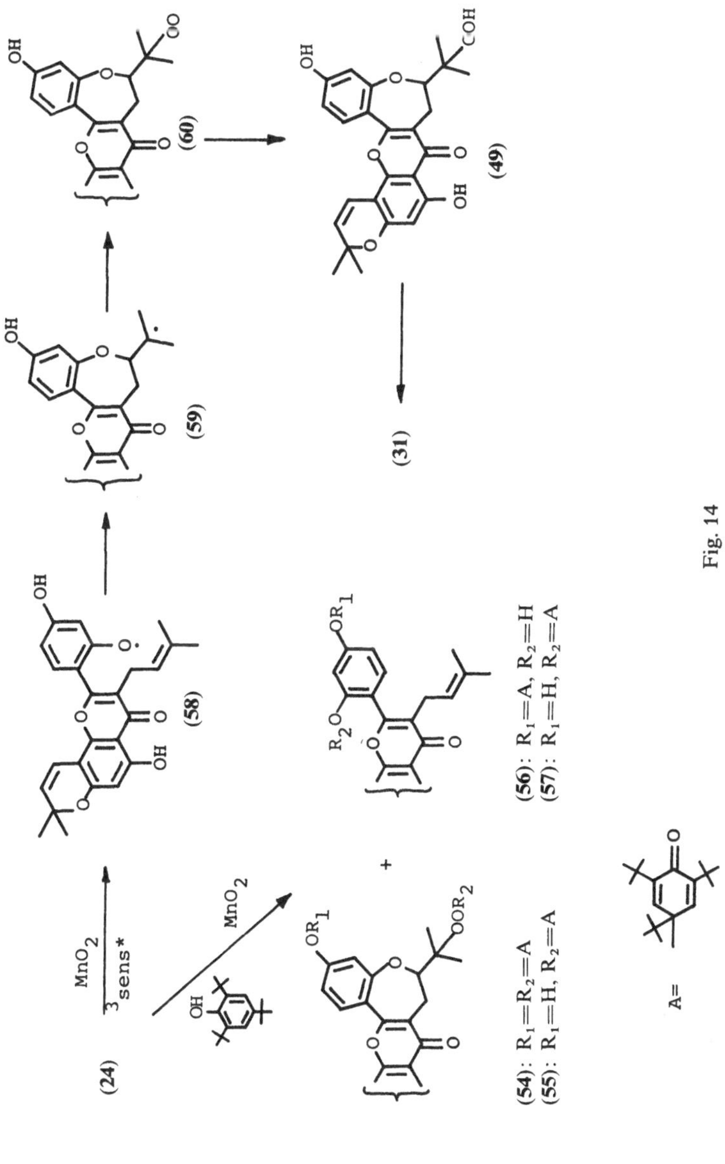

Fig. 14

These results suggest that the photo-oxidations may be models for the biogenesis of certein uncommon flavonoids, for example, compound A (**31**) (*32*), chaplasin (**18**) (*30*), and oxyisocyclointegrin (**23**) (*31*), which are supposedly derived *in vivo* from a 3-prenyl-2'-hydroxy

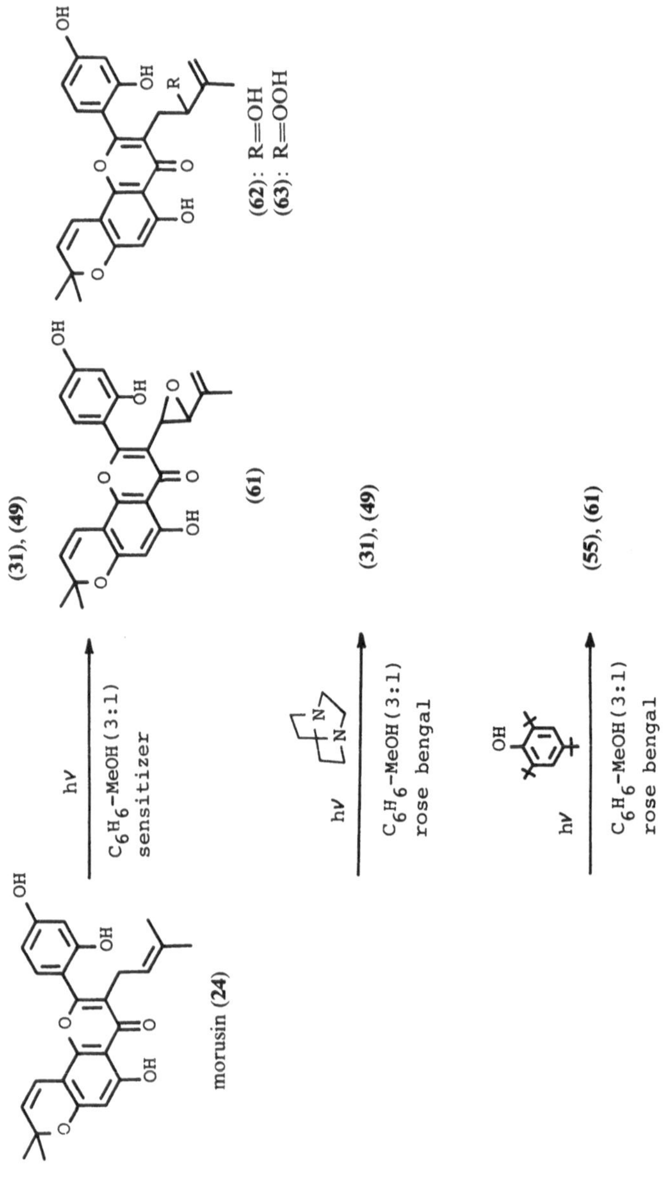

Fig. 15

flavone precursor. The formation of **49** from **24** by dye-sensitized photo-oxidation suggests that formation of the above uncommon flavones can be initiated by a light-dye-oxygen system in the plant (*67*).

References, pp. 191–201

IV. Phytoalexins and Antifungal Substances in the Mulberry Tree

Mulberry leaves are indispensable as food for silkworms. Recently, a shoot-rearing technique, wherein mulberry leaves are harvested along with the shoots, has been used in sericulture in Japan as a labor-saving device. However, this technique often causes browning and death of the shoot in the field, since the resulting wound provides an opening through which pathogens can invade. It therefore became important to clarify the defense mechanisms of the mulberry shoot. For this purpose, TAKASUGI, SHIRATA, MASAMUNE, and co-workers have studied the antifungal substances in mulberry shoots following wounding or infection by pathogenic fungi. In their reports (69, 70), the antifungal substances were classified into two groups: prohibitins (71) and phytoalexins. The prohibitins are antifungal substances obtained from the "intact" epidermis of the shoot, while the phytoalexins are antimicrobial compounds produced by plants in response to microbial infection.

Prohibitins isolated so far include eleven flavonoid derivatives, such as albanins A-E (64–68) (71–73), F (25) (74) [=kuwanon G (33)=moracenin B (75)], G (26) (74) [(=kuwanon H (34)=moracenin A (76)], H (69) (77), morusin (24) (72), kuwanons C (33) (72), and E (42) (73), and three 2-arylbenzofuran derivatives such as albafurans A-C (70–72) (78, 79) (Fig. 16). These prohibitins are obtained from the epidermis and not from the xylem tissue of the shoot, which suggests that the epidermis of the mulberry shoot plays an important role not only as a physical defense structure but also as a chemical defense against pathogens. Details concerned with the Diels-Alder type adducts albanins F (25), G (26), and H (69), will be presented later in this article.

About thirty phytoalexins have so far been isolated from the mulberry tree. A few differences are observed in the structures of phytoalexins isolated from different parts of the mulberry tree. From the fungus-infected xylem tissue of the shoot, two stilbene phytoalexins were isolated and characterized as oxyresveratrol (73) and 4'-prenyloxyresveratrol (74) (80). From the acetone extracts of cortex and phloem tissues of the shoot infected with *Fusarium solani* f. sp. *mori,* the following phytoalexins have been isolated, twenty-six 2-arylbenzofuran derivatives, moracins A-Z (75–100) (71, 81–83), as well as dimoracin (101) (84), a Diels-Alder type adduct composed of two 2-arylbenzofuran derivatives with isoprenoid substituents. Biogenetically these phytoalexins seem to be derived from moracin M (87) by hydroxylation, methylation, and isoprenylation (Fig. 17). Structures were determined on the

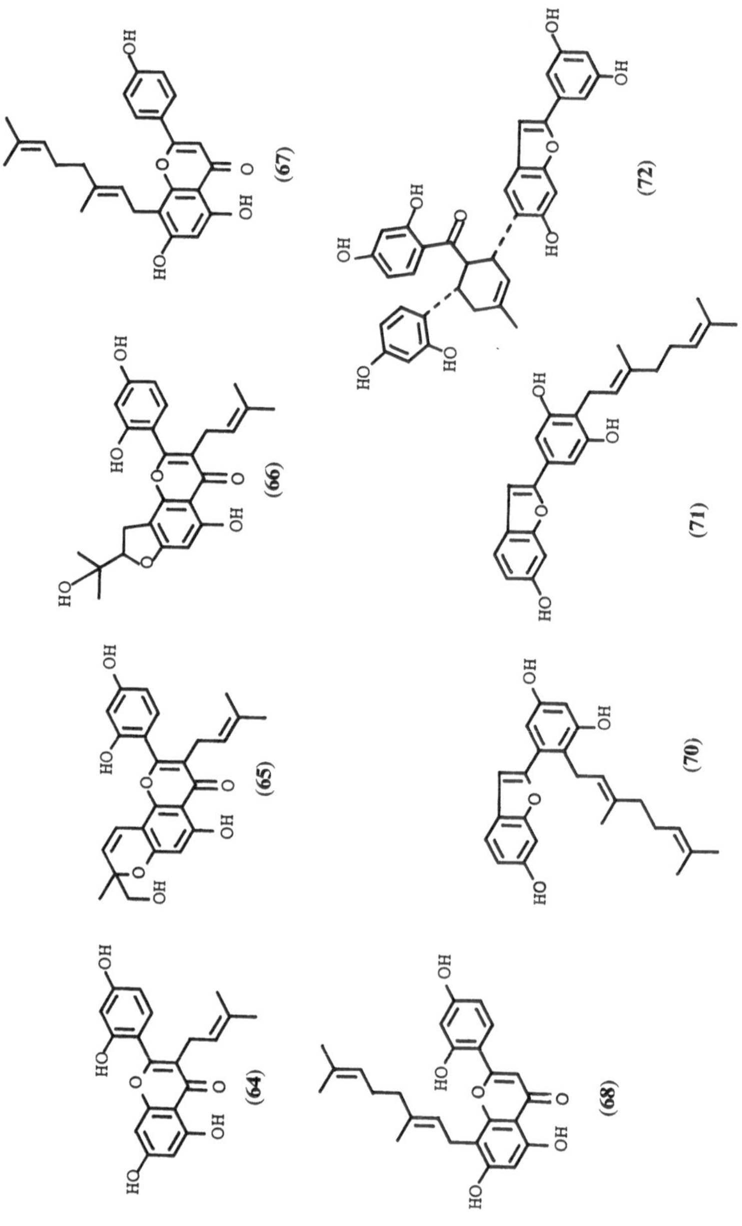

Fig. 16

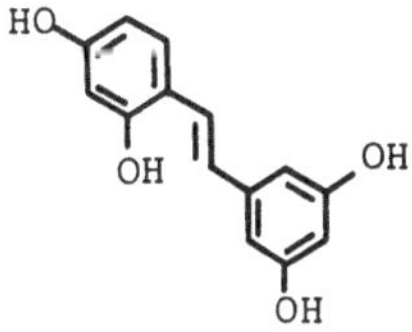

(73)

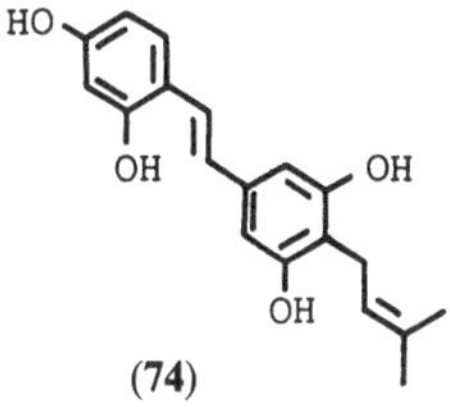

(74)

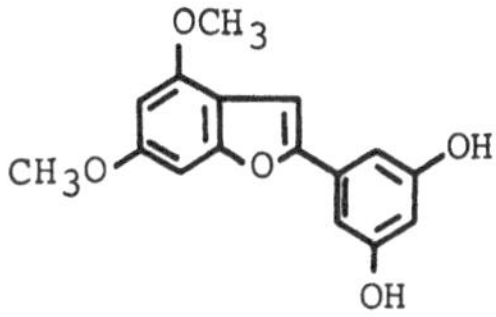

(75)

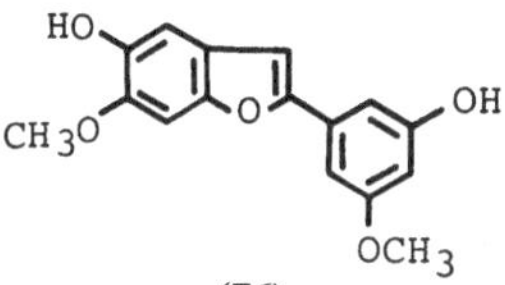

(76)

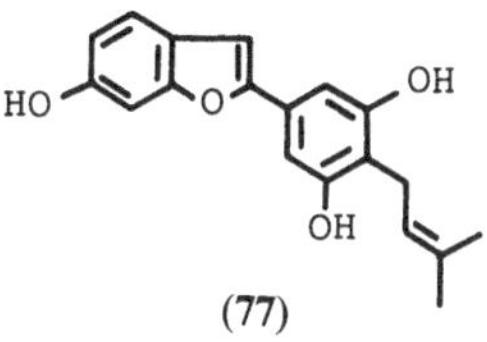

(77)

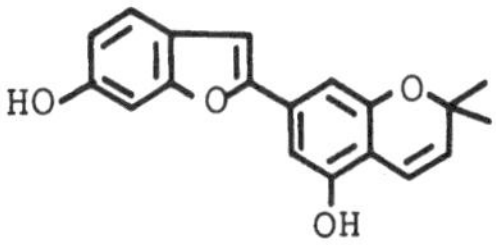

(78)

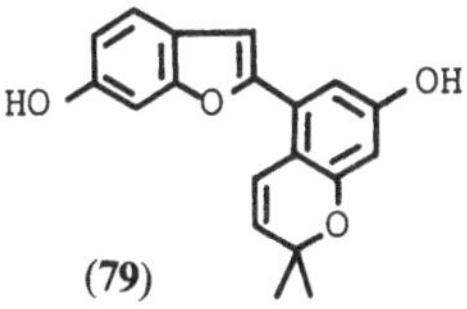

(79)

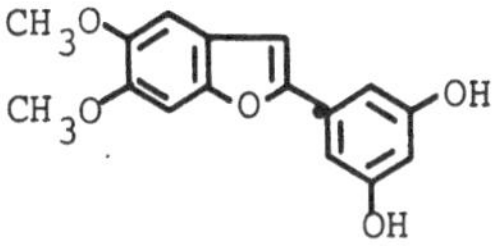

(80)

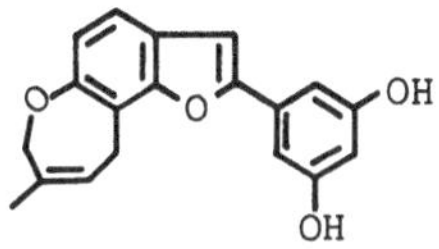

(81)

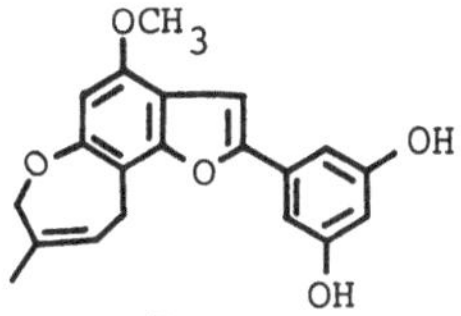

(82)

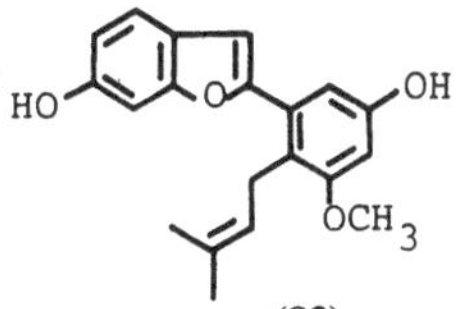

(83)

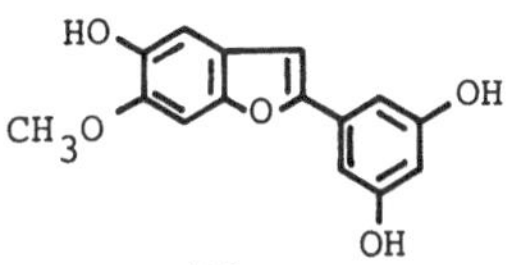

(84)

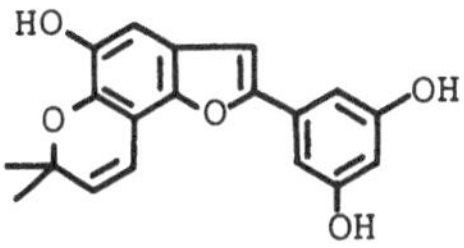

(85)

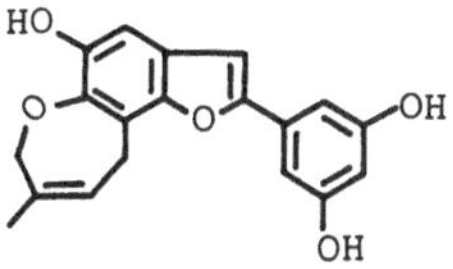

(86)

Fig. 17 (No. 1)

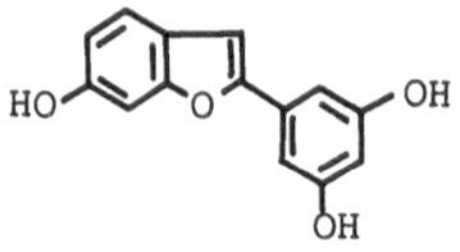

(87)

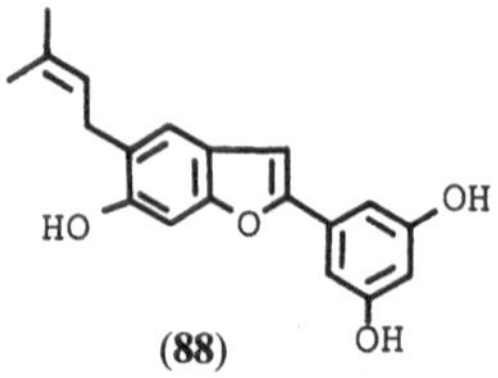

(88)

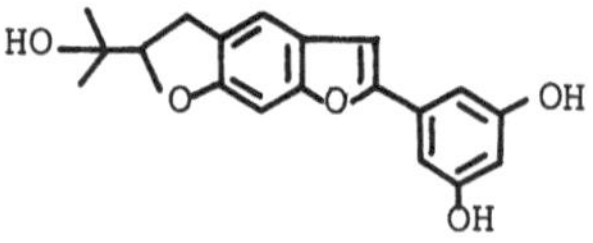

(89)

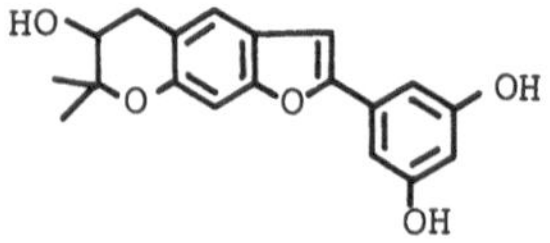

(90)

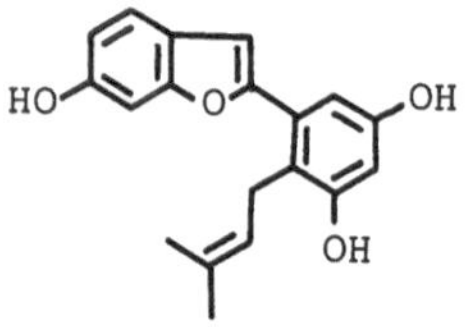

(91)

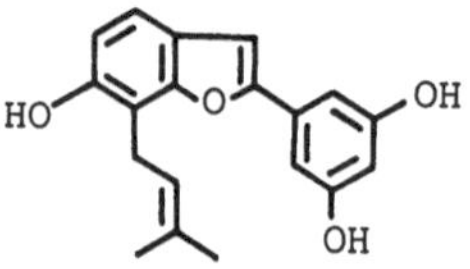

(92)

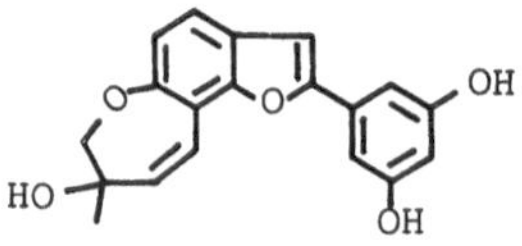

(93)

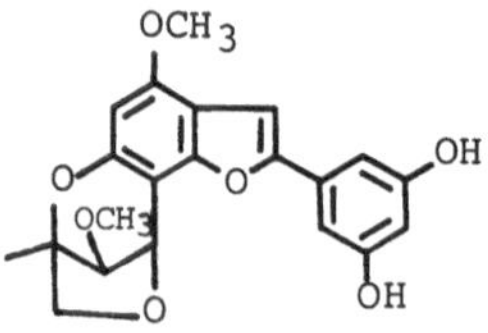

(94)

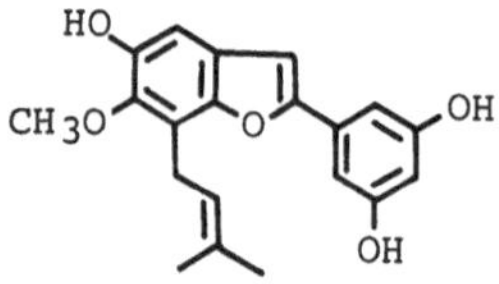

(95)

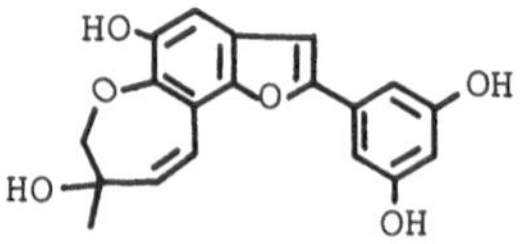

(96)

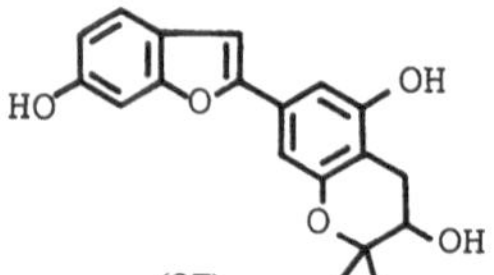

(97)

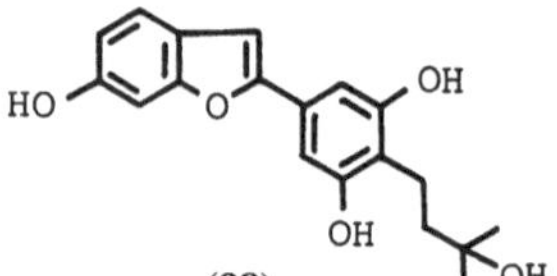

(98)

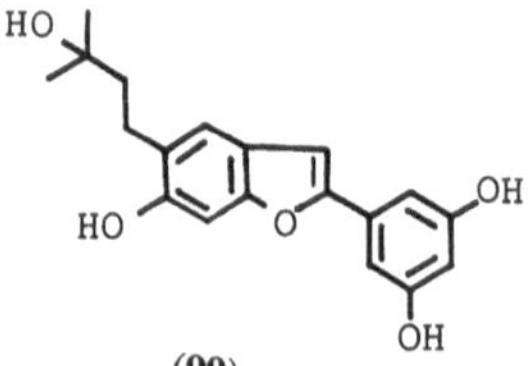

(99)

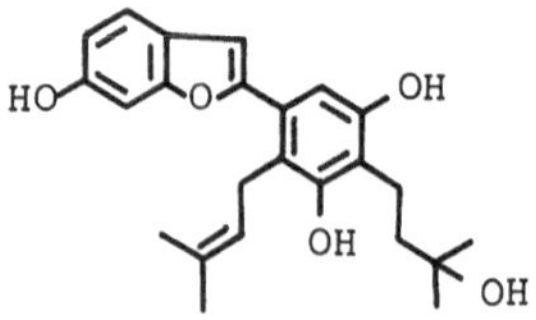

(100)

Fig. 17 (No. 2)

basis of spectral data and chemical evidence. Thus Moracin A (**75**), for example, was characterized in the following manner (*81*).

Moracin A (**75**), mp 83–85 °C, $C_{16}H_{14}O_5$, formed a diacetate (**75a**). The UV spectrum had absorption maxima at 217 nm (26 700), 304 (24 800), 313 (29 400), and 326 (20 900). The ^{1}H-NMR spectrum in CD_3COCD_3 had the following signals: δ 3.88, 3.96 (each 3 H, s, 2 × OCH_3), 6.44 (1 H, t, $J = 2$ Hz), 6.45 (1 H, d, $J = 2$ Hz), 6.82 (1 H, dd, $J = 0.8$ and 2 Hz), 6.94 (2 H, d, $J = 2$ Hz), 7.13 (1 H, d, $J = 0.8$ Hz), and 8.62 (2 H, br s, 2 × OH). The ^{13}C-NMR spectrum in CD_3SOCD_3 had signals as follows: δ 55.7 (q), 55.8 (q), 88.5 (d), 94.5 (d), 98.9 (d), 102.3 (d, intense), 102.7 (d), 112.3 (s), 131.4 (s), 152.9 (s), 153.1 (s), 155.6 (s), 158.6 (s, intense), and 158.9 (s). These spectral data suggested presence of a 2-arylbenzofuran skeleton. Ozonolysis of **75a** afforded aldehyde ester (**102**) which on saponification gave 2-hydroxy-4,6-dimethoxybenzaldehyde and 3,5-dihydroxybenzoic acid. Hence, the structure of moracin A was **75** (Fig. 18). Moracins A (**75**) and B (**76**) were synthesized by short routes which incorporate an intramolecular Wittig reaction to form the heterocycle as key step (*85*). Moracin M (**87**) was also synthesised in good yield by cross coupling of 6-(*tert*-butyldiphenylsiloxy)-2-trimethylstannylbenzofuran with 5-iodoresorcinol-bis(triisopropylsilyl) ether and fluoride ion deprotection of the product (*86*).

From an acetone extract of infected mulberry leaves a new phytoalexin, chalcomoracin (**103**) (*87*), was isolated. Compound **103** can be viewed biogenetically as a Diels-Alder type adduct between a chalcone and a dehydroprenyl-2-arylbenzofuran derivative. The isolation of **103** was the first report on a series of optically active Diels-Alder type adducts obtained from the mulberry tree (*72*). Details of its structure determination will be presented later.

On the other hand, our group has isolated the 2-arylbenzofuran derivatives mulberrofurans A (**104**) (*88*), B (**105**) (*89*), D (**106**) (*90*), L (**107**) (*89*), N (**108**) (*91*), and mulberroside C (**109**) (*92*). Compound **109** is the first example of a glycoside of a 2-arylbenzofuran derivative.

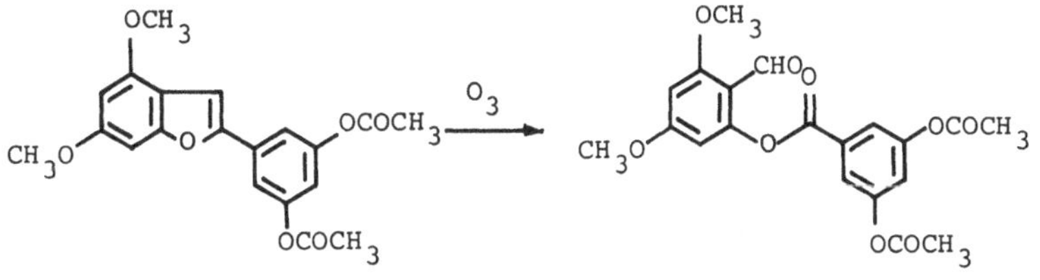

(**75a**)   (**102**)

Fig. 18

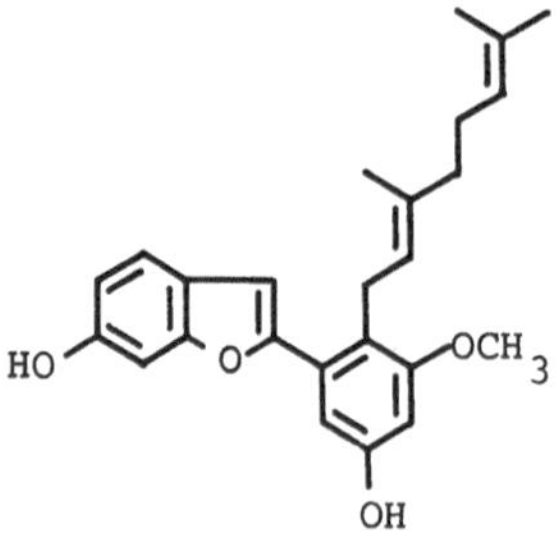

(104)

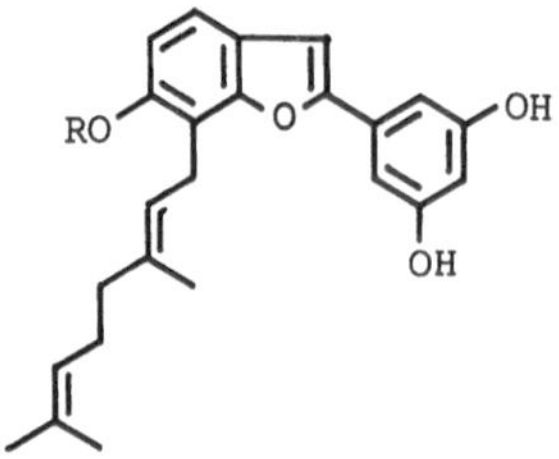

(105): R=CH₃
(107): R=H

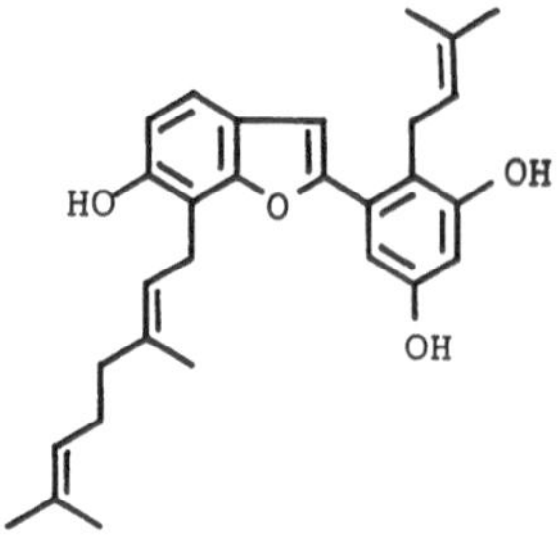

(106)

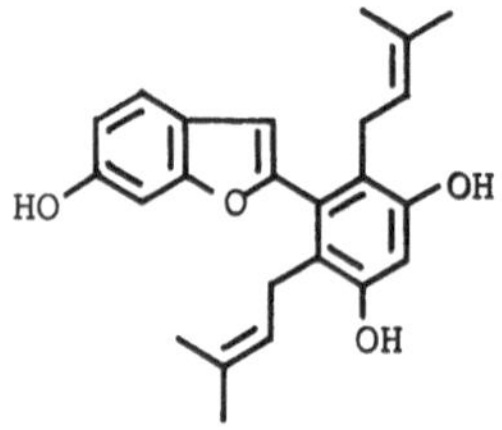

(108) 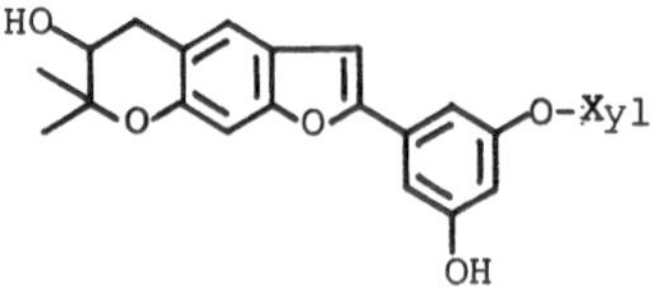   (109) 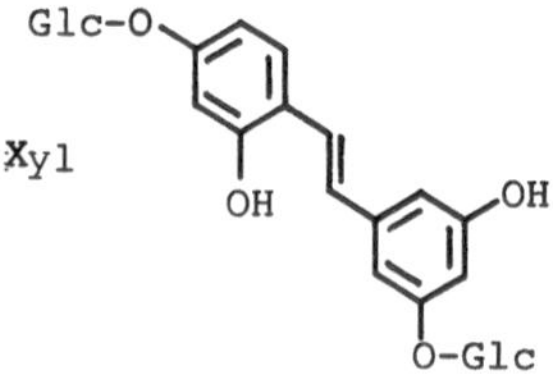   (110)

Fig. 19

Mulberroside A (110) is a glycoside of a stilbene (92). Mulberrofuran A (104) was effective against gram-positive bacteria but inactive against gram-negative bacteria, and showed weak activity against fungi (88) (Fig. 19).

Comparison of the structures of the 2-arylbenzofuran and stilbene derivatives suggests that the latter compounds are biogenetic precursors of the former. The skeletons of the 2-arylbenzofuran and stilbene derivatives could formally be regarded as a 1,2-diphenylethane; biogenetically the skeleton would be derived from cinnamoylpolyketide produced from cinnamic acid and acetate (70).

V. Diels-Alder Type Adducts of the Cultivated Mulberry Tree

1. Hypotensive Constituents, Kuwanons G and H (48, 49, 93)

The root bark of the cultivated mulberry tree was extracted successively with n-hexane, benzene, and methanol. Intravenous injection of the methanol extract, 1 mg–20 mg, showed a dose-dependent decrease in arterial blood pressure in pentobarbital-anesthetized rabbits (94). The extract was fractionated successively by silica gel column chromatography, polyamide column chromatography, silica gel preparative TLC, and HPLC leading to isolation of kuwanons G (25) (33) and H (26) (34) (Fig. 4). Intravenous injections of both compounds (0.1–3.0 mg/Kg) showed an almost equally transient dose-dependent decrease in arterial blood pressure in anesthetized rabbits (95).

In parallel with our findings in the preceding paragraph, HIKINO and co-workers isolated as two hypotensive compounds from *Morus* root bark which they named moracenins B (25) (75) and A (26) (76), while MASAMUNE and co-workers isolated two prohibitins of mulberry shoot which they named albanins F (25) and G (26) (74). By direct comparison kuwanon G was found to be identical with moracenin B and albanin F, and kuwanon H with moracenin A and albanin G (96). Structure determinations of these compounds were carried out by the three research groups independently.

Structure elucidation of these two compounds by our group will be discussed first (33, 34, 97).

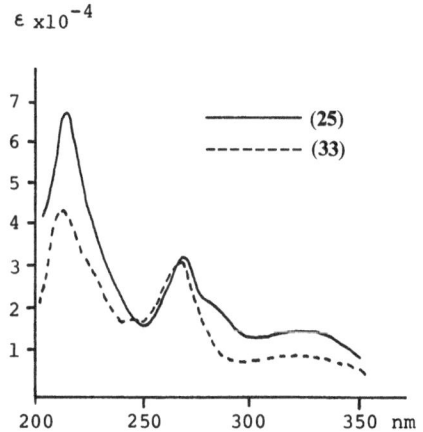

Fig. 20. UV Spectra of Kuwanons C (33) and G (25)

Kuwanon G (**25**), an amorphous powder, $[\alpha]_D$ −534°, had a molecular formula $C_{40}H_{36}O_{11}$, gave positive Mg-HCl and Zn-HCl tests, and characteristically has a large optical rotation. The UV spectrum was similar to that of kuwanon C (**33**) except for a shoulder at 280 nm, which suggested that kuwanon G possesses a kuwanon C partial structure (Fig. 20). Treatment with dimethyl sulfate gave four methyl ethers, a hexamethyl ether (**25a**), a heptamethyl ether (**25b**) which gave a negative Gibbs test, a heptamethyl ether (**25c**) which gave a positive Gibbs test, and an octamethyl ether (**25d**). The EI-MS of **25** had fragment ion peaks at m/z 582, 555, 420, 137, and 110. It is especially noteworthy that the fragment at m/z 420 (**111**) is the same as the molecular ion of morusin (**24**) (Fig. 21). Analysis of the ¹H-NMR spectrum of **25** using decoupling experiments is shown in Fig. 22. The spectrum contained all the signals in the ¹H-NMR spectrum of kuwanon C (**33**), except those of the prenyl group of **33** located at C-8; and suggested presence of the following moieties and groups: a 2,4-dihydroxybenzoyl and a 2,4-dihydroxyphenyl moiety, a methyl group, two methylene protons, three methine protons, and an olefinic proton. In the ¹³C-NMR spectrum of **25**, the chemical shifts of all carbon atoms were in good agreement with those of kuwanon C (**33**) except for C-8 and the carbon atoms in the C-8 substituent of **33** (Tables 2 and 3). Hence, kuwanon G was assumed to be a derivative of kuwanon C (**33**) in which because of the molecular formula a fifteen carbon unit was combined with the prenyl group at C-8.

The location of the prenyl group of **25** at C-3 was supported by formation of kuwanon G hydroperoxide (**112**) on photo-oxidative cyclization of **25**, in analogy with photo-oxidative cyclizations of morusin (**24**) and other prenylflavones (*40, 41*). Location of the substituent of **25** at C-6 was excluded by the results of the Gibbs test of the heptamethyl ethers **25b** and **25c** and by the fact that the chemical shifts of the C-6 and C-8 signals of **25** were in good agreement with those of C-8 substituted flavones with an isoprenoid substituent (*49, 53*). A solution of kuwanon G hexamethyl ether (**25a**) in xylene was pyrolyzed at 450 °C in a sealed tube to give 2'-hydroxy-2,4,4'-trimethoxychalcone (**113a**) identical with an authentic sample synthesized by reaction of 2'-hydroxy-4'-methoxyacetophenone with 2,4-dimethoxybenzaldehyde. These findings indicated that kuwanon G (**25**) contained a dihydrochalcone partial structure as well as a kuwanon C partial structure and led to two possible planar structures **25** and **25'** (Fig. 21). The two can be regarded as Diels-Alder type adducts of dehydrokuwanon C and chalcone derivatives, **25** and **25'** being isomers due to different regioselectivity of the Diels-Alder reaction.

The arrangement of substituents in the methylcyclohexene ring was

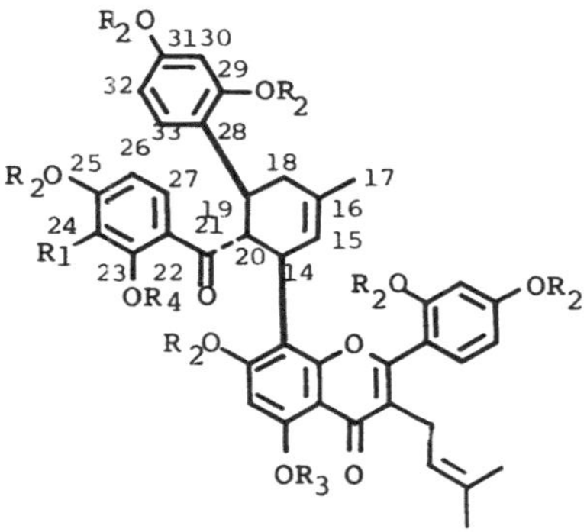

m/z 205   (119)

m/z 420   (111)

(113):   R₁=R₂=R₃=H
(113a): R₁=H, R₂=R₃=CH₃, R₃=H
(113b): R₁=H, R₂=R₃=R₃=CH₃

(120):   R₁= , R₂=R₃=H

(120a): R₁= , R₂=CH₃, R₃=H

(120b): R₁= , R₂=R₃=CH₃

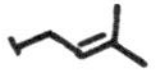

(25):   R₁=R₂=R₃=R₄=H
(25a): R₁=R₃=R₄=H, R₂=CH₃
(25b): R₁=R₃=H, R₂=R₄=CH₃
(25c): R₁=R₄=H, R₂=R₃=CH₃
(25d): R₁=H, R₂=R₃=R₄=CH₃
(25e): R₁=H, R₂=R₃=R₄=CD₃

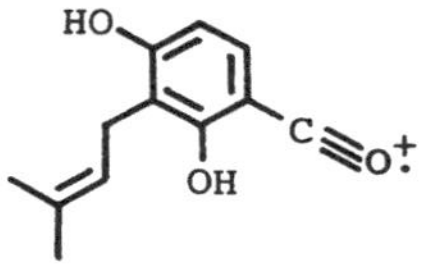

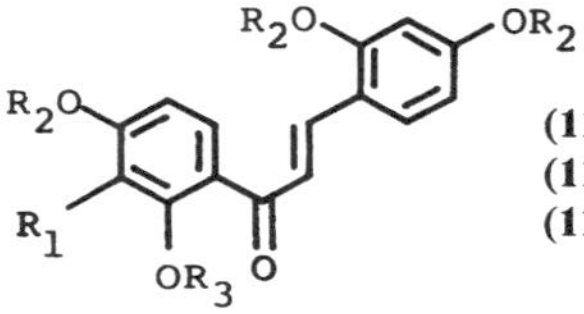

(26):   R₁= , R₂=R₃=R₄=H

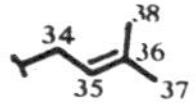

(26a): R₁= , R₂=CH₃, R₃=R₄=H

(26b): R₁= , R₂=R₃=R₄=CH₃

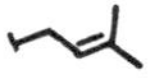

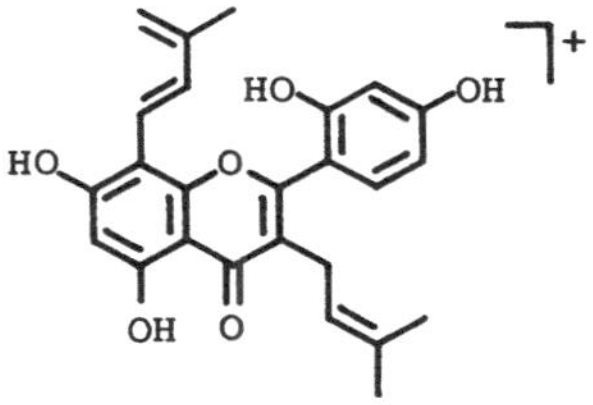

Fig. 21 (No. 1)

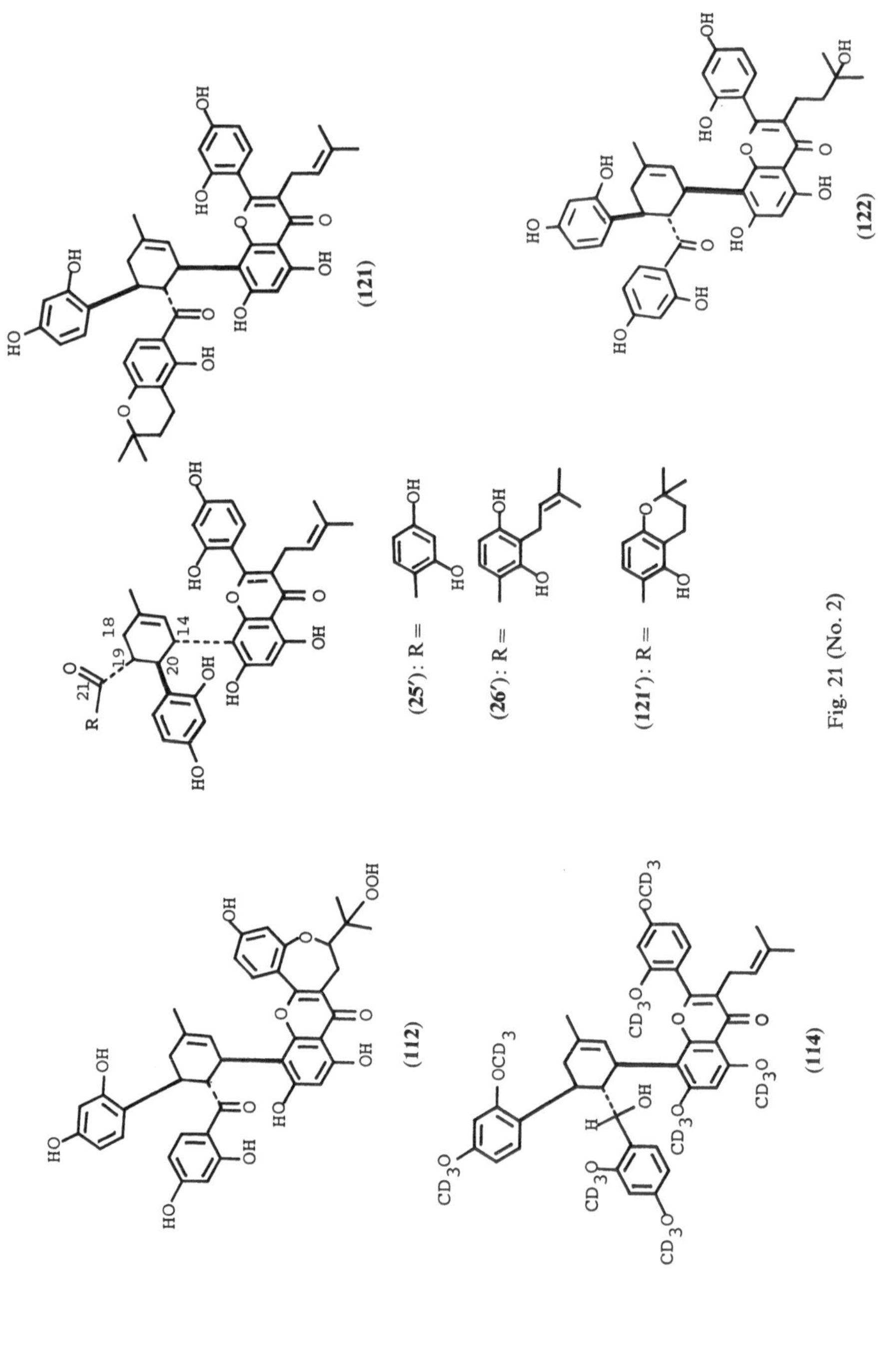

(121)

(122)

(25'): R =

(26'): R =

(121'): R =

(112)

(114)

Fig. 21 (No. 2)

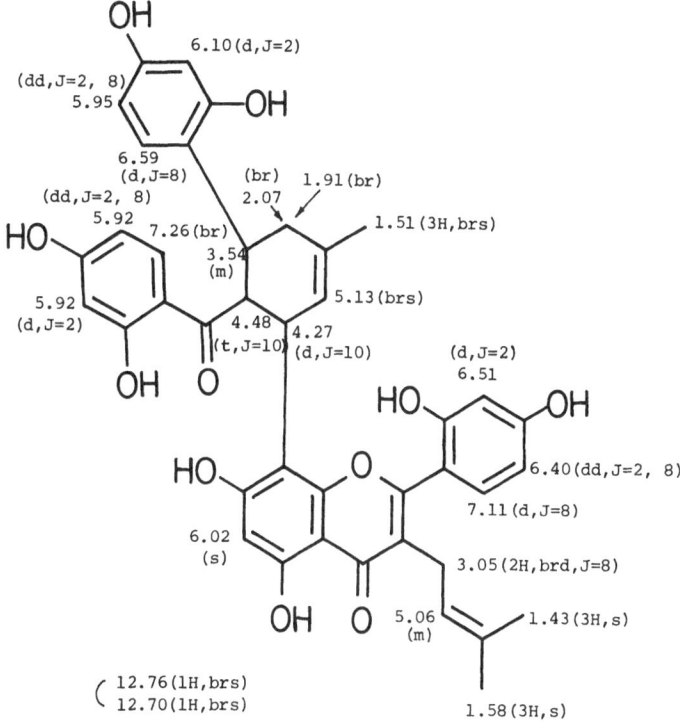

δ in dmso-d$_6$ 120°C

Fig. 22. ^{1}H-NMR chemical shifts (ppm) and coupling constants (Hz) of kuwanon G (**25**)

deduced from the ^{1}H-NMR spectrum of the methylcyclohexene portion of kuwanon G octadeuteromethyl ether **25e** (Fig. 23), all signals having been assigned by extensive decoupling. Assignments of the H-19 and H-20 signals were confirmed by comparison with the ^{1}H-NMR spectrum of alcohol **114** obtained by sodium borohydride reduction of **25e**. The relevant proton signals of **114** are shown in Fig. 23. The H-20 signal of **114** is shifted upfield by about 1.5 ppm when compared with H-20 of **25e**. If kuwanon G were **25′**, three signals (H-14, H-20, and H-21) would appear at 4.0–5.0 ppm in the ^{1}H-NMR spectrum of the NaBH$_4$ reduction product. Hence structure **25** (without assignment of stereochemistry) was preferred to structure **25′**. On the basis of the coupling constants ($J_{14,20} = J_{19,20} = 10$ Hz) of **25**, H-14, H-19 and H-20 are quasiaxial to each other.

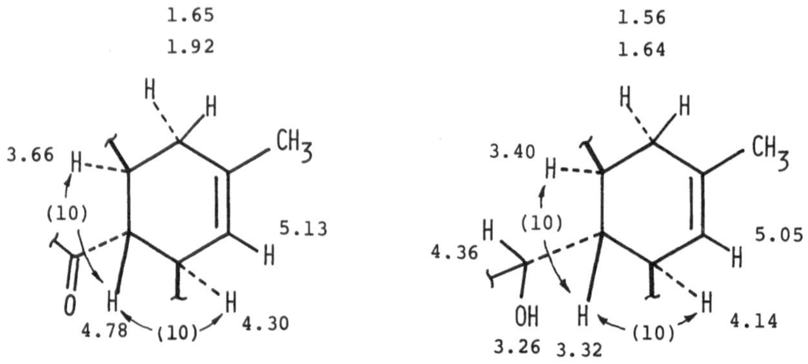

(25e) (114)

δ in $CDCl_3$ δ in $CDCl_3$

Fig. 23. Chemical shifts (ppm) and coupling constants (Hz) of methylcyclohexene ring protons of **25e** and **114**

In order to confirm this deduction and the proposed regioselectivity of the Diels-Alder type reaction between a chalcone and a dehydroprenylphenol, two model compounds **115** and **116** were prepared in 46% resp. 25% yield by a Diels-Alder reaction between *trans*-chalcone and 3-methyl-1-phenyl-1,3-butadiene (250 °C, 5 h). Their structures were established by X-ray crystallography and are shown in Fig. 24. The ^{1}H-NMR spectra of **115** and **116** were compared with those of alcohols **117** and **118** obtained by reduction. Signals of the protons on the relevant carbon atoms of alcohols **117** and **118** were shifted by 1.16 ppm and 1.61 ppm to higher field than the corresponding signals in **115** and **116**, as shown in Fig. 25.

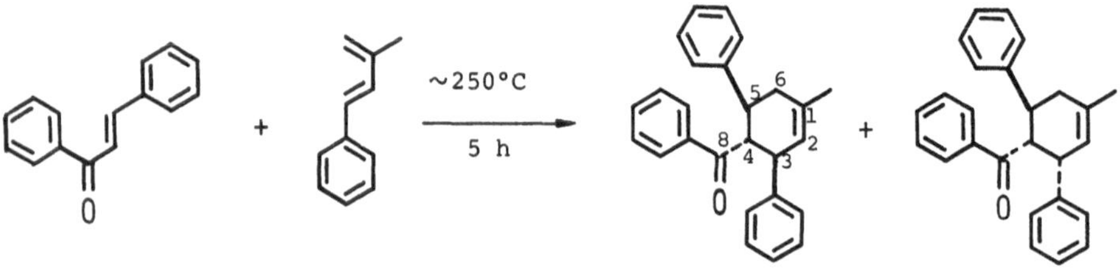

(115) (116)

Fig. 24

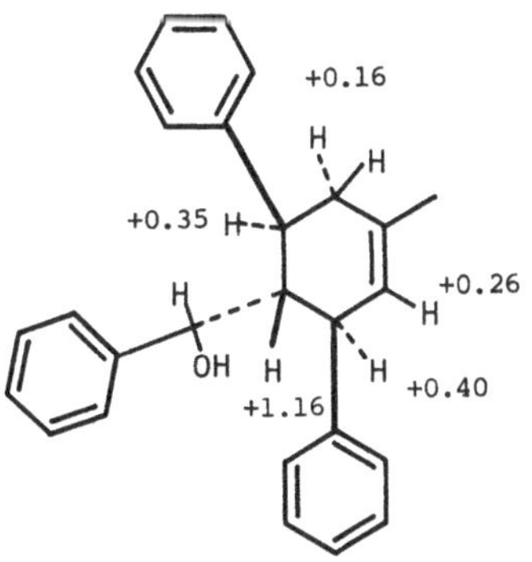

(117) 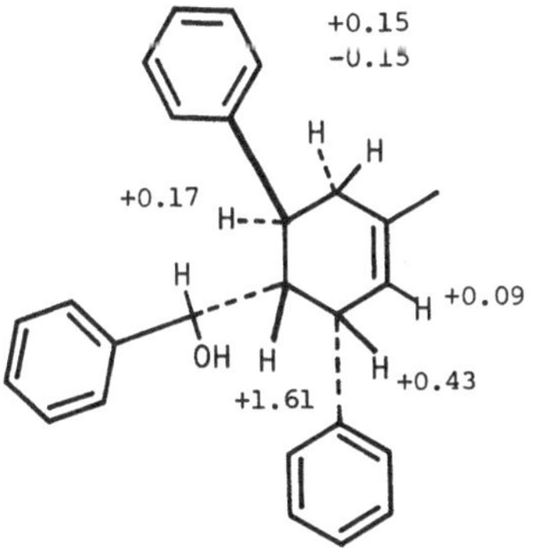   (118)

Fig. 25. Chemical shifts (ppm) of protons in the methylcyclohexene rings of alcohols
117 and 118 on comparison with 115 and 116, respectively

Kuwanon H (26), an amorphous powder, $[\alpha]_D$ $-536°$, had a molecular formula $C_{45}H_{44}O_{11}$. The UV spectrum was similar to that of kuwanon C (33). The MS of 26 showed fragments at m/z 420 (111) and 205 (119). Treatment of 26 with dimethyl sulfate gave a hexamethyl ether 26a and octamethyl ether 26b. The hexamethyl ether was pyrolyzed to give 2′-hydroxy-3′-γ,γ-dimethylallyl-2,4,4′-trimethoxychalcone (=morachalcone A trimethyl ether, 120a), a sample of which was synthesized by reaction of 2′-hydroxy-3′-γ,γ-dimethylallyl-4′-methoxyacetophenone with 2,4-dimethoxybenzaldehyde. The ^{13}C-NMR spectrum is listed in Table 3. From the spectral data, the difference between kuwanons G (25) and H (26) is the presence of a prenyl group at the C-24 position of 26.

It should be noted that kuwanons G (25) and H (26) are optically active and are presumably formed by a Diels-Alder type reaction between a chalcone and dehydrokuwanon C or its equivalent under enzymatic control. This hypothesis is supported by the co-occurrence of morachalcone A (120) (87, 98), morusin (24), equivalent to dehydrokuwanon C, and kuwanon H (26) in the Morus root bark. Although some natural products such as thamnosin (99), heliocides (100), alflabene (101), and paraensidimerins (102) are assumed to be formed by Diels-Alder type reactions, these compounds show no optical activity in contrast to kuwanons G (25) and H (26).

As has been mentioned previously, HIKINO and co-workers also isolated four hypotensive compounds from Morus root bark, named then moracenins A-D and proposed structures 26′ (76), 25′ (75), 121′

(*103*) and **122** (*104, 105*) for them. Allocation of the benzoyl moieties to C-19 was based on results of the long-range selective proton decoupling (LSPD) technique. In the case of moracenin B (**25′**) (*75*), ^{13}C-^1H spin couplings were observed between the signal at δ 209.8 (C-21) of the carbonyl carbon and that at δ 1.92 (H-18 a, b) for the methylene hydrogens, and between that at δ 115.5 (C-22) for the aromatic carbon α to the carbonyl and that at δ 3.66 (H-19) for the methine hydrogen. These results seemed to eliminate C-20 position as the locus of the benzoyl moiety. Similar ^{13}C-^1H couplings were observed in the case of the moracenins A and C and coupled with other spectral data, led to formulae **25′**, **26′**, and **121′** for moracenins B, A, and C (Fig. 21).

Similarly, MASAMUNE and co-workers reported on the prohibitins, albanins F and G (*74*), for which they proposed structures **25** and **26** based on the following results. Pyrolysis of albanin F octamethyl ether (**25d**) afforded two fragmentation products (**113b** and **123**). Compound **123**, $C_{29}H_{32}O_6$, was presumed to be a prenylflavone containing a conjugated diene moiety. The diene **123** on catalytic hydrogenation formed a hexahydro derivative which was identified as tetrahydrokuwanon C tetramethyl ether (**33a**). Discrimination between the two possible formulae **25** and **25′** was accomplished by comparing the ^1H-NMR spectra of albanin F (**25**) and the model compounds **124** and **125**, which were prepared by cyclization of a 3:2 mixture of *trans*- and *cis*-dienes **126** and **127** and the *trans*-chalcone **113b** in toluene (160 °C, 61 h), with **124** the major and **125** the minor product. In consideration of the stereospecificity and regioselectivity due to substituents in the Diels-Alder reaction (*106, 107*), formulae **124** and **125**, different only in (relative) configuration of a single carbon atom were assigned to the major and the minor product, respectively (Fig. 26). The spectral data were also in good accordance with the formulae. In the ^1H-NMR spectra of **124** and **125**, the signals of the H-4 methine protons [δ 4.82 (**124**) and 4.7 (**125**)] adjacent to the benzoyl moiety appeared at lower fields than those of the H-5 methine protons [δ 3.7 (**124**) and 4.24 (**125**)]. In the case of albanin F octamethyl ether (**25d**), spin-decoupling permitted the demonstration that the signal of the methine proton (H-20, δ 4.77) adjacent to the benzoyl moiety appeared at lower field than that of the methine proton (H-19, δ 3.70). These observations, combined with determination of the coupling constants between adjacent protons on the methylcyclohexene ring, led to formula (**25**) for albanin F. In the same manner formula (**26**) was assigned to albanin G essentially. Albanin G octamethyl ether (**26b**) gave morachalcone A tetramethyl ether (**120b**) and the diene (**123**) on pyrolysis.

Final proof for structures **25** and **26** of kuwanons G and H was obtained as the result of joint work between our group and that of

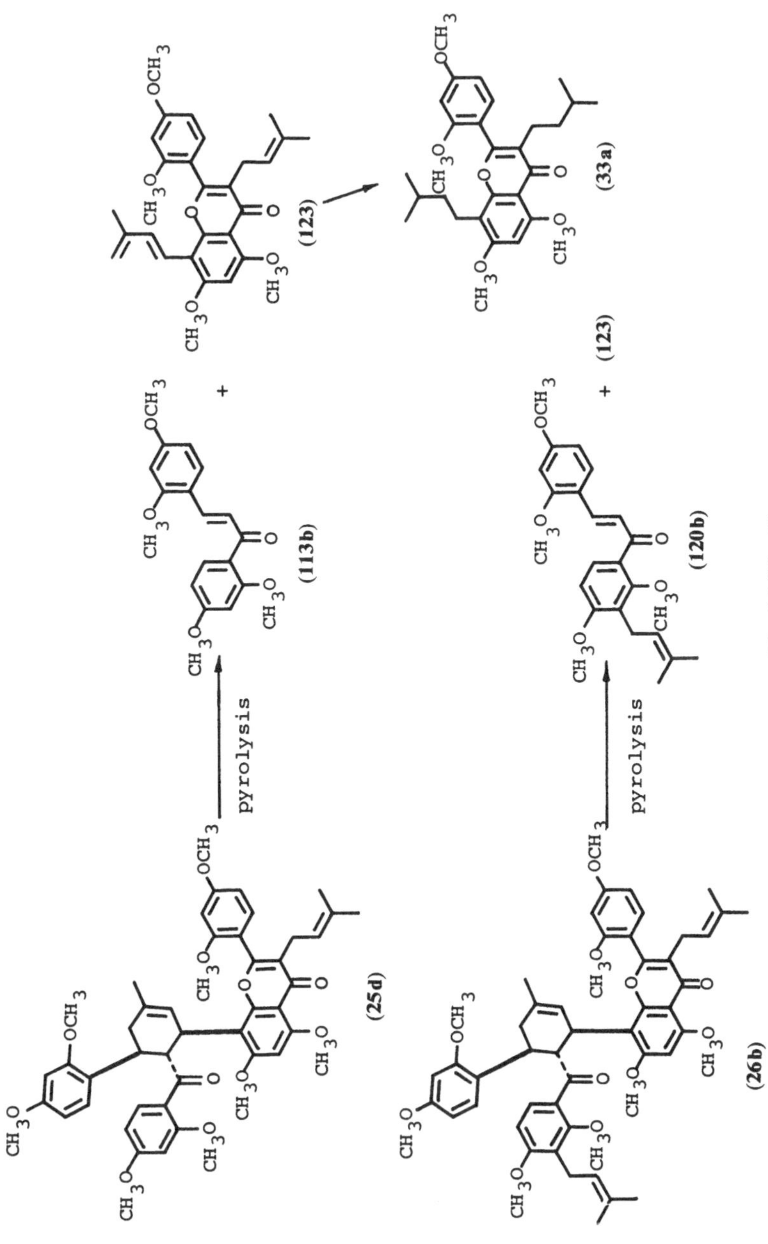

Fig. 26 (No. 1)

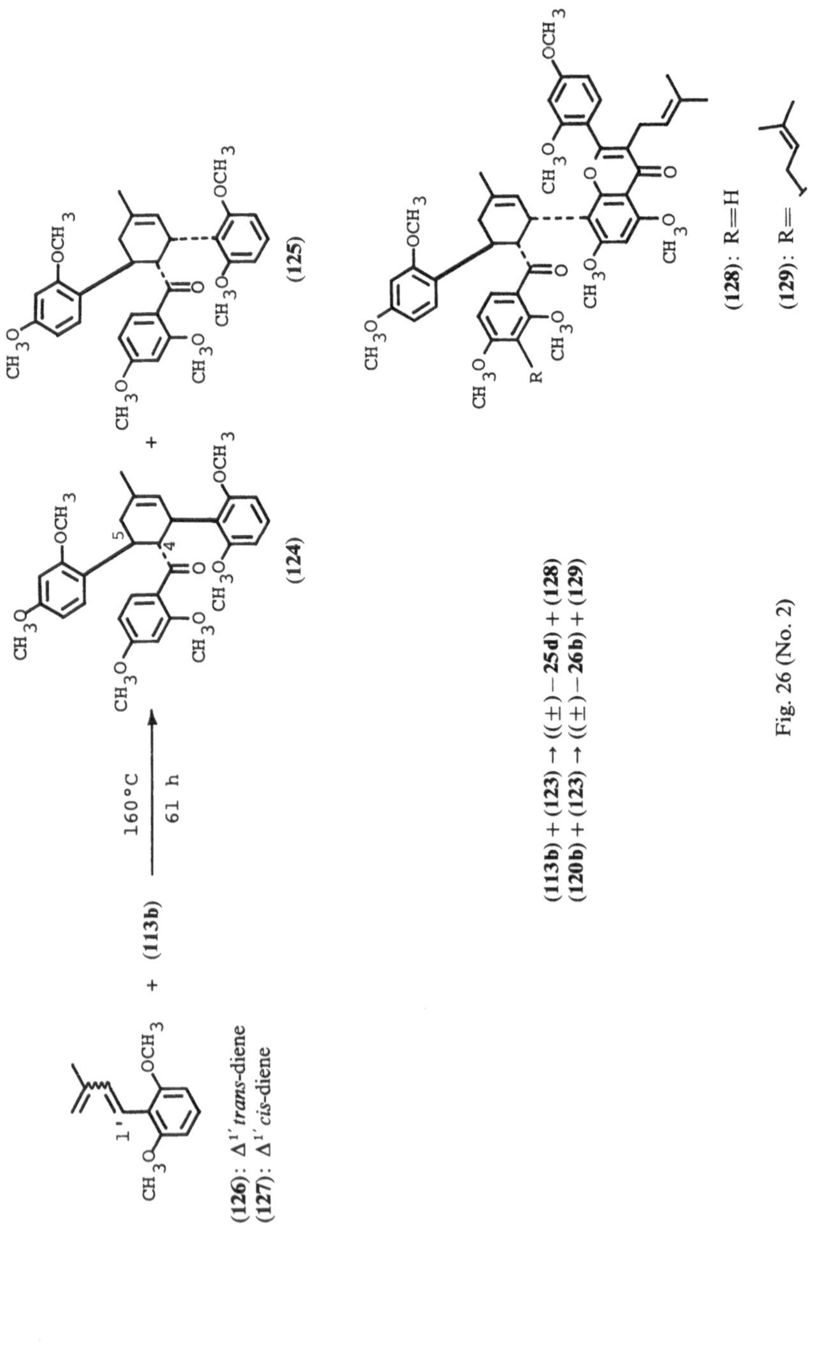

Fig. 26 (No. 2)

MASAMUNE (96). The two fragmentation products **113b** and **123**, formed on pyrolysis of kuwanon G octamethyl ether (**25d**), gave when heated in toluene in the presence of 2,6-di-*tert*-butyl-*p*-cresol at 160 °C for 61 h in a sealed tube, and chromatographyly two amorphous cycloadducts (±) **25d** and **128**, in 35% and 32% yield, respectively. Analogous treatment of the fragmentation products **120b** and **123** from kuwanon H octamethyl ether (**26b**) afforded two amorphous cycloadducts (±) **26b** and **129**, both in 24% yield. No other cycloadducts could be detected. The two Diels-Alder adducts (±) **25d** and (±) **26b** were identified as the octamethyl ethers of (±) kuwanons G and H by direct comparison with samples derived from the natural products. In view of the formation of the model Diels-Alder adducts **115** and **116** cited earlier, structures (±) **25d** and **128**, (±) **26b** and **129** can be considered as proved. Hence kuwanons G and H are **25** and **26** (whose absolute configuration remains uncertain) (Fig. 26).

To buttress these conclusions, we investigated the ^{13}C-^1H spin couplings between the C-21 carbonyl carbon and H-18a, b of kuwanon G (**25**) by the LSPD technique (*108, 109*). The signal at δ208.1 (in DMSO-d_6) due to the C-21 carbonyl carbon remained essentially unchanged on irradiation at δ1.80 (H-18a), 1.90 (H-18b), and 3.60 (H-19), while the area of the C-21 signal increased by 30% on weak irradiation at δ4.30 (H-14 and H-20 both) using a selective ^{13}C-$\{^1H\}$ NOE technique. The same irradiations were also carried out on model compounds **115** and **116** (*110*). In the case of **115**, the signal of the carbonyl carbon atom (C-8) remained unchanged on irradiation at the frequency of H-6 while the signal of C-8 increased in area on weak irradiation at the frequencies of H-3 and H-4 both. In the case of **116**, ^{13}C-^1H spin couplings were observed between the carbonyl carbon atoms and H-3, H-4, and the phenyl protons. These results support formulae **25** and **26** for kuwanons G and H, and exclude attachment of the benzoyl groups to C-19.

More or less simultaneously, HIKINO's group revised their original structures **25'**, **26'** and **121'** for their moracenins B, A, and C, to **25**, **26** and **121**, after analysis of the ^1H-NMR spectra of the reduction products of moracenin B octadeuteromethyl ether (**25e**) (*104*). They attributed their erroneous conclusions drawn from NMR evidence on the locations of the benzoyl and phenyl moieties in the methylcyclohexene rings of moracenins B, A, and C, to inappropriate irradiation power in the ^{13}C-^1H decoupling experiments.

Table 3. ^{13}C-NMR Data of Diels-Alder Type Adducts of Flavonoids Derivatives (Kuwanons G, H, I, J, L, N, O, Q, R, and W: KG, KH, KI, KJ, KL, KN, KO, KQ, KR, and KW)

C	KG (25)	KH (26)	KW (157)	KN (155)	(C)	KL (154)	KO (156)	(C)	KI (152)	KJ (147)	KQ (148)	KR (149)
2	159.2[a]	159.3[a]	158.5[a]	161.0[a]		76.6	75.3	(β)	139.8	141.2	140.9	145.0
3	119.7	119.8	120.0	120.3		42.6	43.2	(α)	116.2	117.3	117.5	118.3
4	181.7	181.8	181.9	181.8		198.5	197.6	(C=O)	192.0	193.4	193.3	192.9
4a	103.7[b]	103.9[b]	103.3[b]	103.6, 103.8		103.3	103.8	(1')	113.5	114.0[a]	114.2[a]	113.9[a]
5	155.2[a]	155.3[a]	155.4[a]	161.5[a]		165.6	165.0	(2')	164.5	165.7	166.0	165.8
6	97.5	97.7	98.0	98.0		96.6	96.8	(3')	114.4	116.2	116.2	117.3
7	161.3[a]	161.6[a]	160.4[a]	163.7[a]		168.2	167.7	(4')	162.2	163.4	163.6	163.7
8	106.8[b]	106.9[b]	106.8[b]	93.2, 93.4		96.3	96.0	(5')	108.1[a]	110.1	109.2	110.2
8a	160.3[a]	160.3[a]	159.7[a]	157.4[a]		165.4	165.3	(6')	128.6	130.6	129.2	130.9
1'	111.4	111.6	111.9	112.9		118.4, 119.0	118.7	(1)	114.1	115.1	115.3[a]	127.4
2'	156.3	156.4	156.1	154.5[a]		153.0[a]	154.0	(2)	159.0	160.0	159.9	131.7
3'	102.6	102.7	103.3	116.6		109.7	108.4	(3)	103.0	103.6	103.6	116.7
4'	160.8[a]	160.9[a]	158.3[a]	158.0[a]		153.3[a]	154.0[a]	(4)	161.4[b]	162.3	162.3[b]	161.0
5'	106.8	106.9	107.0	105.8		107.5	107.2	(5)	107.6[a]	109.1	109.2	116.7
6'	131.2	131.3	130.8	128.2		130.7	130.6	(6)	129.9	131.8	130.8	131.7
9	23.5	23.8	23.5	24.1								
10	121.8	121.9	122.1	121.3								
11	131.2	131.3	131.4	129.8								
12	25.4	25.5	25.2	25.3								
13	17.3	17.5	17.3	17.5								
14	38.3*	39.8**	37.7	38.5	(3')	37.6	38.9	(3'')	39.4***[c]	32.4	33.3	32.5
15	123.3	123.4	123.6	124.8	(2')	124.2, 125.3	126.2	(2'')	124.3	123.2[b]	123.2	123.2[b]
16	132.8	132.9	132.5	131.2	(1')	133.8	133.2	(1'')	131.8	134.7	135.2	134.8
17	22.5	22.6	22.4	22.9	(7')	23.5	23.5	(7'')	22.5	23.8	23.8	23.8
18	38.3*	39.2**	37.5	38.4	(6')	39.1	38.9	(6'')	38.2***[c]	32.4	36.1	32.4
19	38.3*	39.8**	37.7	38.5	(5')	37.8	38.9	(5'')	39.4***[c]	36.4	40.4	36.4
20	45.8	45.7	47.0	44.3	(4')	—	46.6	(4'')	44.8	47.4	50.0	47.4

No.	A	A	A (120°)	A		B	C		A (90°)	C	C	C
21	208.1	208.4	208.7	209.1	(8'')	209.4	208.5	(8')	208.7	209.5	207.6	209.5
22	114.0	114.0	114.8	113.9	(9'')	116.1, 117.2	116.7	(9')	112.8	113.3ᵃ	114.1ᵃ	113.3ᵃ
23	164.2	162.1ᵃ	161.0ᵃ	162.5ᵃ	(10'')	165.4ᵃ	162.3ᵃ	(10')	161.8ᵇ	164.5	164.2	164.6
24	102.1	113.7	107.9	112.9	(11'')	103.9	115.7	(11')	116.2	115.8	115.5ᵃ	115.8
25	164.2	161.6ᵃ	161.4ᵃ	161.3ᵃ	(12'')	165.6ᵃ	163.7ᵃ	(12')	161.4ᵇ	163.4	162.7ᵇ	163.7
26	107.2	106.9	107.1	106.2	(13'')	108.3	107.4	(13')	106.2	108.2	108.1	108.1
27	132.4	129.5	128.1	130.6	(14'')	134.6	132.8	(14')	129.9	128.6	129.2	128.6
28	120.7	121.1	121.0	120.5	(15'')	122.0, 122.5	123.9	(15')	120.7	121.7	136.7	121.7
29	155.8ᵃ	155.9ᵃ	156.0ᵃ	154.5ᵃ	(16'')	157.1ᵃ	156.6ᵃ	(16')	156.0	156.5	131.8	156.4
30	102.0	102.7	103.3	102.6	(17'')	102.7	103.1	(17')	103.0	103.6	116.0	103.6
31	155.8ᵃ	155.9ᵃ	155.4ᵃ	155.9ᵃ	(18'')	158.4ᵃ	157.4ᵃ	(18')	155.7	157.9	156.6	157.9
32	106.8	106.9	107.0	106.5	(19'')	107.5	108.4	(19')	106.2	107.4	116.0	107.5
33	130.8	131.3	130.8	130.8	(20'')	130.7	131.2	(20')	129.5	132.1	131.8	132.3
34		21.4	114.7	21.1	(21'')		22.2	(21')	21.1	22.1	22.2	22.2
35		122.4	128.3	122.4	(22'')		123.4	(22')	122.5	123.5ᵇ	123.2	123.4ᵇ
36		130.5	77.4	130.0	(23'')		131.1	(23')	129.9	131.4	131.4	131.4
37		25.5	28.0	25.3	(24'')		25.8	(24')	25.2	25.8	25.8	25.8
38		17.5	27.8	16.9, 17.0	(25'')		17.8	(25')	17.5	17.8	17.9	17.8
Solvent	A	A	A (120°)	A		B	C		A (90°)	C	C	C

* In pyridine-d_5. ** In CD$_3$OD. A dmso-d_6, B CD$_3$OD, C acetone-d_6. ᵃ⁻ᶜ Assignments may be interchangeable in vertical column.

2. Five Novel 2-Arylbenzofuran Derivatives, Chalcomoracin, Mulberrofurans C, F, G, and H

As described in Chapter IV, a series of 2-arylbenzofuran derivatives have been isolated from the mulberry tree. TAKASUGI et al. isolated a phytoalexin from diseased mulberry leaves and designated it as chalcomoracin (**103**) (*87*). In the course of our studies, three novel 2-arylbenzofuran derivatives, mulberrofurans C (**130**), F (**131**), and G (**27**), were isolated as hypotensive components. Single intravenous injection of **130**, **131**, or **27** caused a marked depressor effect in rabbits (*111*, *112*).

Chalcomoracin (**103**), mp 183 °C (decomp.), $[\alpha]_D$ +194°, had the molecular formula of $C_{39}H_{36}O_9$. The UV spectrum showed absorption maxima at 218 nm (58 600), 294 (shoulder, 33 900), 329 (50 500), and 334 (41 300). Pyrolysis of the heptamethyl ether (**103a**) gave two fragmentation products, morachalcone A tetramethyl ether (**120b**) and the diene **132a**. Hydrogenation of **132a** gave a tetrahydroderivative (**133**) which was identified as dihydromoracin C trimethyl ether. The facile formation of **120b** and **132a** by pyrolysis of **103a** suggested that chalco-

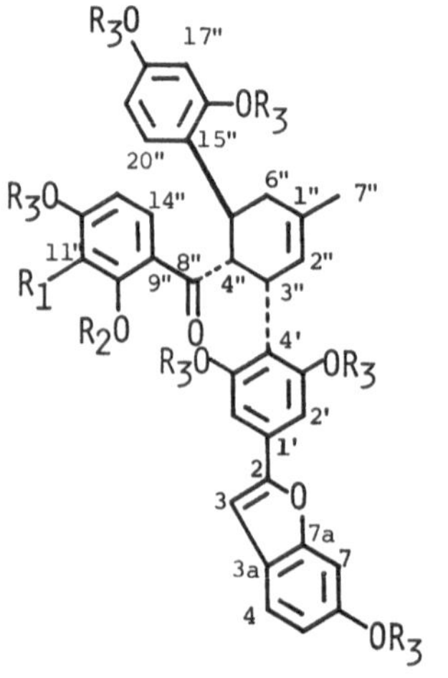

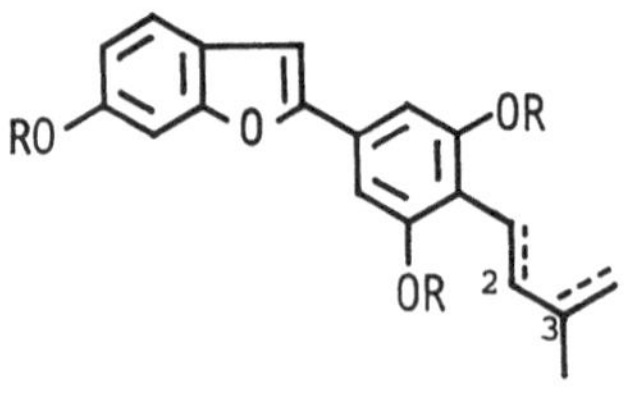

(132): R=H, $\Delta^{1,3}$
(132a): R=CH₃, $\Delta^{1,3}$
(133): R=CH₃, no Δ

(103): R₁= ⟨structure⟩ , R₂=R₃=H
(103a): R₁= ⟨structure⟩ , R₂=R₃=CH₃
(103b): R₁= ⟨structure⟩ , R₂=H, R₃=CH₃
(130): R₁=R₂=R₃=H

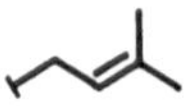

Fig. 27

moracin was probably a Diels-Alder type adduct between the demethylated products **120** (Fig. 21) and **132**.

Location of the benzoyl moiety on the methylcyclohexene ring was established by analysis of the ^{1}H-NMR spectrum of the alcohol obtained by hydride reduction (LiAlH$_4$) of the hexamethyl ether (**103b**). Chalcomoracin (**103**) is optically active and is presumably formed by a Diels-Alder type reaction between morachalcone A (**120**) and dehydromoracin C (**132**) or their equivalents under enzymatic control. This biogenetic hypothesis is supported by the co-occurence of morachalcone A (**120**), moracin D (**78**) (*82*) equivalent to dehydromoracin C (**132**), and chalcomoracin (**103**) as minor phytoalexins in the infected cortical tissues of mulberry shoots.

Mulberrofuran C (**130**) (*111*), an amorphous powder, [α]$_D$ +153°, had molecular formula C$_{34}$H$_{28}$O$_9$. The UV spectrum was similar to that of chalcomoracin (**103**) and had absorption maxima at 215 (56200), 281 (26900), 320 (41700), and 333 nm (36300). Presence of aluminum chloride induced a bathochromic shift of the 281 nm maximum. SHERIF *et al.* had reported earlier that the aluminum chloride-induced shift was not observed in the UV spectra of compounds containing a prenyl substituent *ortho* to a chelated hydroxyl group (*113*). These data, together with the fact that chalcomoracin (**103**) coexists with mulberrofuran C (**130**), suggested that **130** was probably a deprenylchalcomoracin. Chemical shifts and coupling constants of the protons of the methylcyclohexene ring in the ^{1}H-NMR spectrum of mulberrofuran C are shown in Fig. 28; the remaining protons are summarized as follows: 2-arylbenzofuran moiety, δ6.96 (2H, s, H-2′ and H-6′), 6.78–6.86 (3H, m, H-3, H-5 at H-7), 7.41 (1H, d, J=8.5, H-4); 2,4-dihydroxybenzoyl moiety, δ6.27 (1H, d, J=2.5, H-11″), 6.40 (1H, dd, J=2.5 and 9, H-13″), 8.56 (1H, d, J=9, H-14″), 12.61 (1H, s, OH-10″); 2,4-dihydroxyphenyl moiety, δ6.34 (1H, dd, J=2.5 and 8.5, H-19″), 6.55 (1H, d, J=2.5, H-17″), 7.04 (1H, d, J=8.5, H-20″). Detailed examination of the ^{1}H-NMR spectra of mulberrofuran C (**130**), chalcomoracin (**103**), and kuwanon G (**25**) revealed that the chemical shifts and coupling constants of protons of the relevant methylcyclohexene ring of **130** closely resembled those of **103** (Fig. 28). In the ^{13}C-NMR spectrum of mulberrofuran C (**130**), the signal of C-11″ appeared at higher field than that of C-11″ of chalcomoracin (**103**), whereas the chemical shifts of all the carbon atoms other than C-11″ were essentially the same as those of the relevant carbon atoms of **103** (Table 4). Hence, mulberrofuran C was formulated as deprenylchalcomoracin (**130**) (Fig. 27).

Mulberrofuran F (**131**) (*112*), an amorphous powder, [α]$_D$ +513°, had molecular formula of C$_{39}$H$_{34}$O$_8$. The IR spectrum indicated the

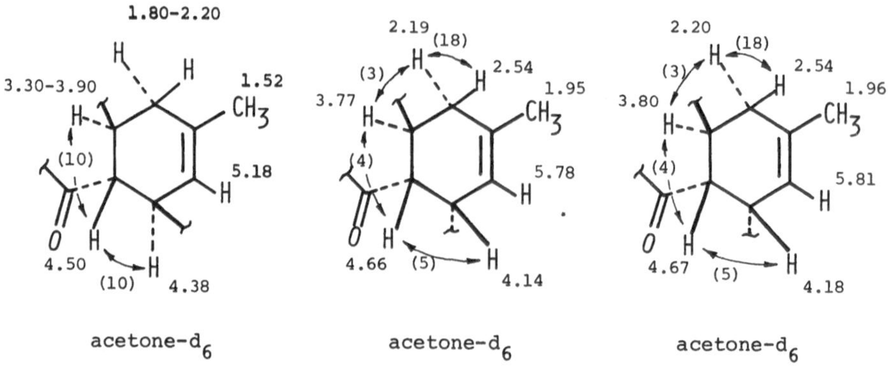

Fig. 28. Chemical shifts (ppm) and coupling constants (Hz) of protons of the methylcyclohexene rings of **25**, **103**, and **130**

absence of carbonyl functions, while the UV spectrum was similar to those of chalcomoracin (**103**) and mulberrofuran C (**130**) and had maxima at 230 (inflection, 32400), 285 (14100), 296 (shoulder, 13200), 306 (inflection, 18200), 321 (28800), and 335 (24500). These results suggested that mulberrofuran F was a 4'-substituted 6,3',5'-trihydroxy-2-arylbenzofuran derivative. The ^{1}H-NMR spectra of mulberrofuran F (**131**) and its derivatives, such as the pentamethyl ether (**131a**) and the pentaacetate (**131b**), given in Fig. 29 suggested that one of the hydroxyl groups in the C-ring is involved in an ether linkage, because of the chemical shifts, of H-2' and H-6' which appeared to be nonequivalent. The presence of only one hydroxyl group in the 2,4-dioxygenated phenyl moiety was suggested in view of the chemical shift changes produced on acetylation. These observations indicated the presence of the following part structures a 4'-substituted-6,3'-dihydroxy-5'-oxygenated 2-arylbenzofuran, a 2,4-dihydroxy-3-prenylphenyl, and a 4-hydroxy-2-oxygenated phenyl group combination of which led to partial structure **131'**. The presence of a trisubstituted methylcyclohexene ring in the C_8H_9 moiety was indicated by the ^{1}H-NMR spectrum of pentamethyl ether (**131a**) and sequential decoupling experiment (Fig. 29).

The ^{13}C-NMR spectrum is presented in Table 4. When the spectrum of mulberrofuran F (**131**) is compared with that of chalcomoracin (**103**), the chemical shifts of the carbon atoms of the 2-arylbenzofuran skeleton, except those of C-3' and C-4', are seen to be similar to those of the corresponding carbon atoms of **103**. Moreover, a characteristic singlet signal at δ 103.3 suggested the presence of a ketal carbon. The

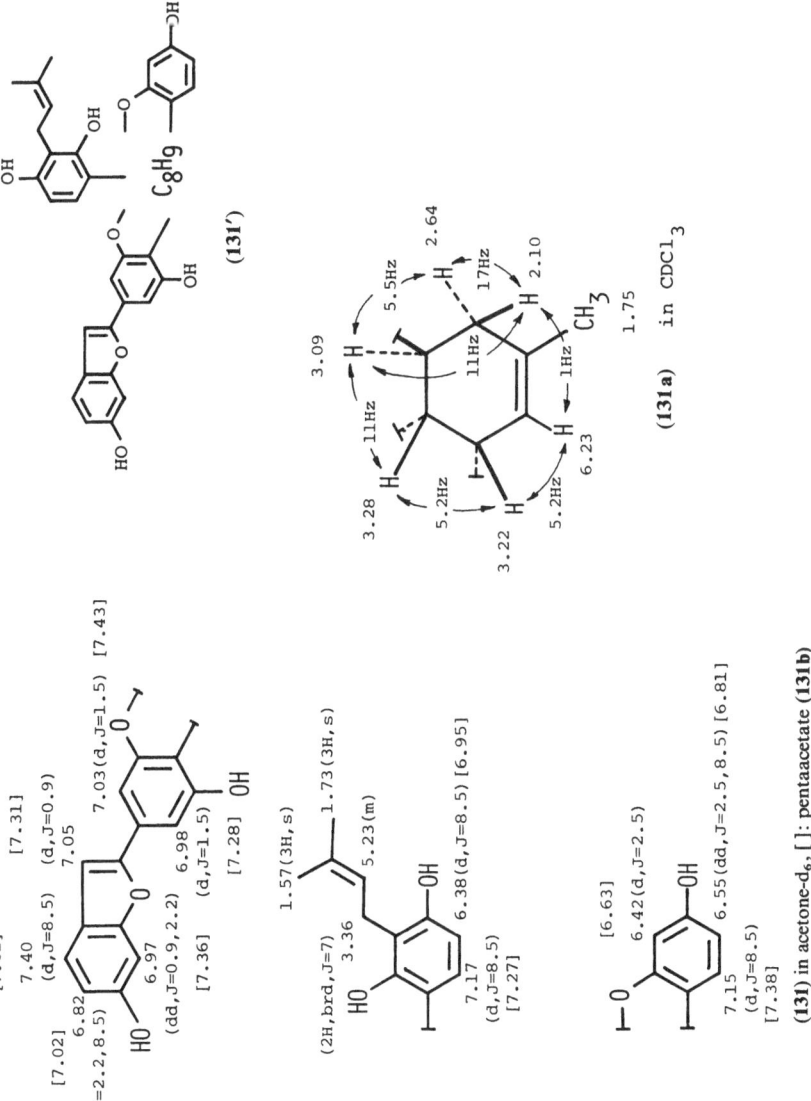

Fig. 29. Chemical shifts (ppm) and coupling constants (Hz) of ¹H-NMR spectra of 131 and its derivatives (131a and 131b)

location of the methylcyclohexene ring on the C-ring was also supported by the shift of the H-2″ signal to higher field (0.47 ppm) on
acetylation of **131** to **131b** (Fig. 30). All of these results indicate that
the structure of mulberrofuran F may be represented by formula **131**.
In order to corroborate this structure, mulberrofuran F was derived
from chalcomoracin (**103**) by means of the following reaction. A solution of **103** in ethanol containing 1.5% sulfuric acid was externally
irradiated in a glass vessel using a high-pressure mercury lamp. From
this reaction mixture, mulberrofuran F was obtained. The configuration of the 3-prenyl-2,4-dihydroxyphenyl moiety with the methine protons on the methylcyclohexene ring was suggested by the coupling
constants of the methylcyclohexene ring protons. The coupling constants of the ring protons were in good agreement with those expected
for the relevant protons on the basis of a Dreiding model. Hence formula **131** can be assigned to mulberrofuran F (Fig. 30).

Mulberrofuran G (**27**) (*112*), an amorphous powder, $[\alpha]_D$ $+546°$,
had a molecular formula $C_{34}H_{26}O_8$. The IR spectrum indicated the

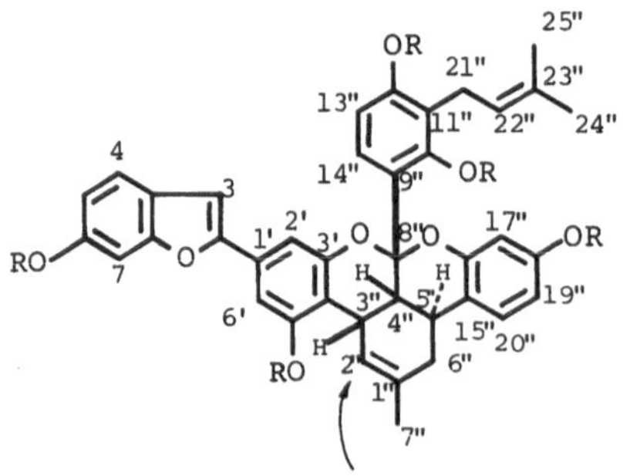

(**131**):   R=H $\delta 6.48$ (**131**)
(**131a**): R=CH$_3$ [6.01 (**131b**)]
(**131b**): R=COCH$_3$

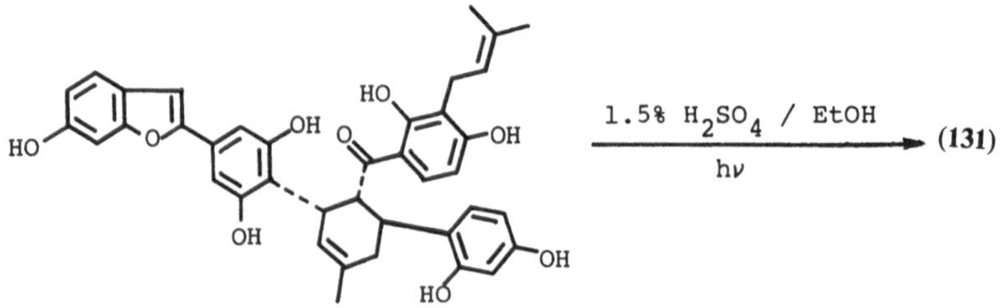

Fig. 30

absence of carbonyl functions, while the UV spectrum had maxima to be at 223 (22900), 285 (19500), 295 (shoulder, 17400), 306 (inflection, 23400), 321 (37200), and 335 nm (31600) similar to those of mulberrofuran F (131). The EI-MS of pentamethyl ether (27a) showed a significant fragment ion at m/z 495 ($M^+ - C_8H_9O_2$, 134) observed in the spectrum of mulberrofuran F pentamethyl ether (131a). These results suggest that mulberrofuran G (27) is a deprenylmulberrofuran F. In the ^{13}C-NMR spectrum of 27, all carbon atoms except those of the F ring exhibit essentially the same chemical shifts as the corresponding carbon atoms of 131 (Table 4). From the spectral evidence, formula 27 was suggested for mulberrofuran G. To confirm the structure, mulberrofuran G was prepared from mulberrofuran C (130), as in the case of mulberrofuran F (131). The ketalization reaction seems to depend on the stereochemistry of the methylcyclohexene ring. When a stereo-

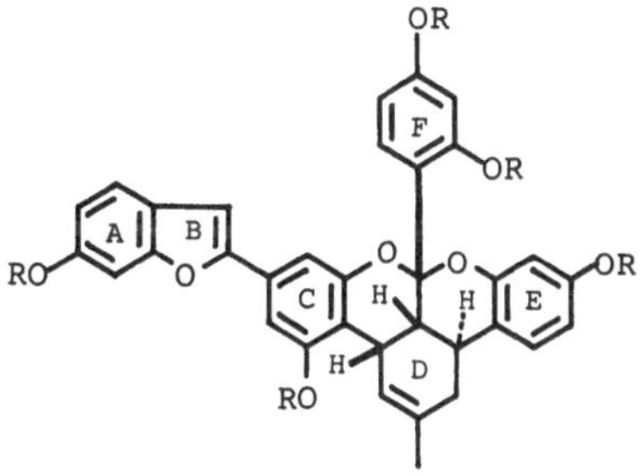

(27):   R=H
(27a): R=CH₃

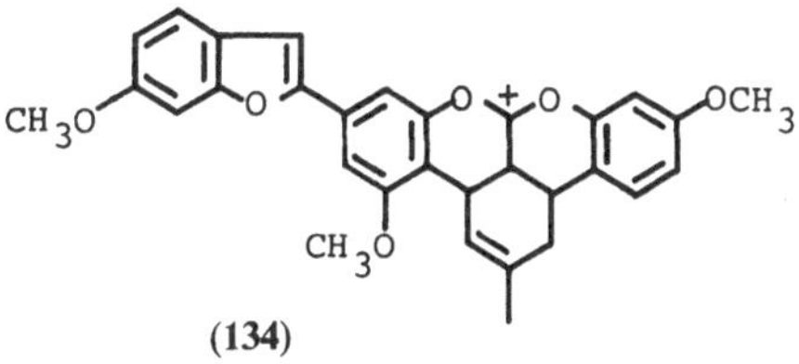

$$(130) \xrightarrow[h\nu]{1.5\%H_2SO_4 \; / \; EtOH} (27)$$

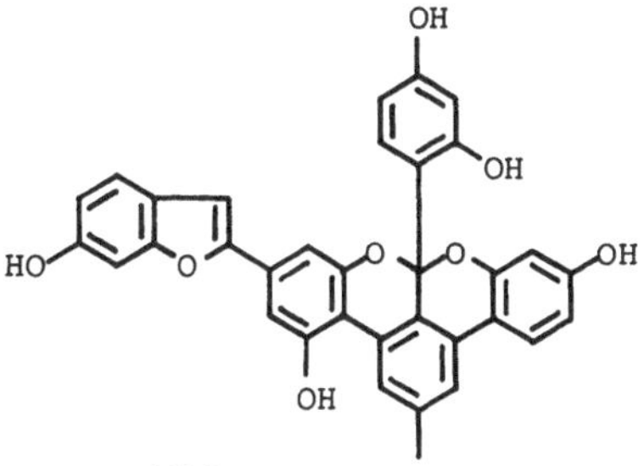

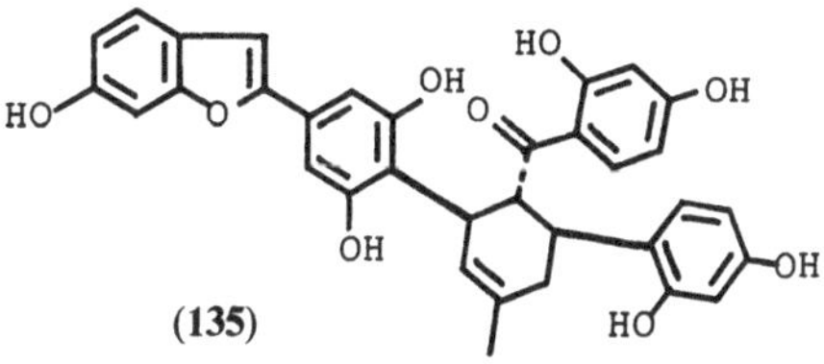

Fig. 31

isomer, mulberrofuran J (135) (*114*) was exposed to acid as in the cases
of chalcomoracin (103) and mulberrofuran C (130), starting material
(135) was recovered unchanged, and no reaction product was obtained.
The location of the benzoyl moieties on the methylcyclohexene rings
of 103 and 130 was reconfirmed on the basis of the results of ketalization reaction.

More or less contemporaneously, RAMA RAO *et al.* reported isolation of two compound from *Morus alba* which they called albanol A
and B and assigned formulae 27 and 136 to them, on the basis of
an X-ray analysis of albanol A pentamethyl ether (27a) (*115*). Mulberrofuran G proved to be identical with albanol A by direct comparison.
Our group has also isolated albanol B (136) from the Japanese cultivated mulberry root bark (*112*).

From a biogenetic point of view, mulberrofurans F (131) and G (27)
seem to be Diels-Alder type adducts derived from chalcomoracin (103)
and mulberrofuran C (130), respectively, by intramolecular ketalization
of the carbonyl group with the two adjoining hydroxyl groups as shown
in Fig. 30.

Mulberrofuran H (137) (*47*), an amorphous powder, $[\alpha]_D$ $+25°$,
had molecular formula of $C_{27}H_{22}O_6$. The UV spectrum was similar
to that of moracin C (77) (*71*), mulberrofurans F (131) and G (27)
and had maxima at 220 nm (shoulder, 30000), 285 (shoulder, 14100),
321 (33100), and 333 (shoulder, 28000). In the ^{13}C-NMR spectrum
of mulberrofuran H (137), the chemical shifts of the carbon atoms
of the 2-arylbenzofuran skeleton were, except for C-4' similar to those
of the equivalent carbon atoms of mulberrofuran C (130) (Table 4).
These results suggested that mulberrofuran H was a 4'-substituted
6,3',5'-trihydroxy-2-arylbenzofuran derivative. Comparison of the ^1H-
NMR spectra of mulberrofuran H (137) and its acetate (137a) also
provided evidence for the presence of a 4-substituted 3,5-dihydroxy-
phenyl and a 4-hydroxy-2-oxygenated phenyl moiety. On this basis
partial structure (137') could be advanced. The remaining C_7H_9 part
of the C-4' side chain was indicated by the ^{13}C-NMR spectrum to
contain seven aliphatic carbons: one methyl, two methylenes, one methine, one tertialy carbon attached to oxygen atom and two olefinic carbons as shown in Table 4. The nature of the C_7H_9 unit was clarified
by sequential decoupling. The results are shown in Fig. 32, along with
the chemical shifts and coupling constants involving the protons of
the C_7H_9 moiety. Supporting data for the structure were obtained
by the LSPD technique; when the proton signal at $\delta 1.56$ (C-1''-CH$_3$)
was weakly irradiated, the carbon signal at $\delta 71.8$ (C-1'') increased
in area (*ca.* $+70\%$). Irradiation of the signal at $\delta 3.16$ (H-5'') increased
the area of the C-1'' signal (*ca.* $+15\%$), and irradiation of the signal

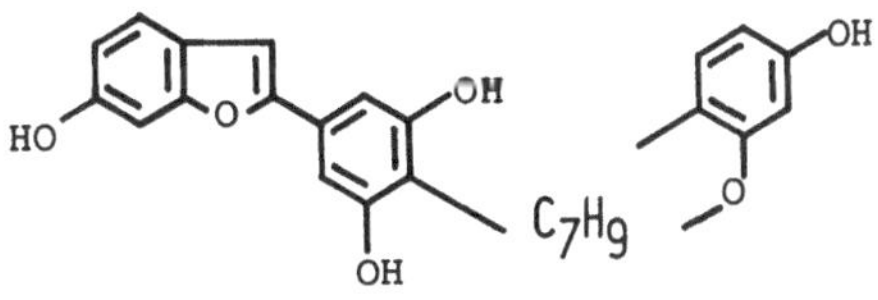

(137')

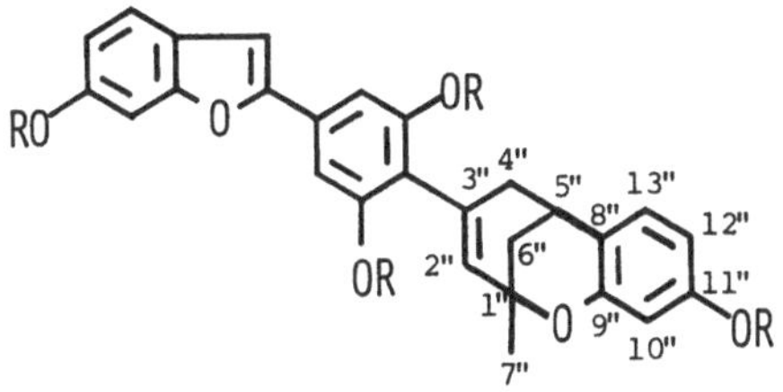

(137):   R=H
(137a): R=COCH₃
(137b): R=CH₃

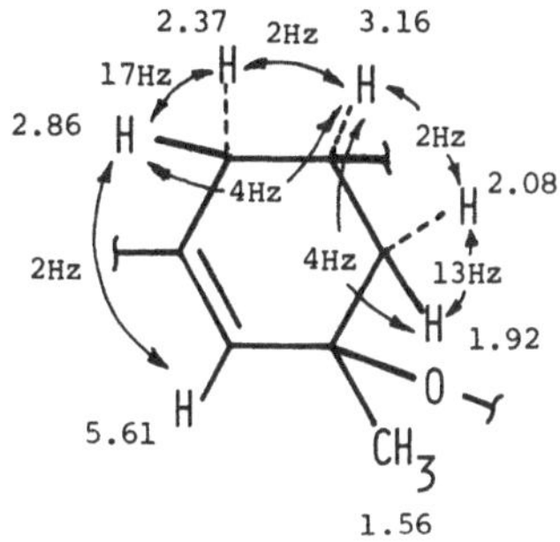

(130) or (135) 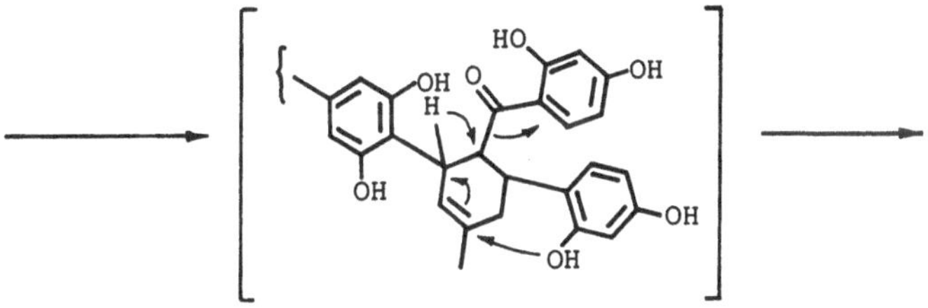 ⟶ [...] ⟶ (137)

Fig. 32

at δ 5.61 (H-2″) also increased the area (*ca.* +30%) of the same carbon signal.

Thus formula **137** could be derived for mulberrofuran H. Biogenetically, mulberrofuran H seems to be a derivative of a Diels-Alder type adduct, like mulberrofurans C (**130**) and J (**135**), possibly through the mechanism illustrated in Fig. 32.

3. Coloring Matter of *Morus* Root Bark, Mulberrofuran I

On the surface of the root bark of the mulberry tree in the lenticel, is found a reddish violet powder whose composition was investigated in our laboratory. Isolated were mulberrofuran I (**138**) (*116, 117*), kuwanols A (**139**) (*118*), B (**140**) (*118*), mulberrofuran P (**141**) (*119*), M (**142**) (*120*), and Q (**143**) (*121*). Compounds **138** and **140** are reddish-violet, and **141** is a blue powder obtained as a minor component.

Table 4. ^{13}C-NMR Data of 2-Arylbenzofuran Derivatives (Diels-Alder type adducts: chalcomoracin: CM, mulberrofurans C, E, F, G, J, and H: MC, ME, MF, MG, MJ, and MH)

C	CM (103)	CM (103)	MC (130)	MF (131)	MG (27)	MJ (135)	ME (151)	MH (137)
2	155.8[a]	156.4	156.5	156.8	156.7	156.5	156.6	156.7
3	102.9	101.9	102.0	101.9	102.2	102.8	101.8	102.2
3a	121.2	122.5	121.9	121.8	122.5	122.0	122.5	122.4
4	121.6	121.9	121.9	121.4	122.0	121.8	121.7	121.9
5	112.5	113.1	113.1	112.7	113.4	113.1	113.0	113.2
6	157.0	157.7	157.8	157.0	157.5	158.2[a]	157.8	156.7
7	97.7	98.4	98.4	97.9	98.4	98.4	98.3	98.4
7a	153.8	155.4	153.3	154.0	153.3	155.3	155.3	155.3[a]
1'	129.2	130.9	130.9	130.4	131.1	130.5	131.1	131.5
2'	103.2	104.8	104.8	104.7	105.0	103.9	104.6	103.7
3'	155.2	157.7	156.5	154.6[a]	155.0[a]	156.5	156.6	155.7
4'	112.8	113.3	113.6	116.6[b]	117.5[b]	115.9	113.8	117.4
5'	155.2	157.7	156.5	155.9[a]	156.7[c]	156.5	156.6	155.7
6'	103.2	104.8	104.8	105.5	105.4	103.9	104.6	103.7
1''	133.2	133.8	134.1	133.3	133.7	134.3	134.2	71.8
2''	127.9	123.1	124.2	121.8	122.9	125.1	123.2	132.4
3''	33.1	33.1	33.2	37.1[d]	37.2[d]	38.8	33.4	135.6
4''	46.9	47.7	47.9	28.5	28.5	46.8	50.6	39.8
5''	33.9	36.4	33.2	35.3[d]	35.1	38.8	40.8	31.8
6''	33.9	32.4	32.4	35.7	36.2[d]	37.7	34.8	34.6
7''	23.2	23.8	23.8	23.7	23.9	23.4	23.8	27.5
8''	207.7	209.2	209.5	103.3	102.6	209.3	107.9	119.0
9''	115.0	116.5	116.4	115.7[b]	113.4	117.2	115.9	155.2[a]
10''	162.4[b]	164.6	166.0[b]	155.9[a]	159.9	166.0[b]	164.0	103.9
11''	114.4	115.8	103.6	112.6	103.9	101.7	115.6	157.8
12''	162.0[b]	163.3	167.0[b]	151.8[c]	157.7[c]	164.8[b]	162.9	108.8
13''	106.2	108.1	108.8	106.8	107.1	107.7	108.2	130.6
14''	131.2	132.1	135.0	125.5	130.3	133.9	130.7	
15''	122.0	121.9	122.5	116.6[b]	116.8[b]	122.5	136.4	
16''	155.4[a]	156.6	157.8	152.8[c]	154.5[a]	157.1[a]	129.0	
17''	102.6	103.5	103.6	103.4	104.6	104.6	115.9	
18''	157.0	156.4	157.8	156.8[a]	157.9[c]	157.1[a]	157.8	
19''	107.5	107.4	107.5	109.8	109.9	107.8	115.9	
20''	131.2	128.9	128.8	127.9	127.9	130.3	129.0	
21''	21.0	22.1		22.9			22.2	
22''	122.0	124.4		123.1			124.1	
23''	130.4	131.4		130.7			131.3	
24''	25.4	25.8		25.7			25.8	
25''	17.6	17.8		17.8			17.8	
Solvent	A	B	B	B	B	B	B	B

A DMSO-d_6, B acetone-d_6. [a-d] Assignments may be interchangeable in vertical column.

Mulberrofuran I (138), $[\alpha]_D$ +212°, had molecular formula of $C_{34}H_{24}O_8$ and furnished a pentaacetate (138a). The UV spectrum of 138 was similar to those of 2-arylbenzofuran derivatives and had maxima at 250 (inflection, 12600), 290 (shoulder, 12600), 330 (shoulder, 20000), and 338 nm (24500). In acid solution the compound showed a red color which faded when neutralized, and exhibited the following bathochromic shifts in the UV spectrum: λ_{max} 296 (shoulder, 48900), 330 (17000), 340 (15800), 456 (6200), 548 nm (12000).

The ^{1}H-NMR spectra of mulberrofuran I (138) and its pentaacetate (138a) indicated the presence of a 4'-substituted-6,5'-dihydroxy-3'-oxygenated-2-arylbenzofuran, 2,4-dihydroxyphenyl, and 4-hydroxy-2-oxygenated phenyl moieties, thus leading to partial structure (138'). Analogy with other Diels-Alder type adducts between 4'-dehydroprenyl-2-arylbenzofuran and chalcone such as mulberrofuran C (130), led to

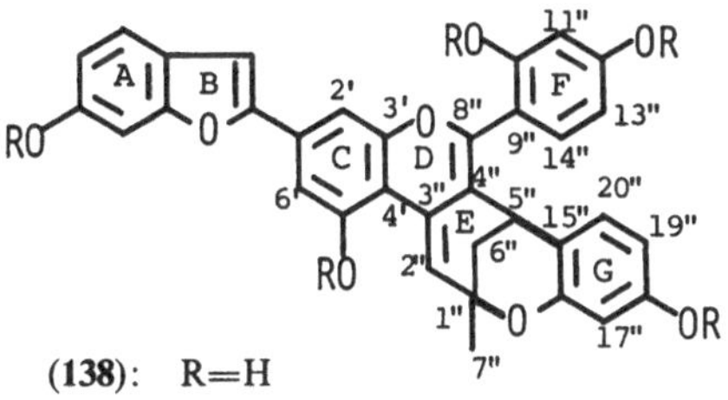

(138): R=H
(138a): R=COCH₃

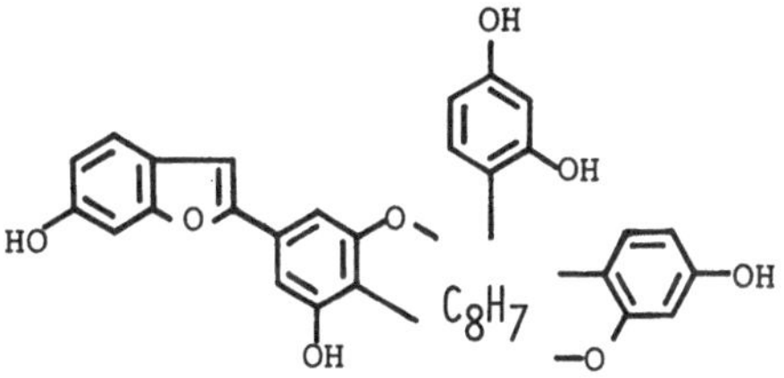

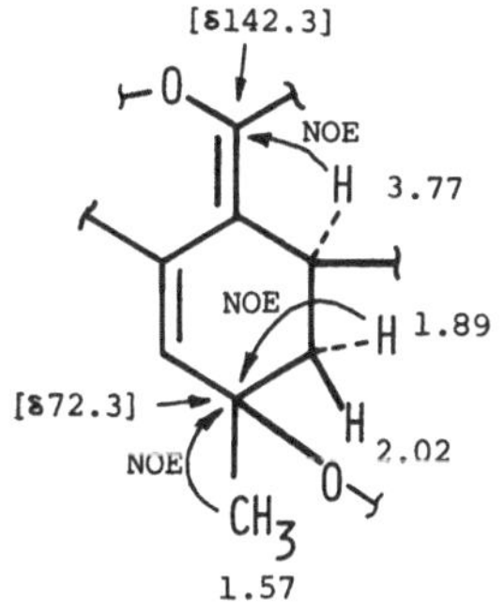

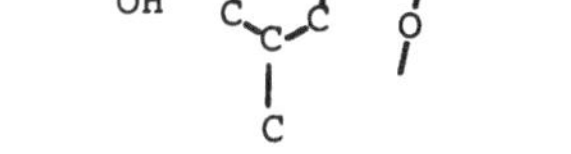

Fig. 33

the assumption that the carbon skeleton of the C_8H_7 unit can be represented as **138″**. According to the ^{13}C-NMR spectrum, the unit contained eight aliphatic carbons: $-CH_3$, $-CH\lessgtr$, $-CH_2-$, $\gtrless C\lessgtr^{O-}$, $\gtrless C=$ $C\lessgtr^{O-}$, $\gtrless C=C\lessgtr^H$ (Table 5). In the ^1H-NMR spectrum of **138**, the C_8H_7 unit gave rise to signals at δ 1.57 (3 H, s, -CH$_3$), 1.89 (1 H, ddd, $J=1.2$, 3.7 and 12.5, $-CH_2-$), 2.02 (1 H, dd, $J=2.1$ and 12.5, $-CH_2-$), 3.77 (1 H, br s, $-CH\lessgtr$), 6.78 (1 H, d, $J=1.2$, $-CH=C\lessgtr$). Further information which permitted discrimination between possible structures were obtained by selective ^{13}C-{^1H} nuclear Overhauser effect (NOE) measurements (*122*). A clear NOE was observed between the methyl protons and the oxygenated carbon at δ 72.3, and a weak NOE between the methylene protons and the same carbon atom. A NOE was also observed between the methine proton and the carbon atom at δ 142.3. The results are summarized in partial structure (**138‴**). In view of the shift ($+0.46$ ppm) experienced by the signal of the olefinic proton (H-2″) on acetylation the olefinic proton must be located at a position near the hydroxy group on the C-ring. This led to structure **138** for mulberrofuran I.

In an attempt to shed light on the color changes exhibited by **138** (*117*) measuring the ^1H-NMR spectrum in $CD_3COCD_3 + CF_3COOD$ solution, showed that the olefinic proton could be exchanged for deuterium. Genesis of the color in acid solution seems to be due to formation of a flavilium ion through the zwitter ion, as shown in Fig. 34.

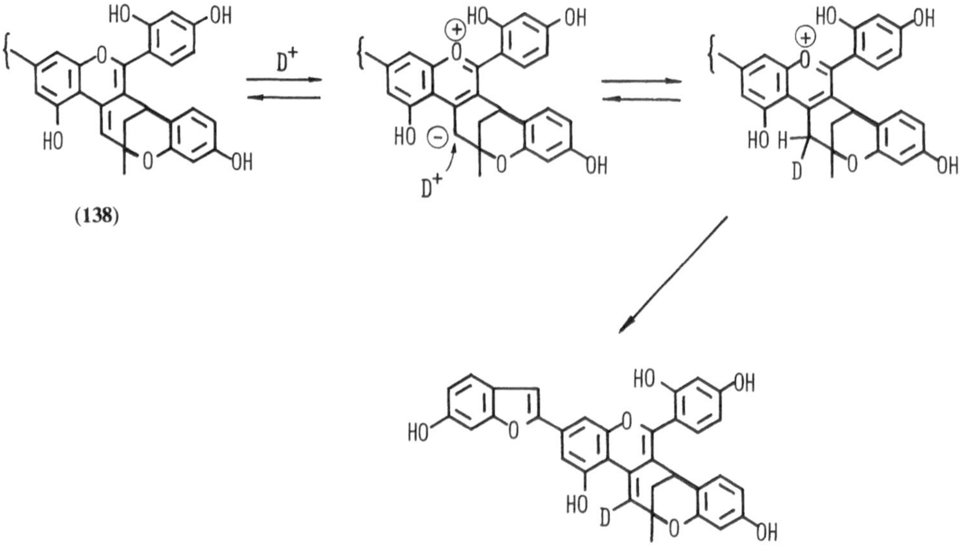

Fig. 34

References, pp. 191–201

(130) or (135) 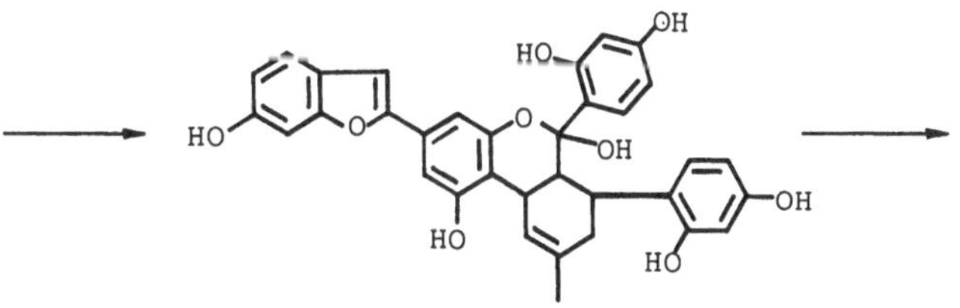 ⟶ ⟶ (138)

Fig. 35

Biogenetically, mulberrofuran I (138) seems to be a derivative of Diels-Alder type adducts like mulberrofurans C (130) and/or J (135) through a hemiketal intermediate, as outlined in Fig. 35.

4. Other Phenolic Compounds Related to Mulberrofuran I

In addition to mulberrofuran I (138) the reddish-violet powder on the surface of cultivated mulberry trees, gave five minor phenolic compounds, namely kuwanols A (139), B (140), mulberrofurans P (141), M (142), and Q (143).

Kuwanol A (139) (118), an amorphous powder, $[\alpha]_D$ +557°, had a molecular formula $C_{34}H_{28}O_8$, and gave a hexaacetate (139a). Formula 139 was proposed for kuwanol A on spectral evidence. The ^{13}C-NMR spectrum of 139 is listed in Table 5. Biogenetically kuwanol A (139) seems to be a Diels-Alder type adduct derived from kuwanon Y (144) (123) by an intramolecular ketalization reaction, as in the case of mulberrofurans F (131) (112) and G (27) (112). Kuwanol A (139) seems to be an interesting intermediate for examining the biogenetic route to mulberrofuran G (27) which is postulated to involve oxidative cyclization of hydroxy stilbenes (124).

Kuwanol B (140) (118), a reddish-violet amorphous powder, $[\alpha]_D$ +103.5°, had molecular formula $C_{34}H_{26}O_8$, and gave a hexamethyl ether (140a). Its acid solution of 140 had a red color which faded when neutralized. The UV spectrum of 140 had absorption maxima at 286 (12000), 310 (inflection, 13000), 326 (shoulder, 17000), and 336 nm (17000), which in acid solution showed bathochromic shifts as follows; λ_{max} 286 (10000), 310 (inflection, 11000), 326 (14000), 336 (14000), 470 (shoulder, 2200), and 555 nm (4900). The same phenomenon was observed in the case of mulberrofuran I (138) (116). Formula 140 was proposed for the structure of kuwanol B, based on the spectral data. The ^{13}C-NMR spectrum is listed in Table 5. It seems to be another derivative of a Diels-Alder type adduct such as kuwanon X (145) (114) and/or Y (144) (123), by the route already outlined in the case of mulberrofuran I (138) (116) (Fig. 35).

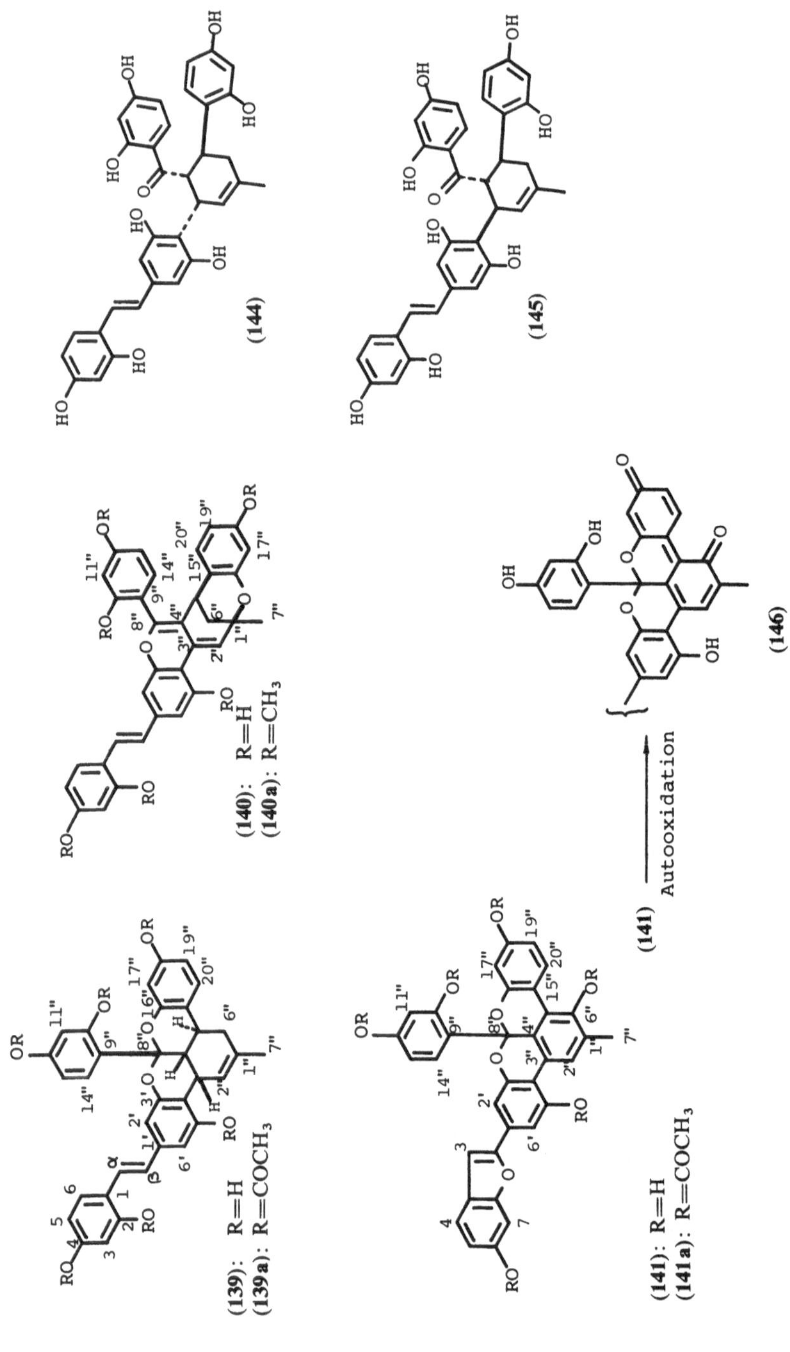

(144)

(145)

(140): R=H
(140a): R=CH$_3$

(139): R=H
(139a): R=COCH$_3$

(141): R=H
(141a): R=COCH$_3$

Autooxidation

(146)

Fig. 36

Mulberrofuran P (141) (*119*), a blue amorphous powder, had molecular formula of $C_{34}H_{22}O_9$, and gave a hexaacetate (141a). Its UV spectrum had absorption maxima at 210 (52000), 284 (15000), 315 (shoulder, 21000), 336 (34000), 351 (40000), and 370 nm (36000); it also showed a bathochromic shift in the presence of sodium methylate. Formula 141 was proposed for mulberrofuran P on the basis of spectral evidence. The ^{13}C-NMR spectrum is listed in Table 5. The alkaline solution of 141 showed a blue color, while in a nitrogen atmosphere coloration did not occur. This coloration seems to be due to the formation of a π-complex between 141 and the quinoid substance (146) produced by autooxidation of 141.

Mulberrofuran M (142) (*120*), an amorphous powder, $[\alpha]_D$ +15°, had molecular formula $C_{34}H_{22}O_{10}$, and gave a tetraacetate (142a). The UV spectrum had maxima at 208 (51000), 225 (inflection, 26000), 261 (20000), 299 (inflection, 18000), 323 (shoulder, 28000), 335 (34000) and 350 nm (shoulder, 28000), and was similar to those of other 4'-substituted 6,3',5'-trioxygenated 2-arylbenzofuran derivatives. The ^{13}C-NMR spectrum of 142 is listed in Table 5. Hydrolysis of mulberrofuran M with 10% H_2SO_4 in methanol gave β-resorcylic acid methyl ester and a compound (142') were obtained. Considering the spectral data of 142, 142a, and 142', formula 142 was proposed for mulberrofuran M.

Mulberrofuran Q (143) (*121*), an amorphous powder, $[\alpha]_D$ +82.4°, had molecular formula $C_{34}H_{24}O_{10}$. The UV spectrum had maxima at 222 (shoulder, 27000), 280 (14000), 284 (shoulder, 14000), 321 (28000), and 335 nm (shoulder, 20000). The ^{13}C-NMR spectrum is shown in Table 5. Considering the ^{13}C and ^{1}H-NMR spectra and the biogenetic analogy to other Diels-Alder type adducts from *Morus* root bark, the presence of 4'-substituted-6,5'-dihydroxy-3'-oxygenated-2-arylbenzofuran, 2,4-dihydroxyphenyl, 4-hydroxy-2-oxygenated phenyl, and $C_8H_7O_2$ moieties was indicated thus leading to structure 138″ (Fig. 33). The $C_8H_7O_2$ unit to contained seven aliphatic carbons and one carbonyl carbon. From the ^{13}C and ^{1}H-NMR spectra and the results obtained by the LSPD technique (*122*), the $C_8H_7O_2$ unit was incorporated as in 143' (Fig. 37). Hydrolysis of mulberrofuran Q with potassium carbonate in acetone to give compound 142' identical with the hydrolysis product of mulberrofuran M (142). Compound (142') seems to be derived from the hemiketal intermediate 143″ through an intramolecular Michael addition described in Fig. 37. Further support for the presence of an epoxide ring in the $C_8H_7O_2$ unit was obtained by examination of the hexamethyl ethers 143a, b formed by reaction of 143 with dimethyl sulfate. Formulas 143a and 143b were supported by the spectral data (Fig. 38). From the above results, the formula 143 was proposed for the structure of mulberrofuran Q. Mul-

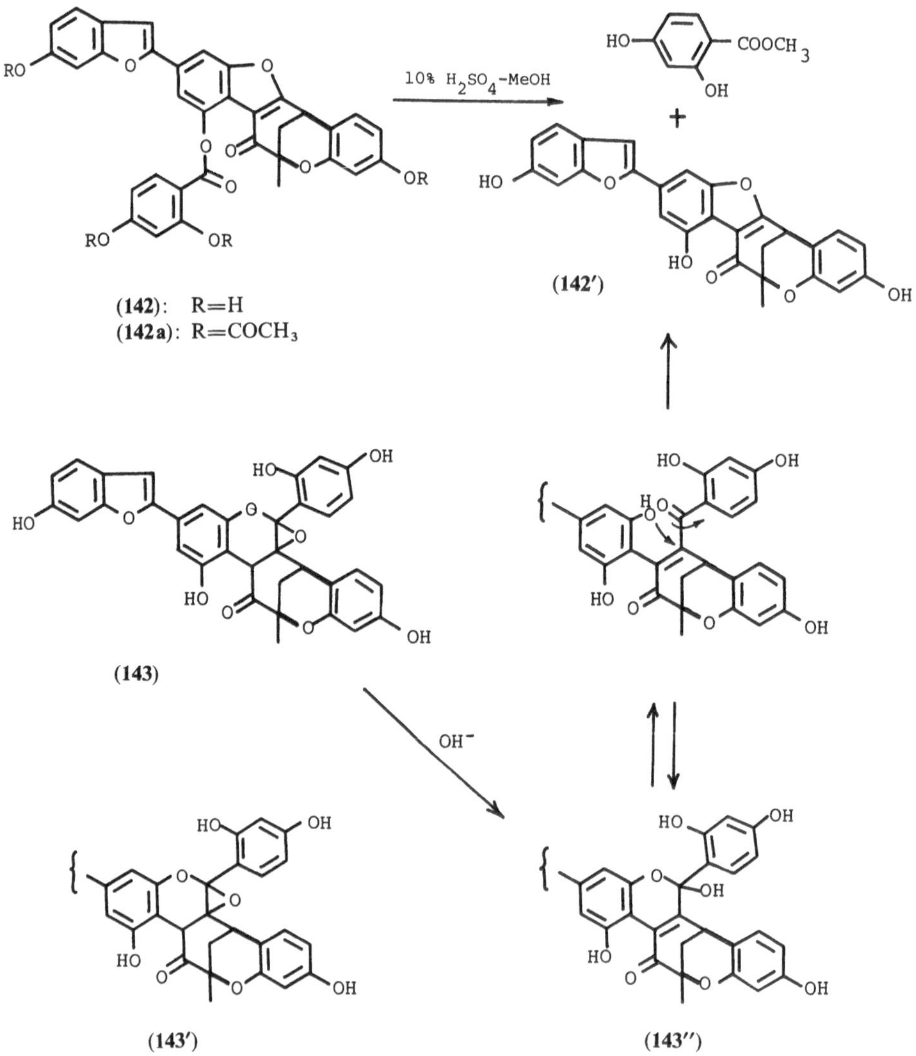

Fig. 37

berrofuran Q (143) seems to be derived biogenetically from a Diels-Alder type adduct, such as mulberrofurans C (130) and/or J (135), through an intermediate hemiketal and mulberrofuran I (138), and also seems to be an intermediate on the way to mulberrofuran M (142). The proposed biogenetic pathway to mulberrofurans F (131), G (27), I (138), M (142), and Q (143) is outlined in Fig. 39.

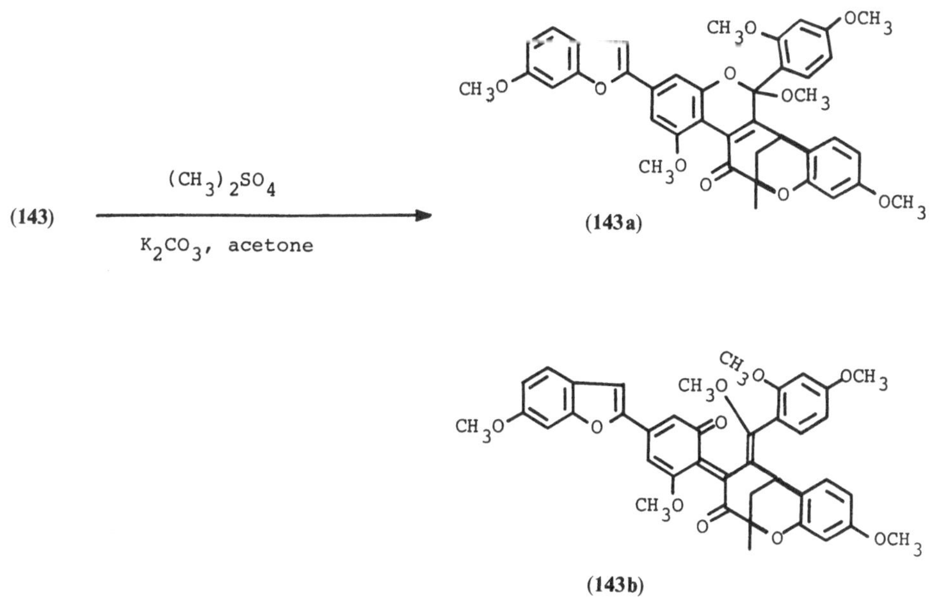

Fig. 38

5. Diels-Alder Type Adducts of *Morus* Cell Cultures

A few reports dealing with mulberry tissue culture have been published during the past decade (*125–127*). Apart from the isolation of β-sitosterol (*126*) and detection of yellow substances which have not yet been characterized at all (*126*), these studies were mainly concerned with the response of the callus growth to culture conditions. To examine the production and biosynthesis of secondary metabolites in plant cell cultures and to identify the constituents of *Morus* root bark, we attempted to obtain pigment-producing callus cultures of *Morus alba* L. Callus tissues induced from seedlings were subcultured under specified conditions (*128*) and subjected to selection over nine years, giving rise to cell strains having a high pigment productivity. From the extract of the callus tissues, six Diels-Alder type adducts, kuwanons J (**147**) (*129, 131*), Q (**148**) (*130, 131*), R (**149**) (*130, 131*), V (**150**) (*130, 131*), mulberrofuran E (**151**) (*130*), and chalcomoracin (**103**) (*87*), along with β-sitosterol and stigmast-5-en-3 β-ol-7-one (*129*) were detected. In addition to these, moracin C (**77**) (*71*), moracin M (**87**) (*71*), and morachalcone A (**120**) (*87*), were also detected on TLC.

The structures of kuwanons J (**147**), Q (**148**), R (**149**), V (**150**), and mulberrofuran E (**151**) were based on spectral evidence. The ^{13}C-NMR

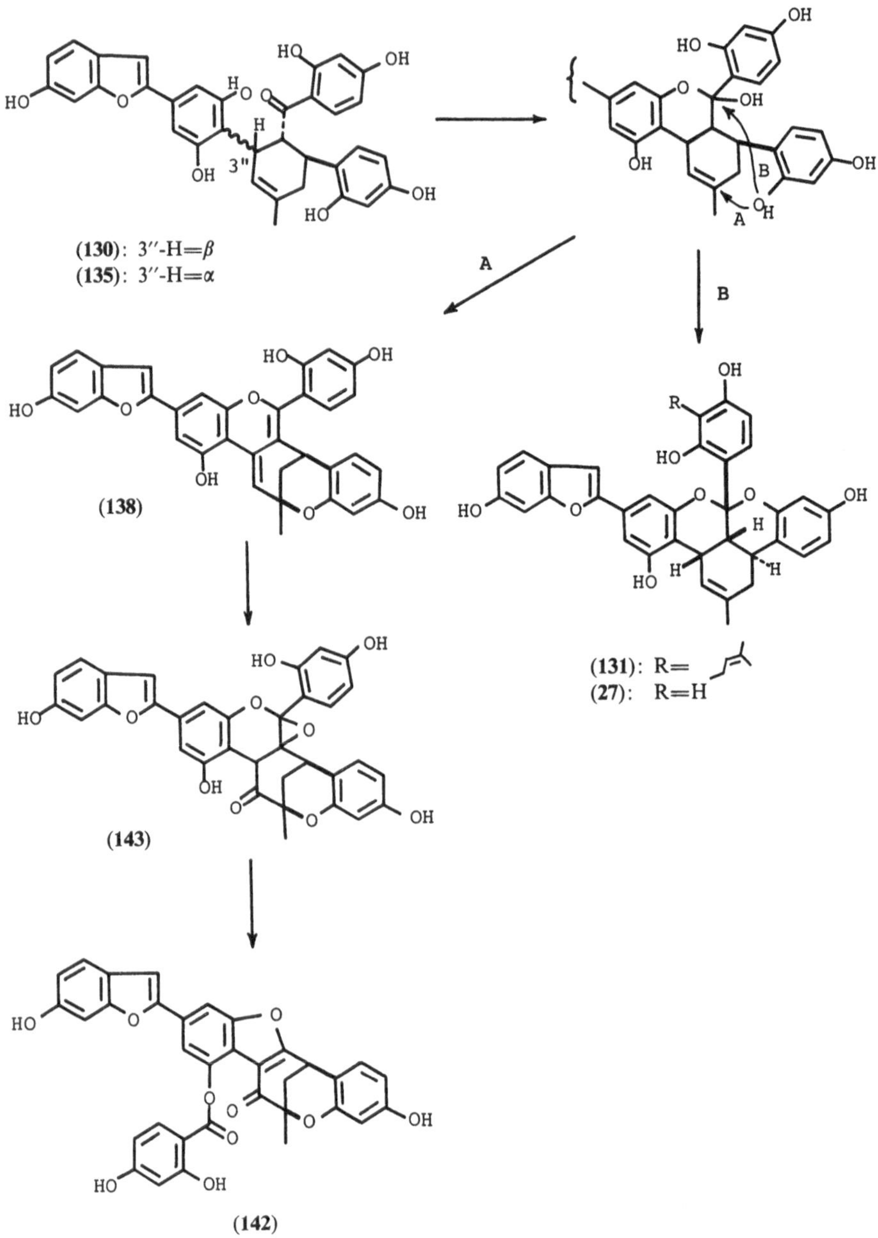

(130): 3″-H=β
(135): 3″-H=α

(138)

(143)

(142)

(131): R=
(27): R=H

Fig. 39

Table 5. ^{13}C NMR Data of Mulberrofuran I (138) and the Related Compounds (Mulberrofurans M, P, Q, Kuwanols A, and B)

C	MI (138)	MP (141)	MM (142)	MQ (143)	(C)	KA (139)	KB (140)
C-2	155.0ᵃ	157.5ᵃ	159.1ᵃ	158.1ᵃ	(C-1)	117.1	117.0
C-3	102.3	102.5	102.8	101.7	(C-2)	156.9ᵃ	156.2ᵃ
C-3a	121.3	120.6	121.8	122.0	(C-3)	103.2	103.7
C-4	121.2	121.5	121.8	121.3	(C-4)	158.4ᵃ	157.2ᵃ
C-5	112.5	112.9	113.1	112.7	(C-5)	108.0	108.2ᵇ
C-6	154.0ᵃ	153.8ᵃ	154.3ᵃ	154.7ᵃ	(C-6)	124.0	124.8
C-7	97.4	98.0	97.9	98.5ᵇ	(C-α)	125.1	125.6
C-7a	156.9ᵃ	155.3ᵃ	156.5ᵃ	155.9ᵃ	(C-β)	127.8	128.5
C-1′	130.5	130.6	129.6	132.1		138.8	140.1
C-2′	106.0	105.4	108.7	104.9		106.4	108.1ᵇ
C-3′	156.6ᵃ	156.3ᵃ	164.4	160.2		156.9ᵃ	159.5ᵃ
C-4′	119.0	114.5ᵇ	113.1	109.9		116.7ᵇ	119.1
C-5′	156.0ᵃ	156.2ᵃ	153.1	154.6		157.2ᵃ	157.5ᵃ
C-6′	103.7	105.9	105.4	107.4		106.7	106.7
C-1″	72.3	126.7	77.4	75.2		132.7	73.2
C-2″	119.9	129.9	186.0	195.4		122.7	120.2
C-3″	126.6	125.3	144.5	49.0		37.2ᶜ	127.9
C-4″	116.9	121.9	175.3	92.2		28.4	114.2
C-5″	32.8	114.4ᵇ	32.3	37.9		35.0ᶜ	33.8
C-6″	33.1	150.9	35.7	31.0		36.1	34.1
C-7″	28.5	17.5	21.5	22.4		23.8	28.5
C-8″	142.3	106.1	169.1	109.1		102.1	142.7
C-9″	112.9	111.1	105.4	114.3ᶜ		111.4	111.5
C-10″	156.7ᵃ	158.1ᵃ	164.4	166.2		157.0ᵃ	157.8ᵃ
C-11″	102.9	104.3	103.4	97.9ᵇ		103.5	103.8ᶜ
C-12″	159.0ᵃ	160.0ᵃ	164.9	172.2		159.3ᵃ	160.6ᵃ
C-13″	107.2	107.0	108.7	102.8		106.7	108.7ᵇ
C-14″	132.2	130.5	133.3	133.7		129.7	133.3
C-15″	118.1	117.0ᵇ	112.4	112.8ᶜ		116.6ᵇ	118.0
C-16″	153.2ᵃ	151.5	156.6ᵃ	155.1		153.4ᵃ	152.8ᵃ
C-17″	102.9	104.7	104.1	98.2ᵇ		104.2	103.9ᶜ
C-18″	155.9ᵃ	151.7	159.1ᵃ	155.9		152.5ᵃ	154.8ᵃ
C-19″	107.2	110.3	115.3	111.8		109.3	108.6ᵇ
C-20″	129.4	128.1	128.8	125.9		127.2	130.3

MI mulberrofuran I, MP mulberrofuran P, MM mulberrofuran M, MQ mulberrofuran Q, KA kuwanol A, KB kuwanol B measured in acetone-d_6. ᵃ⁻ᶜ Assignments may be interchangeable in vertical column.

spectra of **147**, **148**, **149**, and **151** are listed in Tables 3 and 4, and $[\alpha]_D$ values were as follows: **147**, $[\alpha]_D$ +85°; **148**, $[\alpha]_D$ +160°; **149**, $[\alpha]_D$ +56°; **150**, $[\alpha]_D$ +145°; **151**, $[\alpha]_D$ +302°. Compounds (**147–150**) are stereochemically identical with Diels-Alder type adducts of two prenylchalcone moieties such as A (morachalcone A, **120**) and B, as

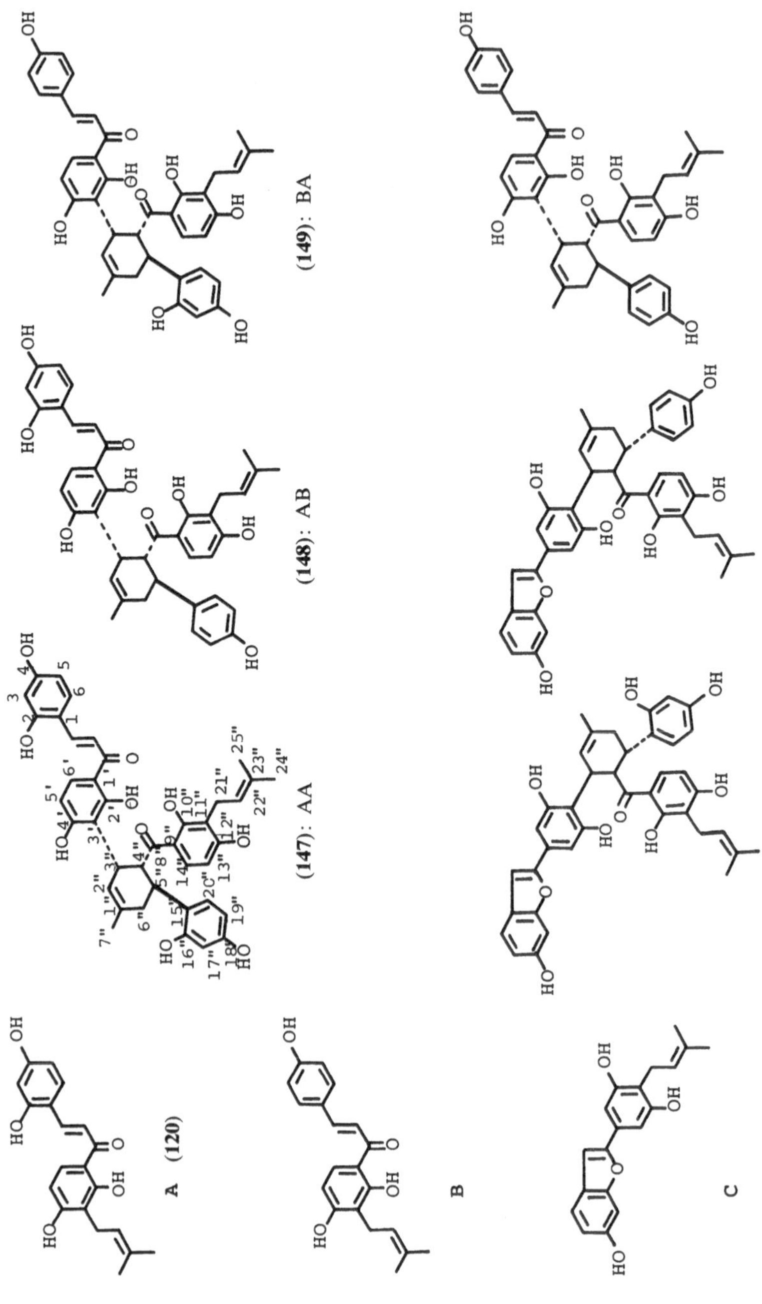

(149): BA

(150): BB

(148): AB

(151): CB

(147): AA

(103): CA

Fig. 40

A (120)

B

C

shown in Fig. 40. Kuwanon J is an adduct of dehydro-A and A (AA type), Q is an adduct of dehydro-A and B (AB), R is an adduct of dehydro-B and A (BA), and V is an adduct of dehydro-B and B (BB). Moreover, chalcomoracin (103) is a Diels-Alder type adduct of dehydro-C (dehydroprenyl moracin M) and A (CA), and mulberrofuran E (151) is an adduct of dehydro-C and B (CB). It is interesting that all possible combinations of the A, B, and C moieties could be isolated from *Morus* callus tissues. These results strongly suggest that kuwanons, chalcomoracin, and mulberrofuran E isolated from tissue cultures are naturally occurring Diels-Alder type adducts.

6. Other Diels-Alder Type Adducts of Cultivated Mulberry Tree

From the root bark of the cultivated mulberry tree and *Morus* cell cultures, thirty-two Diels-Alder type adducts have so far been isolated and their structures characterized. In this section, the Diels-Alder type adducts, other than those discussed previously are described.

Kuwanon I (152) (*132*), $[\alpha]_D$ −454°, seems to be a Diels-Alder type adduct of dehydromorachalcone A and morachalcone A (120); its stereoisomer kuwanon J (147) (*129*) was also isolated from the same *Morus* root bark. Kuwanons K (153) (*133*), L (154) (*133*), N (155) (*134*), O (156) (*134*), and W (157) (*135*) seem to be Diels-Alder type adducts of dehydroprenylflavonoids and chalcone derivatives. Kuwanons P (158) (*114*), Y (144) (*123*), and X (145) (*114*), are regarded as Diels-Alder type adducts of a chalcone derivative and dehydroprenylstilbene derivatives, while kuwanon Z (159) (*123*) seems to be derived from Diels-Alder type adducts, such as kuwanons X (145) or Y (144) through a hemiketal intermediate, as described in the case of mulberrofuran Q (143) (*121*). Comparison of the structures of such 2-arylbenzofuran derivatives as mulberrofurans C (130), G (27), I (138), J (135), Q (143), and albafuran C (72) with those of the stilbene derivatives kuwanon Y (144), kuwanols A (139) and B (140), kuwanons X (145), Z (159), and P (158), seems to indicate that these 2-arylbenzofuran derivatives are biogenetically derived from the respective stilbene derivatives.

Kuwanon M (160) (*136*), mp 252–254 °C, $[\alpha]_D$ −2.0°, $C_{50}H_{48}O_{12}$, was found to have a hypotensive action in hypertensive rats (2 mg/Kg, i.v.). Its structure was derived on the basis of spectral evidence. Kuwanon M (160) is optically active and is presumably formed by an enzymatically-controlled Diels-Alder type reaction between kuwanon C (33) and a coexisting dehydro derivative and subsequent cyclodehydrogenation. TAKASUGI *et al.* have isolated a similar compound, dimoracin (101) (*84*), mp 238–240 °C, $[\alpha]_D$ −5°, as a phytoalexin from diseased

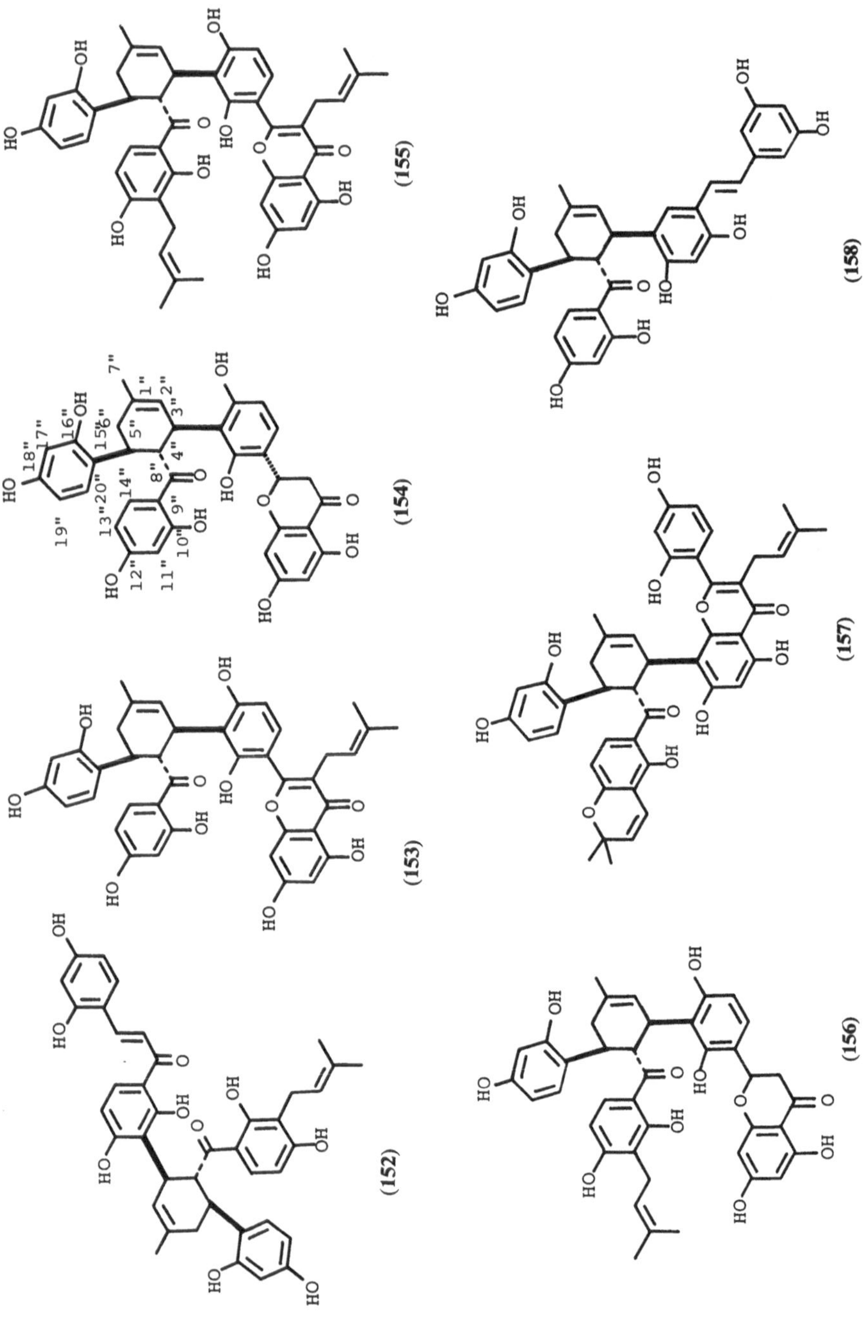

T. Nomura:

(155)

(158)

(154)

(157)

Fig. 41 (No. 1)

(153)

(156)

(152)

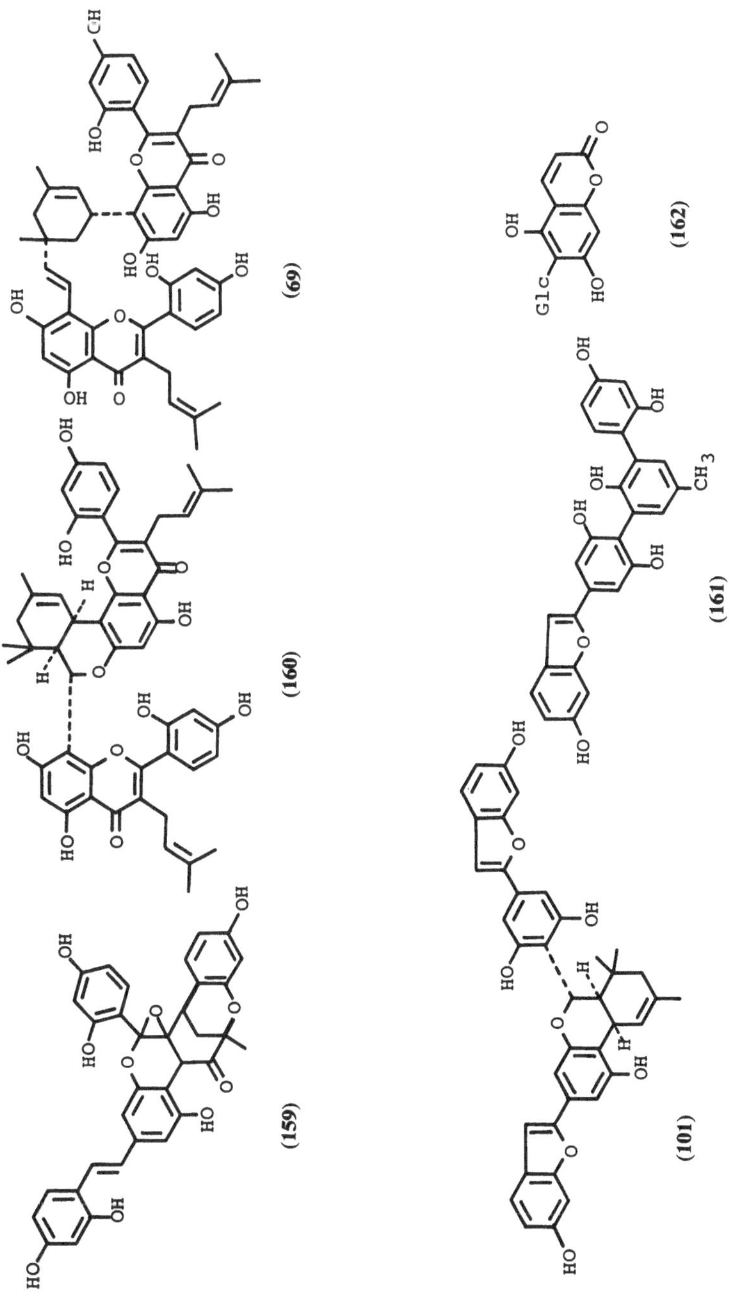

(69)

(160)

(159)

(162)

(161)

(101)

Fig. 41 (No. 2)

mulberry shoots and proposed formula **101** on the basis of spectral evidence. Takasugi's group also isolated albanin H **(69)** *(77)*, mp 215 °C (decomp.), $[α]_D$ 0°, as an antifungal compound from the epidermis of mulberry shoots and proposed the formula **69**, based on the finding that the octamethyl ether of **69** was pyrolyzed to give dehydrokuwanon C tetramethyl ether **(123)** *(96)*.

A new 2-arylbenzofuran derivative, mulberrofuran R **(161)** *(137)*, can regarded as a variation of Diels-Alder type adducts such as mulberrofurans C **(130)** and J **(135)**. To a new glucosylcoumarin, mulberroside B **(162)** *(138)*, formula **162** was attributed.

VI. Phenolic Constituents of the Crude Drug "Sang-Bai-Pi"

In Japan, the crude drug "Sang-Bai-Pi" imported from the People's Republic of China has been used as a herbal medicine; hence a study of the components of this crude drug was undertaken. Its phenolic constituents were isolated by a method similar to that shown in Fig. 4 and are different from the phenolic components of the Japanese cultivated mulberry tree. For example, morusin **(24)** and kuwanon G **(25)** are the main phenolic components of Japanese mulberry trees, while in the case of "Sang-Bai-Pi", these compounds are minor ones while sanggenons A **(163)** *(139)*, C **(164)** *(140)*, and D **(165)** *(141)* are the main components. Recently, the origin of the "Sang-Bai-Pi" studied in our laboratory was found to be the root bark of *Morus mongolica* Schneider by analysis of the extract of the crude drug and the root bark of authentic *M. mongolica* *(142)*.

1. Structures of the Flavonoids Carrying Isoprenoid Substituents Isolated from "Sang-Bai-Pi"

Sanggenon A **(163)** *(139)*, an amorphous powder, $C_{25}H_{24}O_7$, $[α]_D$ +175°, gave positive Mg-HCl and $NaBH_4$ tests. The UV spectrum had maxima at 208 (25700), 228 (15500), 235 (inflection, 14500), 270 (shoulder, 28800), 279 (31600), 315 (12000), and 377 nm (2300). The color reactions and the UV spectrum suggested that sanggenon A **(163)** was a flavanone derivative. In the UV spectra of **163** and its monoacetate **(163a)**, no red shift was observed upon addition of aluminium chloride. From the above results, combined with ^{1}H-NMR and mass spectra studies, formula **(163)** was proposed for sanggenon A.

The case for a linear structure of sanggenon A was supported by the changes in the chemical shifts of the olefinic protons of the chromene portion when the ^{1}H-NMR spectrum of the monoacetate (163a) is compared with that of the diacetate (163b) as described earlier in the case of morusin (24) (32). The hemiketal structure of 163 was confirmed by the following results. Sanggenon A (163) was easily isomerized in alkaline solution to give a 2:5 equilibrium mixture of 163 and its isomer sanggenon M (166). The latter (166) was also isolated from "Sang-Bai-Pi", and its structure deduced on the basis of spectral evidence (143). Treatment of 163 with dimethyl sulfate following the usual procedure gave a trimethyl ether 163d which was identical with

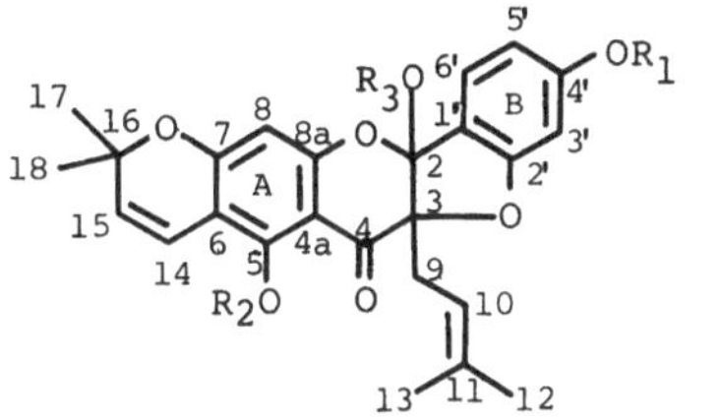

(163): $R_1 = R_2 = R_3 = H$
(163a): $R_1 = COCH_3$, $R_2 = R_3 = H$
(163b): $R_1 = R_2 = COCH_3$, $R_3 = H$
(163c): $R_1 = R_2 = R_3 = COCH_3$

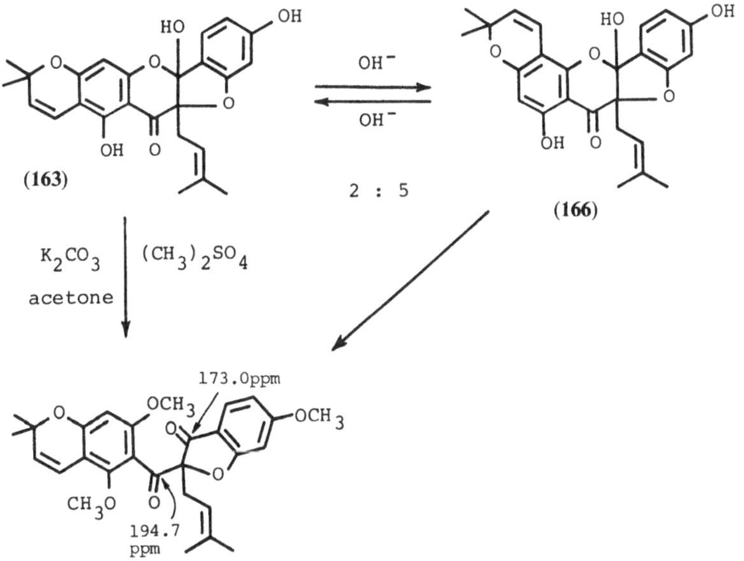

(163)

2 : 5

(166)

K_2CO_3 $(CH_3)_2SO_4$
acetone

173.0ppm

194.7 ppm

(163d)   Fig. 42

a trimethyl ether from **166**. In the ^{13}C-NMR spectrum of trimethyl ether **163d**, signals of two carbonyl carbons appeared at 173.0 and 194.7 ppm and were attributed to C-2 and C-4 carbons, respectively. The ^{13}C-NMR spectra of sanggenon A (**163**), its trimethyl ether (**163d**), and sanggenon M (**166**) are listed in Table 6. The relative configuration of the substituents on C-2 and C-3 positions was based on comparison of the ^{1}H-NMR spectra of diacetate (**163b**) and triacetate (**163c**) which showed that one of the H-9 protons shifted to higher field as follows: **163b**, δ2.74 (1H, dd, $J=7$ and 15 Hz, H-9a) and 3.13 (1H, dd, $J=8$ and 15 Hz, H-9b); **163c**, δ2.77 (2H, m, H-9a, b). This suggests that the hydroxy group at C-2 and the prenyl group at C-3 position *cis* to each other. Sanggenon A (**163**) is the first example of an isoprene substituted flavanone derivative which has a hemiketal partial structure (Fig. 42).

Table 6. ^{13}C-NMR Data of Sanggenons A (**163**), M (**166**) and Sanggenon A Trimethyl Ether (**163d**)

C	SA (**163**)[a]	SM (**166**)[a]	SAMe (**163d**)[b]
C-2	102.5	102.1	173.0
C-3	92.6	92.5	89.0
C-4	188.6	188.8	194.7
C-4a	100.5	100.8	110.3
C-5	163.3	157.5	161.7
C-6	103.3	97.6	109.8
C-7	164.4	165.5	164.0
C-8	96.5	102.6	99.5
C-8a	161.4	164.4	161.9
C-1'	121.2	121.0	121.7
C-2'	159.5	161.2	157.9
C-3'	99.6	99.6	101.0
C-4'	161.4	161.5	157.9
C-5'	109.9	110.0	105.0
C-6'	125.9	125.6	127.8
C-9	32.1	32.2	36.0
C-10	118.7	118.5	116.5
C-11	136.9	137.0	137.3
C-12	25.9	25.9	25.8
C-13	18.1	18.2	17.9
C-14	115.5	115.9	115.8
C-15	127.5	127.2	128.9
C-16	79.1	79.3	78.3
C-17	28.5	28.5	28.5
C-18	28.5	28.4	28.5

Solvent; [a] acetone-d_6. [b] CDCl$_3$. SA sanggenon A, SM sanggenon M, SAMe sanggenon A trimethyl ether.

Structures of sanggenons F (**167**) (*144*), H (**168**) (*145*), I (**169**) (*145*) J (**170**) (*145*), K (**171**) (*145*), L (**172**) (*143*), and N (**173**) (*143*) are shown in Fig. 43. In addition to these compounds, morusin (**24**) and cyclomorusin (**30**) were also isolated from the crude drug as minor components.

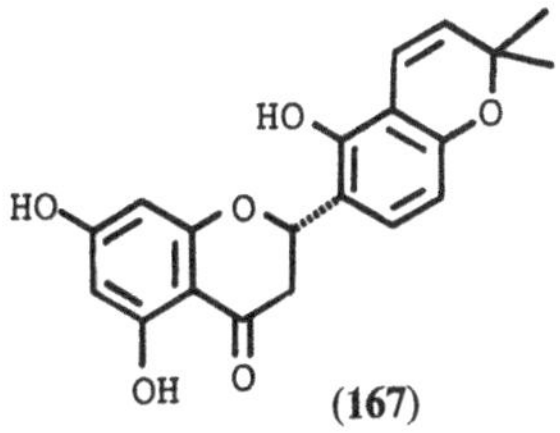

(167)

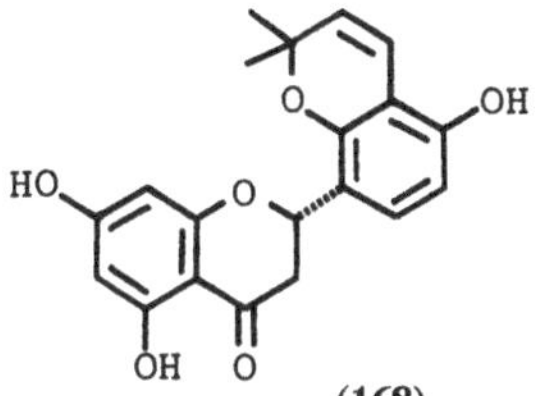

(168)

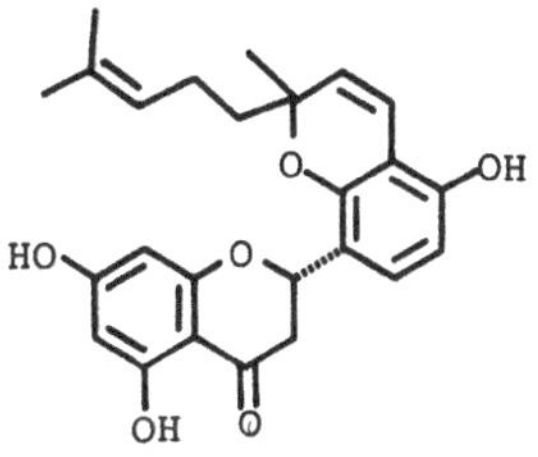

(169)

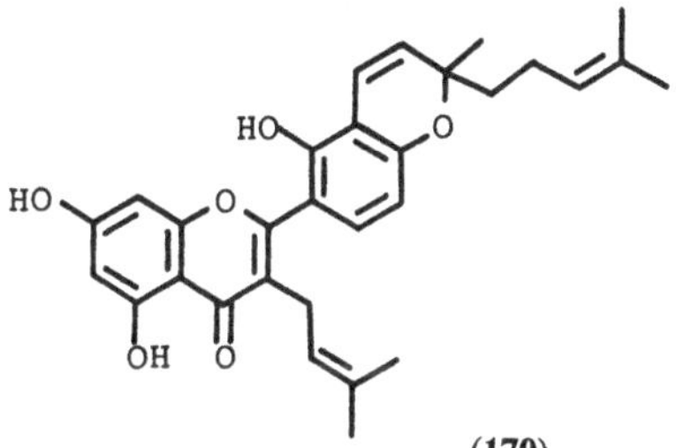

(170)

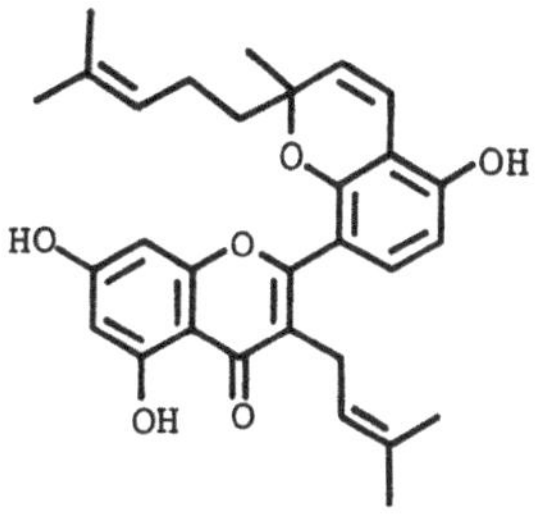

(171)

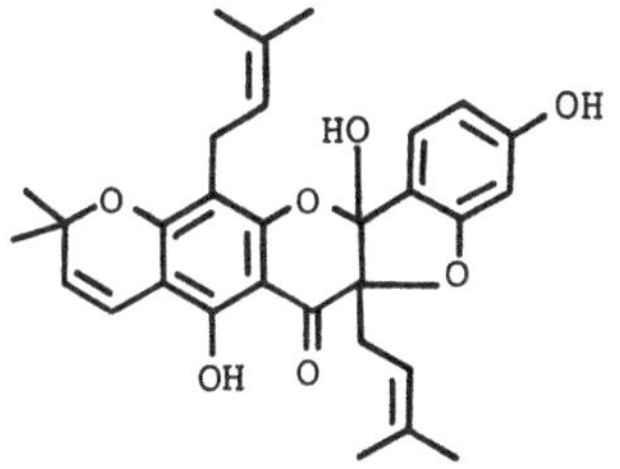

(172)

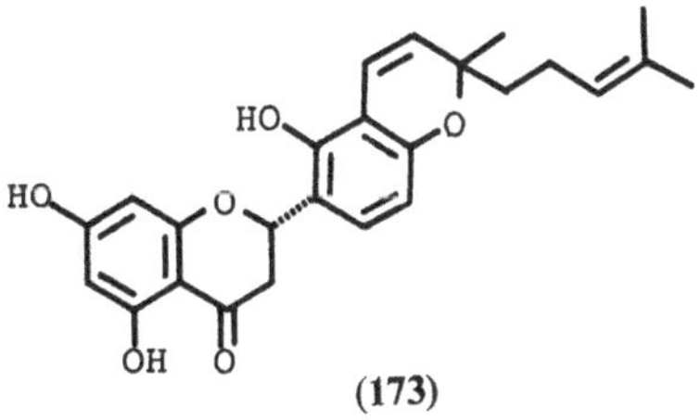

(173)

Fig. 43

2. Structures of Diels-Alder Type Adducts of "Sang-Bai-Pi"

From the methanol extract of the crude drug "Sang-Bai-Pi" a series of Diels-Alder type adducts, sanggenons B (174) (146), C (164) (140), D (165) (141), E (175) (147), G (176) (148), O (177) (149), and P (178) (147), mulberrofurans K (179) (91) and O (180) (91) were isolated along with kuwanons G (25) (33) and L (154) (133) which had also been isolated from Japanese cultivated mulberry trees. Sanggenons C (164) and D (165) showed a marked hypotensive effect in rabbits.

Sanggenon C (164) (140), an amorphous powder, $[\alpha]_D$ + 304°, $C_{40}H_{36}O_{12}$ had a UV spectrum, similar to that of sanggenon A (163) with maxima at 220 (inflection, 43700), 230 (inflection, 35500), 283 (25100), 288 (shoulder, 24500) and 309 nm (22400). In the presence of aluminium chloride, part of the absorption at about 285 nm showed a bathochromic shift, and the absorption at 290 nm was observed as a shoulder as follows: λ_{max} 225 (42700), 290 (shoulder, 22400), 305 (25100), 350 (shoulder, 10200), and 420 nm (1500). The ^{1}H-NMR spectrum showed two hydrogen-bonded hydroxyl groups (δ 12.23, 12.60 ppm). If the ^{1}H-NMR spectrum is taken into account, the absorption at about 285 nm can be ascribed to two conjugated carbonyl groups which are hydrogen bonded. These results led to the assumption that a prenyl unit is substituted on one of the positions adjacent to the two hydrogen bonded hydroxyl groups (113). From the spectral evidence, sanggenon C (164) seemed to be a Diels-Alder type adduct having a sanggenon A partial structure.

Treatment of 164 with dimethyl sulfate gave two octamethyl ethers 164a and 164b. In the ^{13}C-NMR spectrum of 164b, signals of three carbonyl carbons appeared at δ 173.0, 194.9, and 201.3 ppm. This suggested that one of the octamethyl ethers (164b) does not have a hemiketal partial structure but is a triketone. The LSPD technique demonstrated that the areas of the signals at δ 173.0 and 194.9 ppm increased when the H-9 signals were weakly irradiated thus indicating the location of the prenyl group, and leading to the suggestion that the prenyl group was located at C-3 and the carbonyl groups of C-2 and C-4. Octamethyl ether 164b was pyrolyzed at 280 °C to give 2,4,2′,4′-tetramethoxychalcone (113b). The disposition of the dihydroxyphenyl and dihydroxybenzoyl moieties on the cyclohexene ring of 164b was supported by the LSPD as described in the case of kuwanon G (Fig. 44).

Comparison of the ^{1}H-NMR spectra of sanggenon C (164), kuwanon G (25), and chalcomoracin (103), revealed that the chemical shifts and coupling constants of protons on the cyclohexene ring of 164 resembled those of 103 more closely than they did those of 25 (Figs. 28 and 45). In the ^{13}C-NMR spectra (Table 7), the chemical shifts of

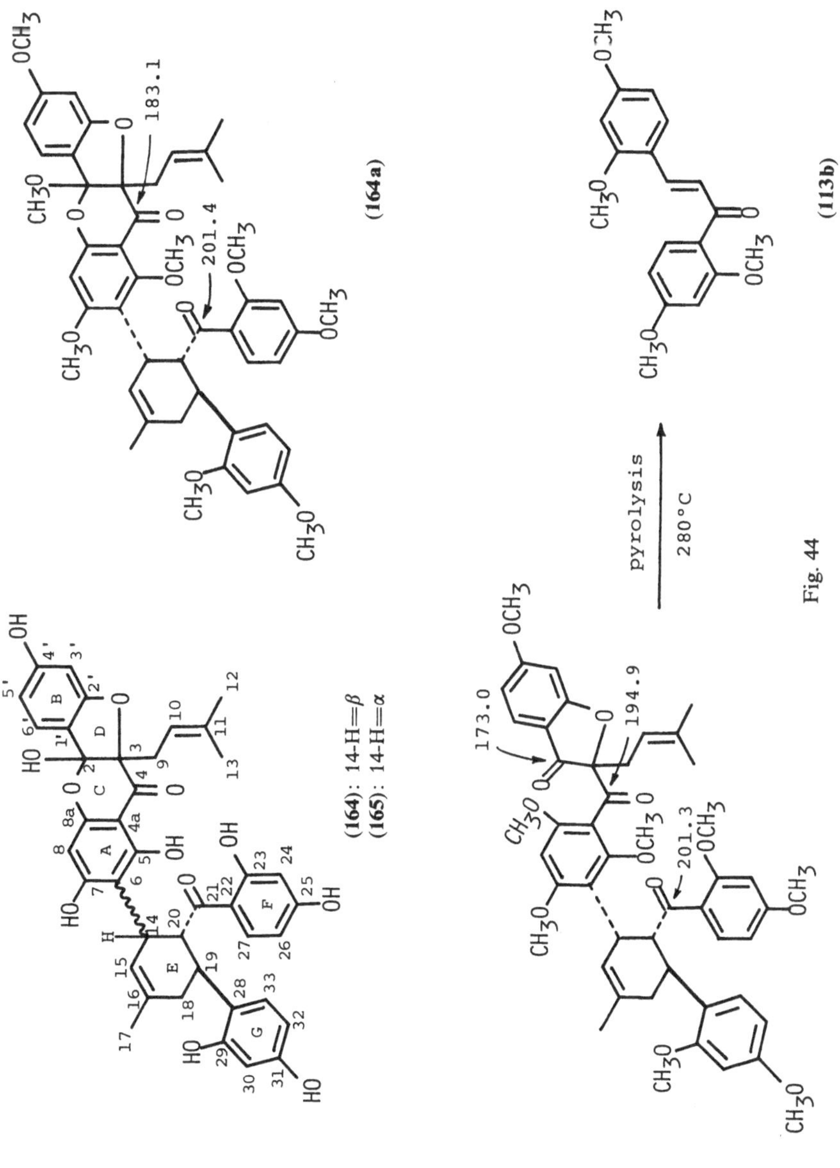
pyrolysis
280°C
183.1
201.4
173.0
194.9
201.3

(164): 14-H=β
(165): 14-H=α

(164a)

(164b)

(113b)

Fig. 44

the carbon atoms of the methylcyclohexene ring of **164** resembled those of **103** (Table 4) more closely than those of **25** (Table 3). From these results, formula **164** was proposed for the structure of sanggenon C.

Final proof for this formulation was obtained as follows. A ketal **181** was derived from sanggenon C (**164**) by treatment with acidic solu-

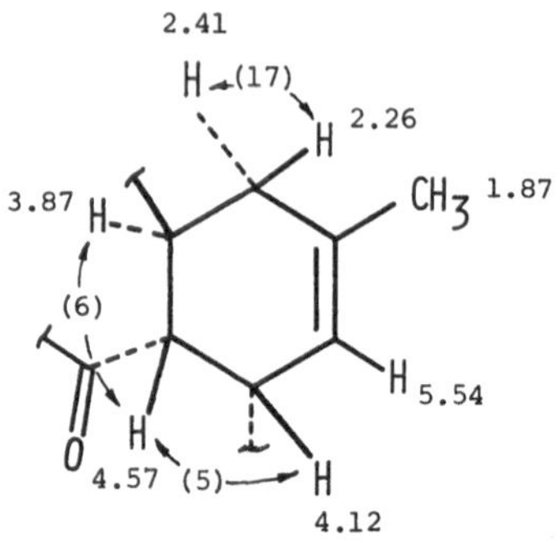

(164) δ in acetone-d$_6$

Fig. 45

tion under irradiation as described in the case of mulberrofurans F
(131) and G (27). Comparison of the ^{1}H-NMR spectra of ketal 181
and its acetate 181a showed a remarkable shift of the olefinic proton
(H-15) to higher field (+0.32 ppm). On the other hand when sanggenon C (164) was exposed to alkali an equilibrium mixture of compounds
164' and 177' was obtained. The IR, ^{1}H-NMR, and UV spectra of
164' and 177' were identical with those of sanggenons C (164) and
O (177), respectively. A similar result was described in the case of sanggenons A (163) and M (166) (143). Compound 177' exposed to acidic
solution gave ketal 182'. In the ^{1}H-NMR spectrum of pentaacetate
182a', the chemical shift of the olefinic proton (H-15) was affected
only insignificantly (−0.05 ppm) (Fig. 46). These results suggested that
sanggenon C was a C-6-substituted compound and the 2,4-dihydroxy-
benzoyl group was located at C-20. Structures 164 and 177 for sanggenons C and O were thus established.

Sanggenon D (165) (141), is an amorphous powder, [α]$_D$ −145°,
$C_{40}H_{36}O_{12}$. Its UV spectrum closely resembled that of sanggenon C
(164). The ^{1}H-NMR and ^{13}C-NMR spectra of 165 showed complex
patterns and broadened signals at room temperature; which were resolved and sharpened at higher temperature. By contrast, the ^{1}H-NMR
spectrum of 164 did not vary significantly with temperature. In the
^{13}C-NMR spectrum of 165, the chemical shifts of the carbon atoms
of the methylcyclohexene ring and the C-21 carbonyl carbon differed
from those of the equivalent carbon atoms of 164, but the shifts of
the other signals were similar to those found for 164. This suggested
that sanggenon D was a stereoisomer of sanggenon C at one or more
concerning of the points of attachment to the methylcyclohexene ring.
Comparison of the shifts of the ring carbon atoms and the carbonyl

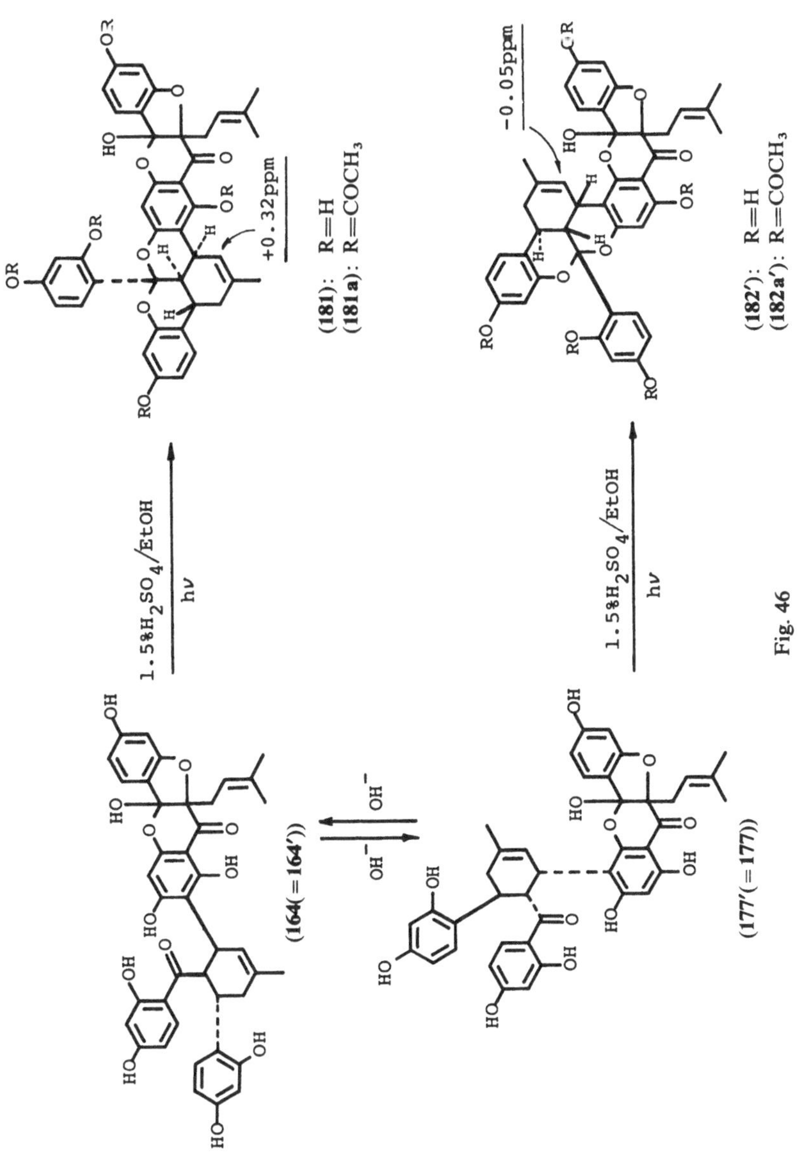

Fig. 46

carbon of **165** with those in the ^{13}C-NMR spectra of kuwanon G (**25**) (Table 3), sanggenon C (**164**) (Table 7), and chalcomoracin (**103**) (Table 4) showed that sanggenon D (**165**) was more similar to **25** than to **164** and **103**. Moreover, the chemical shifts and coupling patterns of the proton signals on the methylcyclohexene ring of **165** were similar to those of **25** (Fig. 47). These observations suggested that the 2,4-dihydroxyphenyl and 2,4-dihydroxybenzoyl substituents on the cyclohexene rings of sanggenon D (**165**) and kuwanon G (**25**) are similarly disposed in the same relative configuration, thus leading to formula **165** for sanggenon D. In view of the formation of the model Diels-Alder adducts **115** and **116** cited earlier, it is interesting that two stereoisomers, sanggenons C (**164**) and D (**165**), coexist in *Morus* root bark. The biogenesis of sanggenons C (**164**) and D (**165**) presumably involves the same type of cycloaddition between a chalcone and sanggenon A type diene compounds invoked previously (Fig. 48).

Sanggenon B (**174**) (*146*) $[\alpha]_D$ +62°, $C_{33}H_{30}O_9$, was obtained as an amorphous powder. Formula **174′** (*150*) originally suggested for sanggenon B (*150*) was revised to **174** on the basis of the ^{1}H-NMR and ^{13}C-NMR spectra, in particular a comparison of the ^{13}C-NMR spectrum of **174** with those of sanggenon A (**163**) and mulberrofuran H (**137**), as shown in Table 7. Sanggenon B (**174**) seems to be derived

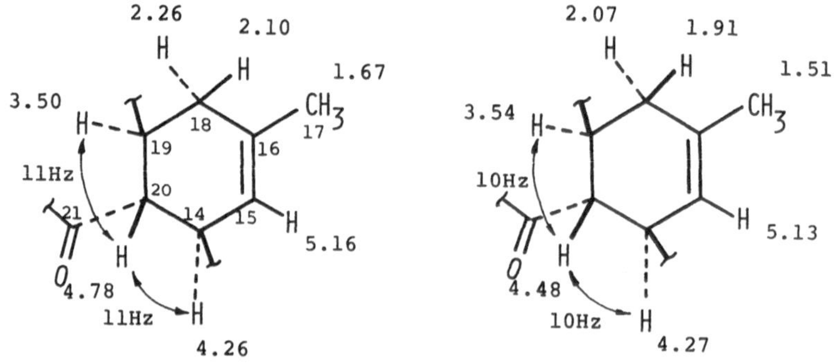

(**165**) dmso-d_6 at 120°C (**25**) dmso-d_6 at 120°C

Fig. 47

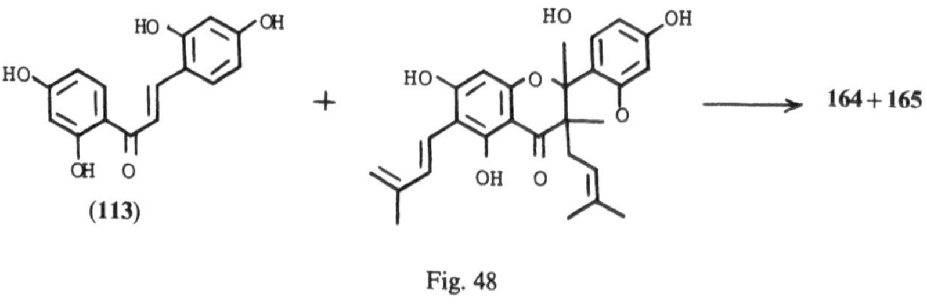

Fig. 48

References, pp. 191–201

from a Diels-Alder type adduct such as sanggenons C (**164**) or D (**165**) in the manner similar to that described in the case of **137** (Fig. 32).

Sanggenons E (**175**) (*147*), [α]$_D$ −86°, C$_{45}$H$_{44}$O$_{12}$, and P (**178**) (*147*), [α]$_D$ +215°, both amorphous powders are C-14 stereoisomers. Structure assignments were based on spectral evidence.

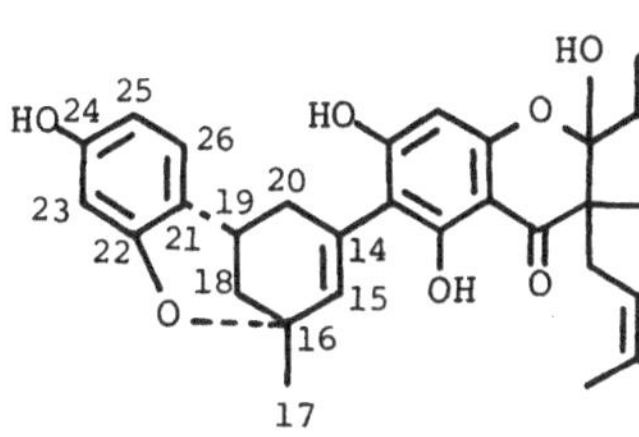

(**174**) 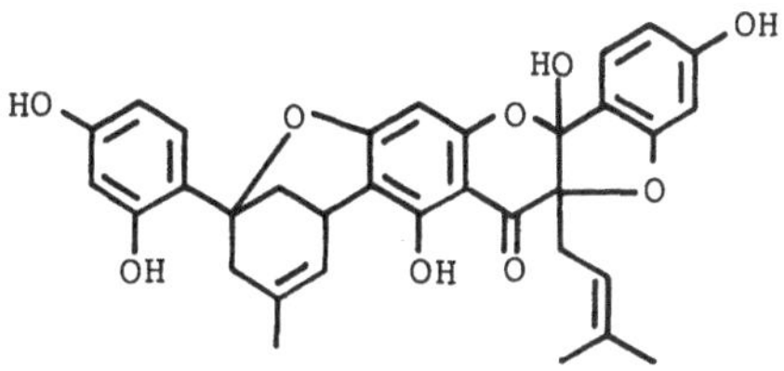   (**174′**)

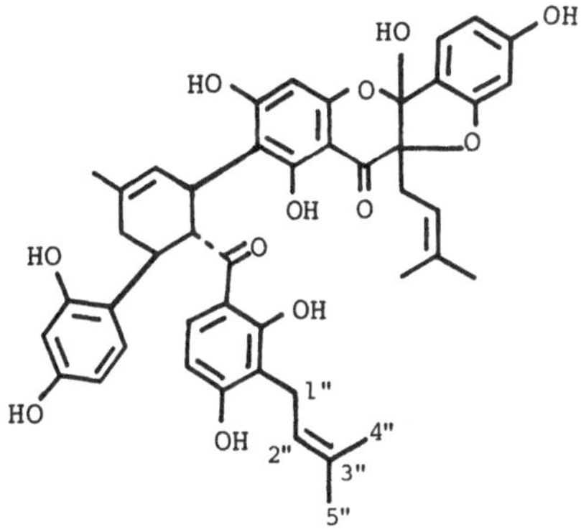

(**175**) 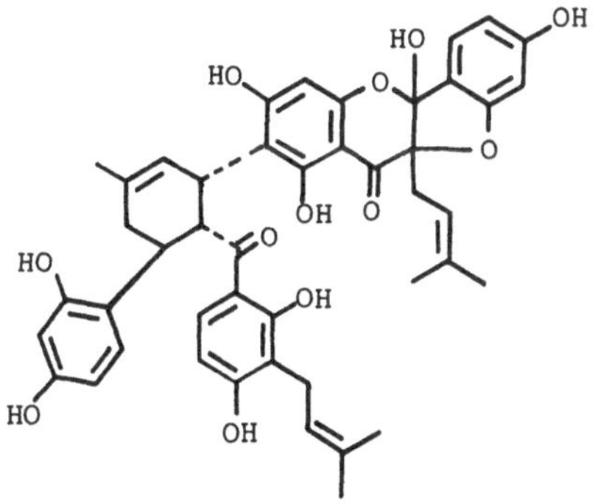   (**178**)

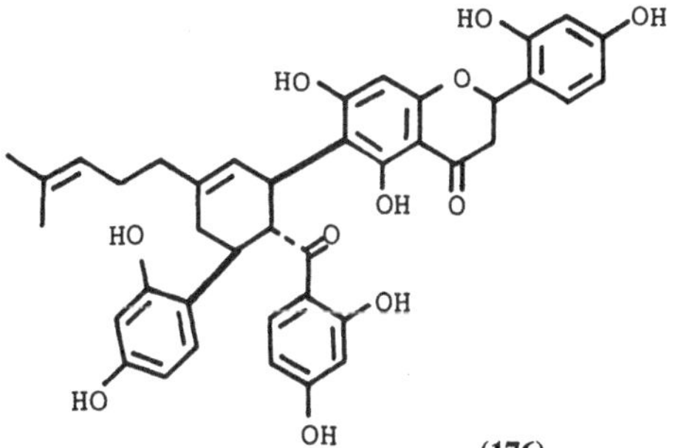

(**176**)

Fig. 49

Sanggenon G (**176**) (*148*), $[\alpha]_D$ −277°, $C_{40}H_{38}O_{11}$, may be regarded as a Diels-Alder type adduct between a dehydrogeranylflavanone derivative and a chalcone; formula **176** was proposed on the basis of spectral evidence (Fig. 49).

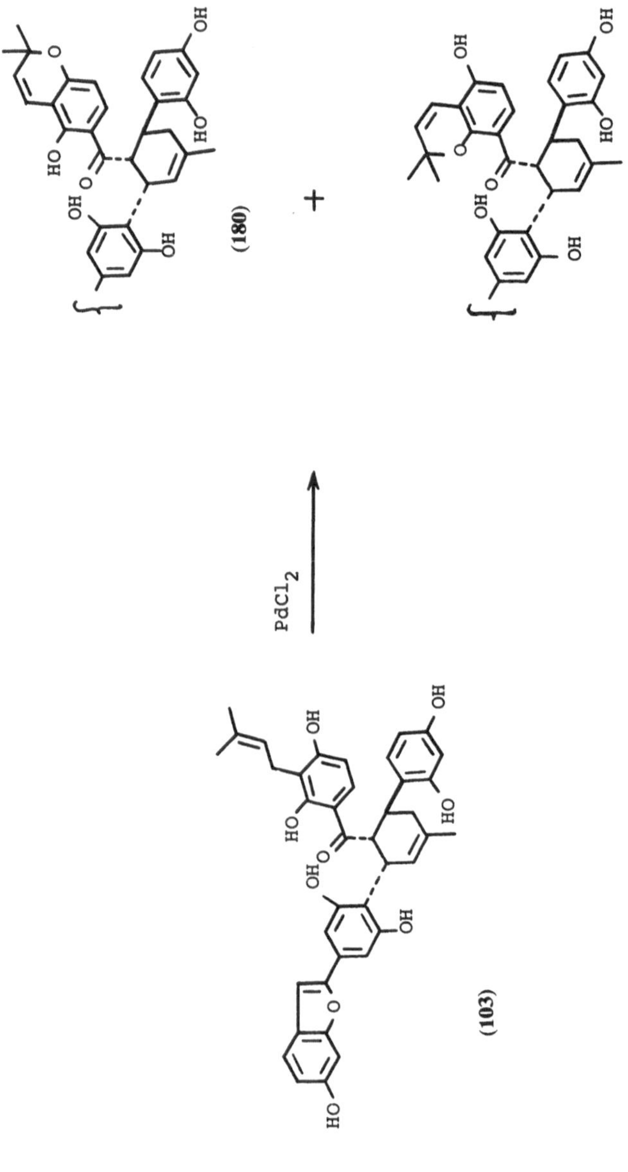

Fig. 50 (No. 1)

Sanggenon O (177) (*149*), [α]_D −64°, is a structural isomer of sang-genon C (164); formula 177 was confirmed by chemical correlation with 164 (Fig. 46).

Mulberrofuran K (179) (*91*), mp 176 °C (decomp.), [α]_D +425°, had a UV spectrum similar to that of 131. Formula 179 was confirmed by its derivation from chalcomoracin (103) shown in Fig. 50.

Mulberrofuran O (180) (*91*), [α]_D +196°, was identical with another compound derived from 103 (Fig. 50). The ^{13}C-NMR spectra of sang-genons B (174) and C (164), sanggenon C octamethyl ether (164b), sanggenons D (165), E (175), and P (178) are listed in Table 7.

(180) $\xrightarrow[\text{h}\nu]{\text{1.5\% H}_2\text{SO}_4\,/\,\text{EtOH}}$

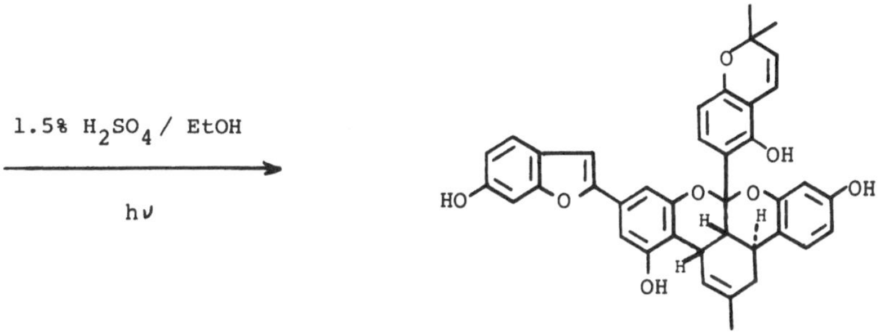

(184)

(183) $\xrightarrow[\text{h}\nu]{\text{1.5\% H}_2\text{SO}_4\,/\,\text{EtOH}}$

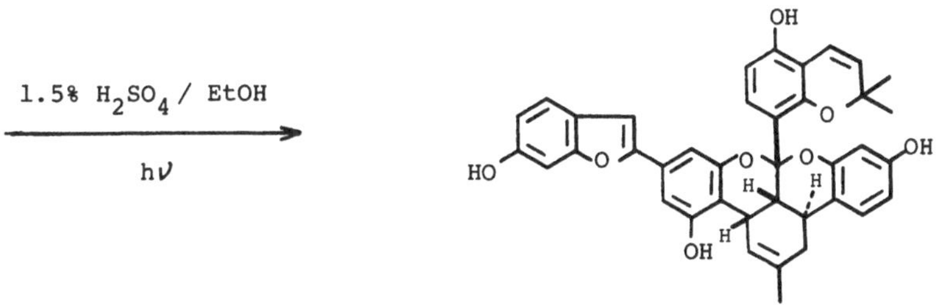

(179)

Fig. 50 (No. 2)

Table 7. ^{13}C-NMR Data of Sanggenons B (**174**), C (**164**), D (**165**), E (**175**), P (**178**) and Sanggenon C Octamethyl Ether (**164b**)

	SB (**174**)[a]	SC (**164**)[a]	SP (**178**)[a]	SCMe (**164b**)[b]	SD (**165**)[c]	SE (**175**)[d]
C-2	101.2	102.4	102.0	173.0	102.1	103.0
C-3	90.8	92.0	91.5	88.5	90.7	90.7
C-4	186.6	188.4	187.5	194.9	187.0	186.2
C-4a	98.7	99.9	99.5	118.7	99.0	98.9
C-5	160.7	163.9	161.1	163.3	164.3	162.3
C-6	109.7	109.0	108.6	110.0	109.1	109.0
C-7	164.3	167.6	166.8	166.6	167.0	167.0
C-8	94.2	96.5	96.2	95.4	94.8	94.9
C-8a	160.7	162.0	162.4	158.7	160.2	160.3
C-1'	119.8	122.2	122.3	121.4	120.5	120.6
C-2'	159.6	161.2	160.4	160.1	159.7	160.0
C-3'	98.3	99.5	99.0	99.5	98.7	98.6
C-4'	159.7	161.2	160.4	161.1	160.2	160.2
C-5'	108.6	109.7	109.2	105.0	109.1	109.1
C-6'	124.2	125.6	125.0	127.9	124.9	124.8
C-9	31.2	32.0	31.7	35.7	31.1	31.2
C-10	117.2	118.6	118.0	116.6	117.5	117.6
C-11	135.3	136.2	136.1	137.0	135.5	135.3
C-12	25.1	25.9	25.8	25.8	25.1	25.3
C-13	17.4	18.1	18.0	17.9	17.6	17.5
C-14	132.0	33.1	33.3	33.1	37.8	37.3
C-15	131.3	121.4	121.6	122.5	124.3	124.6
C-16	70.6	132.6	131.2	132.5	131.9	129.8
C-17	27.0	23.3	23.6	23.7	22.8	22.8
C-18	33.9	32.8	31.9	37.3	37.8	37.3
C-19	30.9	33.1	32.5	33.7	37.8	37.6
C-20	38.5	47.2	48.0	52.3	45.2	45.0
C-21	117.9	206.2	208.6	201.3	208.6	208.7
C-22	154.0	113.8	113.2	123.5	114.3	114.3
C-23	102.5	164.0	163.7	161.1	164.3	162.0
C-24	156.0	102.6	115.3	98.0	102.1	113.6
C-25	107.0	164.3	163.1	163.3	164.3	162.1
C-26	129.1	105.9	107.2	104.2	107.3	106.6
C-27		132.6	134.3	132.5	132.5	130.0
C-28		119.5	120.8	122.5	119.7	119.7
C-29		155.5	155.7	157.9	155.7	155.7
C-30		102.2	103.3	98.7	103.1	103.1
C-31		155.8	157.1	158.3	156.1	·156.2
C-32		107.5	107.6	104.6	106.2	106.6
C-33		128.2	128.3	128.7	129.0	130.0
C-1''			22.1			21.3
C-2''			122.6	OCH$_3$ 55.3 × 7		122.7
C-3''			130.8	61.5		130.0
C-4''			25.7			25.4
C-5''			17.8			17.5

SB sanggenon B, SC sanggenon C, SD sanggenon D, SE sanggenon E, SP sanggenon P, SCMe sanggenon C octamethyl ether.

Solvent; a acetone-d_6. b CDCl$_3$. c dmso-d_6 at 80 °C. d dmso-d_6 at 90 °C.

VII. Absolute Configuration of Diels-Alder Type Adducts from *Morus* Sp.

1. Classification of Diels-Alder Type Adducts

The Diels-Alder type adducts from *Morus* root bark and tissue cultures are the first examples of naturally-occurring optically active Diels-Alder type adducts and may be divided into five types as follows: a) adducts of chalcones and dehydroprenylflavonoids [e.g. kuwanon G **(25)** (*33*), sanggenon C **(164)** (*140*)], b) adducts of chalcones and dehydroprenyl-2-arylbenzofurans [e.g. mulberrofurans C **(130)** (*111*), F **(131)** (*112*)], c) adducts of chalcones and dehydroprenylstilbenes [e.g. kuwanon P **(158)** (*114*), kuwanol A **(139)** (*118*)], d) adducts of chalcones and dehydroprenylchalcones [e.g. kuwanons I **(152)** (*132*), J **(147)** (*129*)], e) adducts of prenylphenols and dehydroprenylphenols [e.g. kuwanon M **(160)** (*136*), dimoracin **(101)** (*84*)].

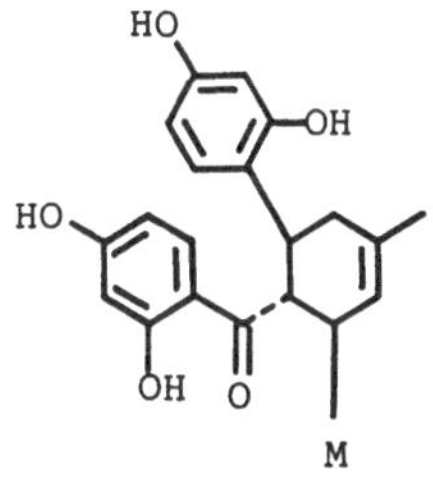

all-<u>trans</u> adduct 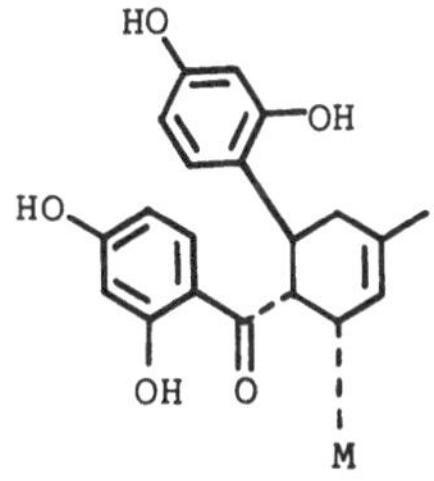   cis-<u>trans</u> adduct

Fig. 51

The relative configuration at the point of attachment of each substituent on the methylcyclohexene ring of these adducts having been clarified by ^{1}H-NMR spectral evidence, the adducts can be divided into two classes, a) all-*trans* type adducts, b) *cis-trans* type adducts (Fig. 51).

This section deals with the absolute configuration of the three chiral centers on the methylcyclohexene ring, based on circular dichroism (CD) spectroscopic studies (*151, 152, 153*), and X-ray analysis (*153*).

2. Stereochemistry of Diels-Alder Type Adducts

The Diels-Alder reaction is a [4+2] cycloaddition of a diene and a dienophile to form a six-membered ring which may involve either *exo-* or *endo*-additions. In the case of the Diels-Alder type adducts

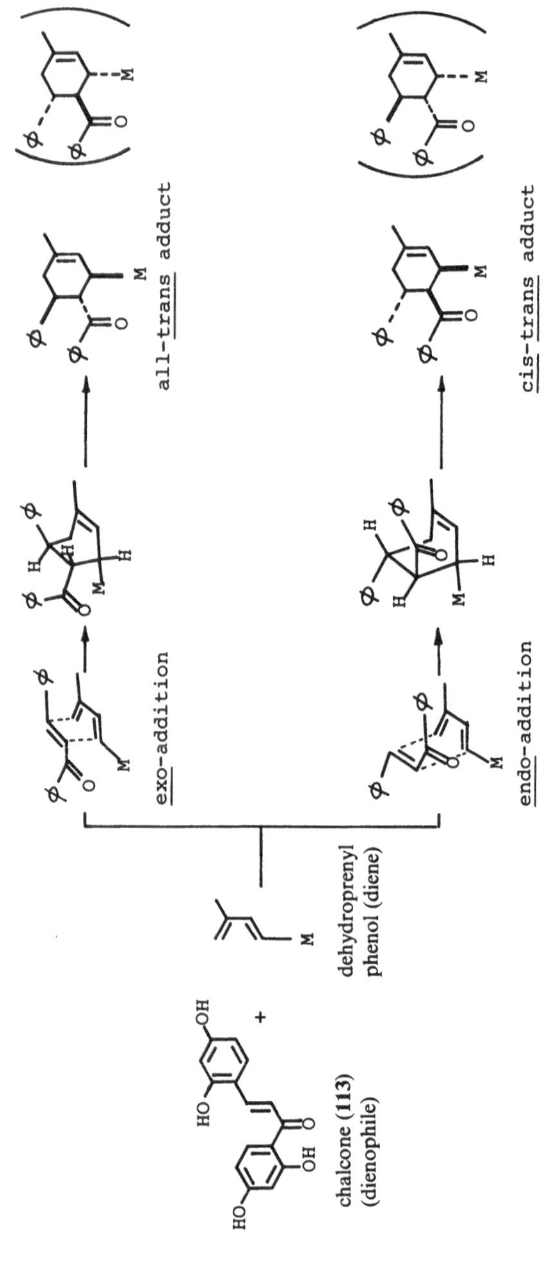

Fig. 52

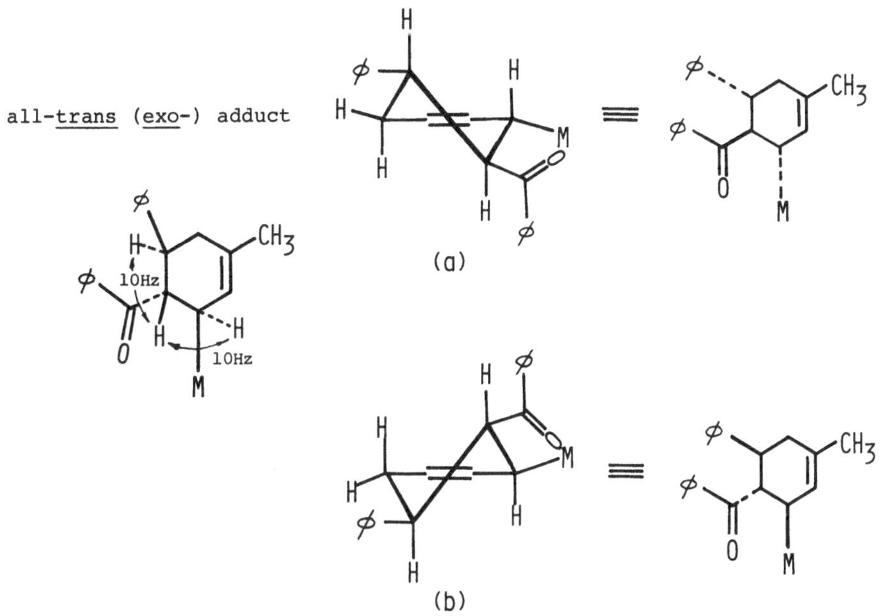
all-trans (exo-) adduct
10Hz
10Hz
(a)
(b)

Fig. 53 (No. 1)

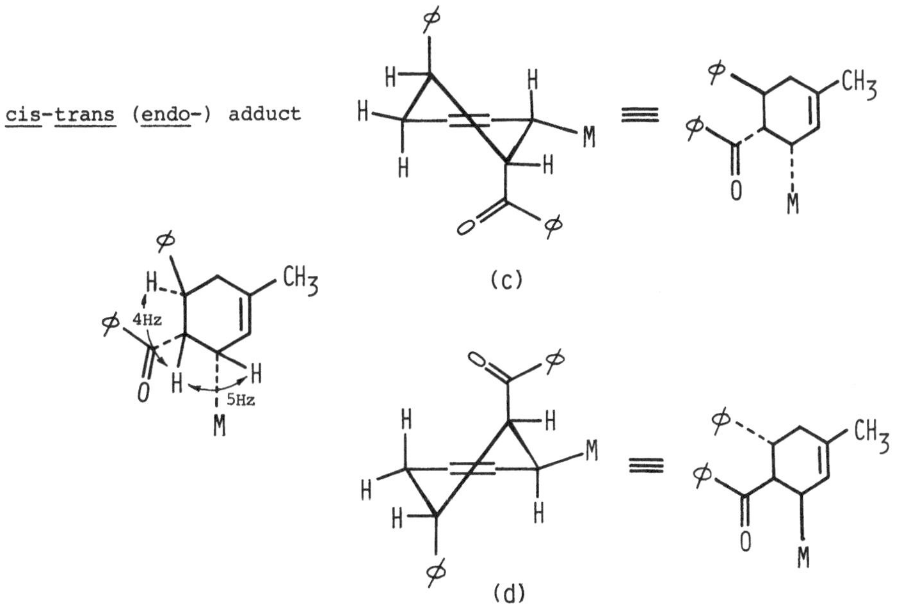
cis-trans (endo-) adduct
4Hz
5Hz
(c)
(d)

Fig. 53 (No. 2)

from *Morus* root bark, the chalcone can be assumed to be the dieno-phile and the dehydroprenylphenol as the diene. *Exo*-addition in the Diels-Alder reaction of a chalcone with a dehydroprenylphenol results in an all-*trans* adduct, while *endo*-addition gives a *cis-trans* adduct (Fig. 52). In view of the magnitude of the coupling constants (10 Hz) between the proton signals on the methylcyclohexene ring in an all-*trans* adduct (*exo*-addition), two possible conformations for the methylcyclo-hexene ring can be written and are shown in Fig. 53. In the case of the *cis-trans* adducts (*endo*-addition) two conformations for the cyclo-hexene ring can again be deduced from the coupling constants (4 and 5 Hz) and are also in Fig. 53. The absolute configuration of an all-*trans* adduct may be represented by either formula (a) or formula (b), and that of a *cis-trans* adduct by formula (c) or (d).

3. Absolute Configuration of Mulberrofurans C and J by CD Spectra

CD spectra of the following Diels-Alder type adducts were mea-sured in EtOH: kuwanons G (**25**), L (**154**), X (**145**), Y (**144**), sanggen-ons C (**164**), D (**165**), mulberrofurans C (**130**), J (**135**). The spectra are shown in Fig. 54 along with the UV spectra. It is notable that the magnitude of θ values in the CD spectra of mulberrofurans C (**130**) and J (**135**) is larger than any others in the spectra of the other com-pounds and that a strongly split Cotton effect is observed in the 280–350 nm region of both of these. This property in the CD spectra seem to indicate exciton coupling (*154*). As the absorption bands at 280–350 nm in the UV spectrum of **130** (or **135**) are supposed to be due to the 2,4-dihydroxybenzoyl and 2-arylbenzofuran chromophores, one may suspect that the split Cotton effect originated from exciton coupling between the two chromophores. In order to ascertain whether this was so, **130** and **135** were reduced with LiAlH$_4$ to give the reduced products, **130a** and **135a**, respectively. In the CD spectra of **130a** and **135a**, the strong split Cotton effect, observed in **130** and **135**, disap-peared and the magnitude of the θ values decreased remarkably. These results clearly indicate that the strong split Cotton effect is due to exciton coupling between the 2,4-dihydroxybenzoyl and the 2-arylben-zofuran chromophores. Additionally, the CD spectra of **130a** and **135a** were mirror images of each other in the region 270–350 nm, as shown in Fig. 55. This suggests that the stereochemistries of **130** and **135** at the C-3″ chiral center bearing the 2-arylbenzofuran chromophore are antipodal to each other (Fig. 55). According to the exciton chirality method (*154*), the absolute configuration may be deduced from the sign of the Cotton effect in the CD spectrum. Now the CD spectra of **130** and **135** show two strong Cotton effects in the 280–350 nm region (Fig. 54). Of these, the strongly positive Cotton effect in the 280–320 nm region may be attributed to exciton coupling between the benzoyl and 2-arylbenzofuran chromophores because in the UV spectra the strong π-π* transitions of the 2,4-dihydroxybenzoyl and 2-arylben- zofuran chromophores appear at about 285 and 320 nm, respectively. In line with theory (154) this means that the transition moments of the 2,4-dihydroxybenzoyl and 2-arylbenzofuran chromophores are twisted to the right side of each other, with the direction of each transi- tion moment in the 2,4-dihydroxybenzoyl and 2-arylbenzofuran chro- mophores being approximately expressed by Fig. 56. However, in the 2,4-dihydroxybenzoyl chromophore, the direction of the transition mo-

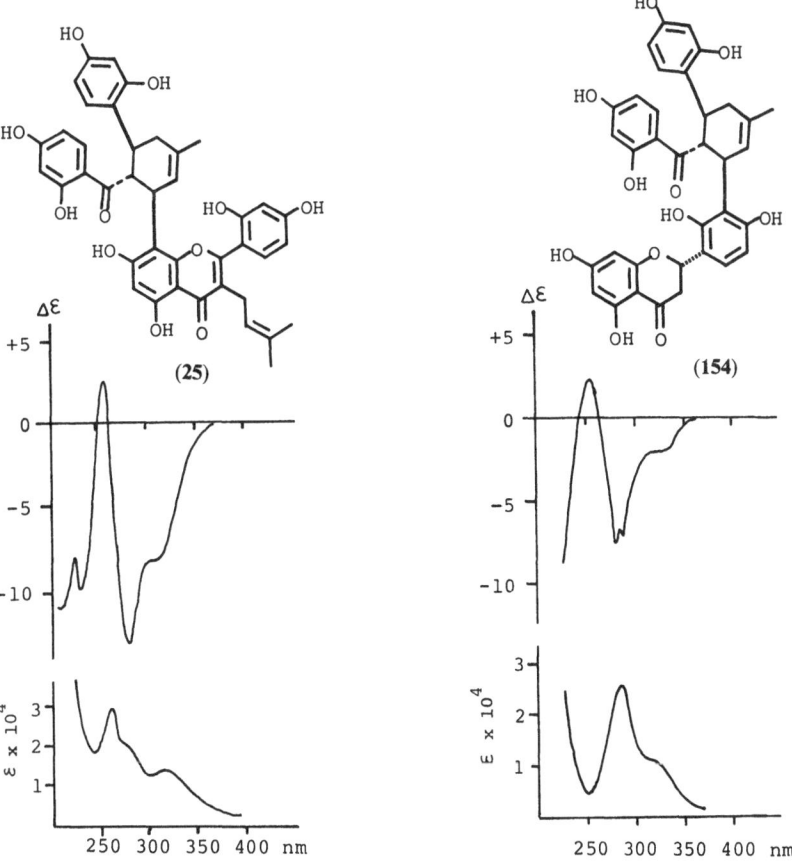

Fig. 54 (No. 1)

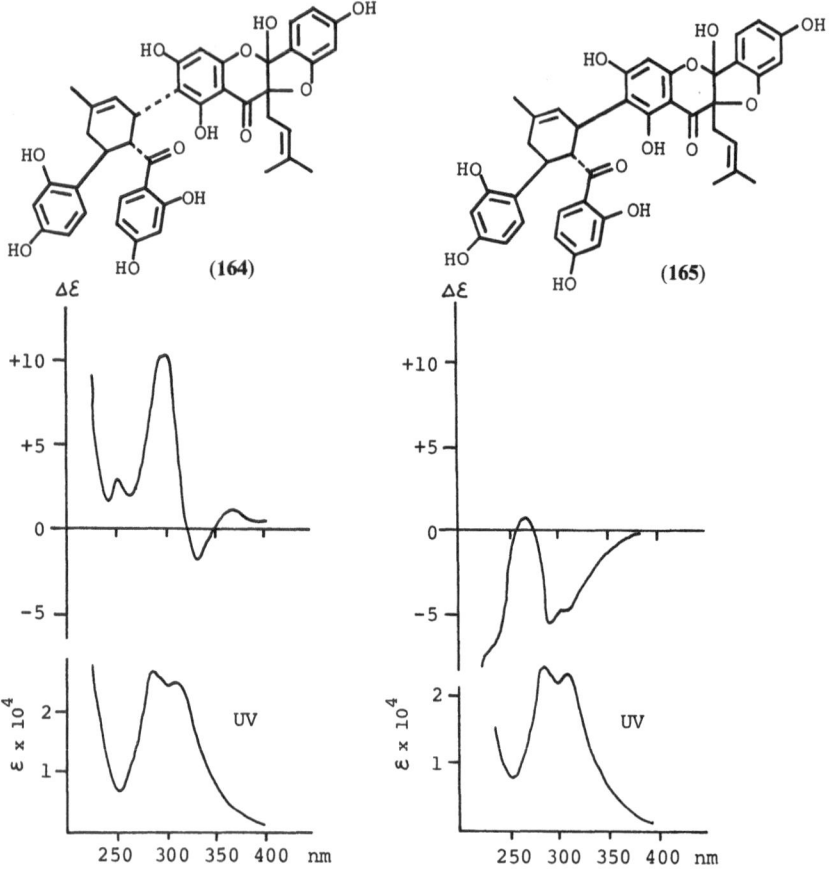

Fig. 54 (No. 2)

ment may be altered by the orientation of the benzoyl group. In order to determine the orientation of the benzoyl group in **130** and **135**, NOE experiments were carried out. In both compounds, the H-4″ signal increased in area (*ca.* 20%) on irradiation at the frequency of the H-14″ signal. This indicates that the C-8″ carbonyl group sites are *anti* to H-4″, and that the orientation of the benzoyl group is the same in both compounds. Therefore, the twist between the 2,4-dihydroxybenzoyl and 2-arylbenzofuran chromophores on the methylcyclohexene rings of **130** and **135** is as illustrated in Fig. 57 and 58, respectively.

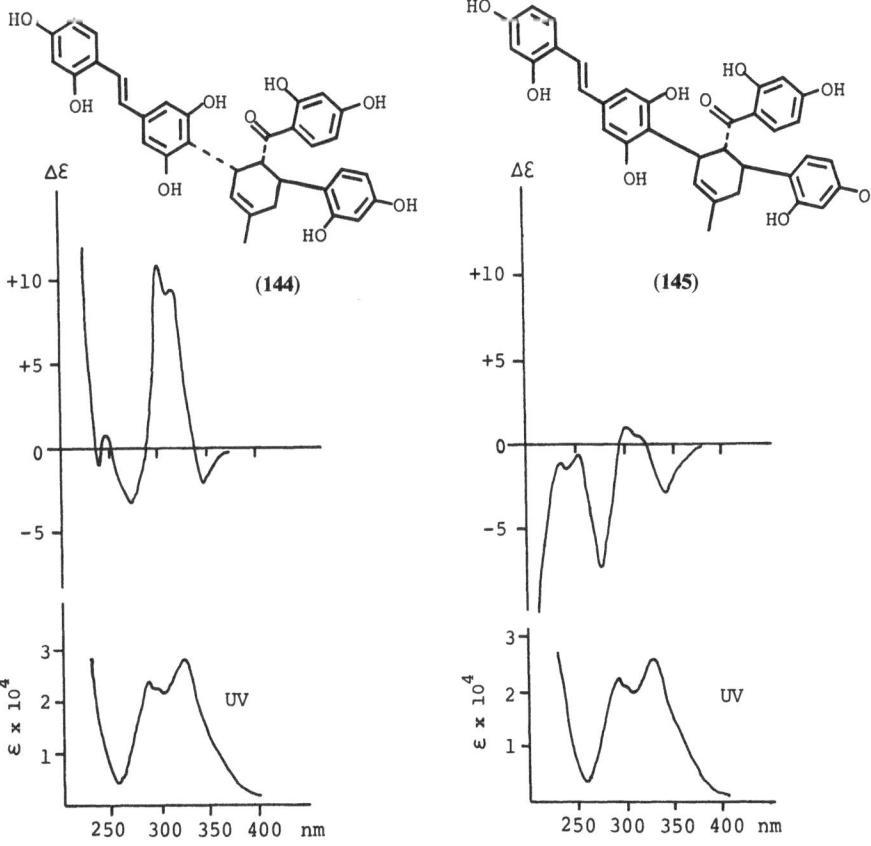

Fig. 54 (No. 3)

Since both **130** and **135** exhibit a positive Cotton effect, as mentioned above, the absolute configuration of mulberrofuran C (**130**) may be represented by formula (c) in Fig. 57, while that of mulberrofuran J (**135**) is shown as formula (b) in Fig. 58, based on the exciton chirality theory. Finally, the absolute configuration of the three chiral centers in mulberrofuran C (**130**) may be written as 3″S, 4″R, 5″S which includes the conformation of the cyclohexene ring, discussed earlier in this section, and that of mulberrofuran J (**135**) as 3″R, 4″R, 5″S (Fig. 59).

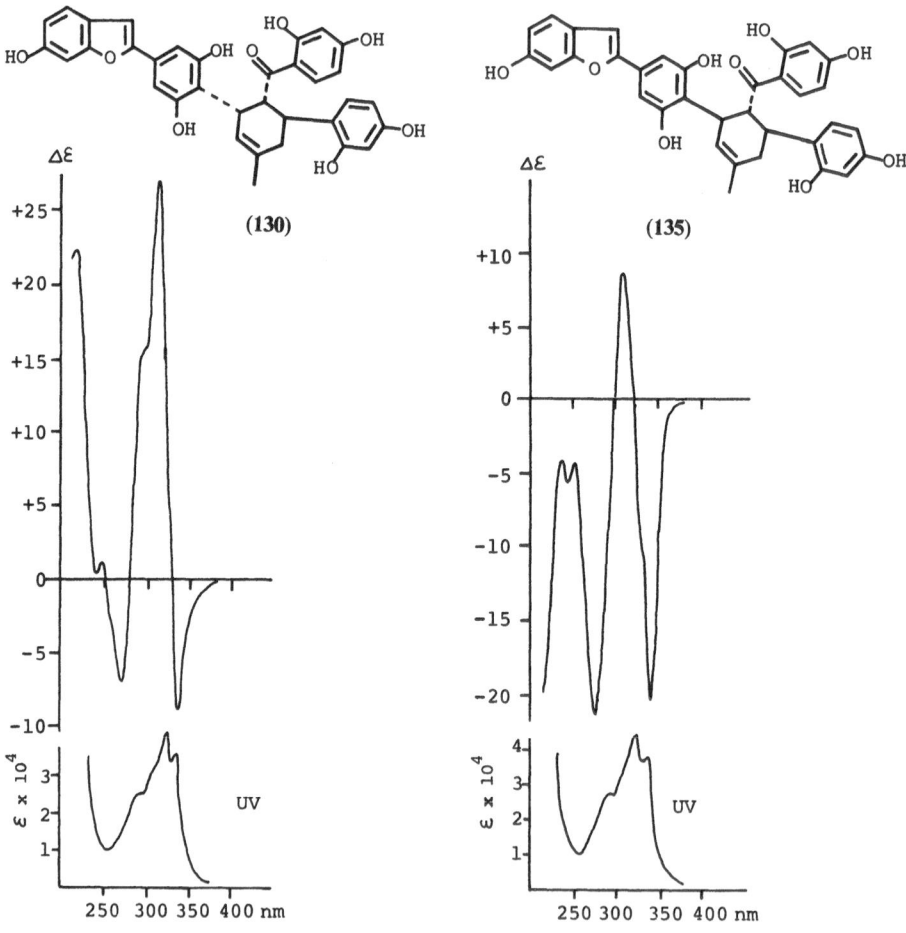

Fig. 54 (No. 4)

4. Absolute Configuration of Other Diels-Alder Type Adducts

Optical rotations of Diels-Alder type adducts from *Morus* root bark are summarized in Table 8. The *cis-trans* adducts exhibit positive optical rotations, while the all-*trans* adducts exhibit negative values. Compared with the absolute configuration of **130** and **135**, only the configuration at C-3″ is different. Consequently, one may conclude that the configuration at the C-3″ chiral center influences the sign of the rotation. Hence, the absolute configurations of the *cis-trans* adducts would be the same as that of mulberrofuran C (**130**), and those of the all-*trans*

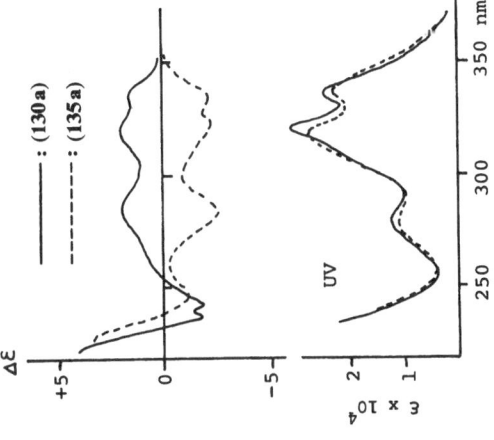

Fig. 55

adducts the same as that of mulberrofuran J (135). Therefore, the absolute configurations of three chiral centers in the methylcyclohexene
ring of *cis-trans* adducts obtained from the mulberry tree may be specified as 3″ S, 4″ R, 5″ S, and those of the all-*trans* adducts as 3″ R,
4″ R, 5″ S.

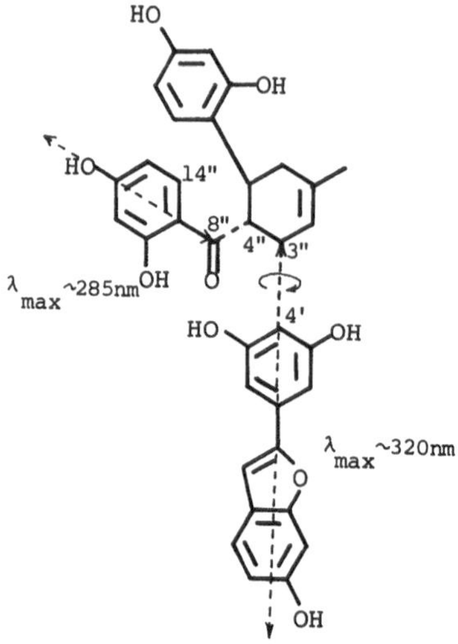

(130) 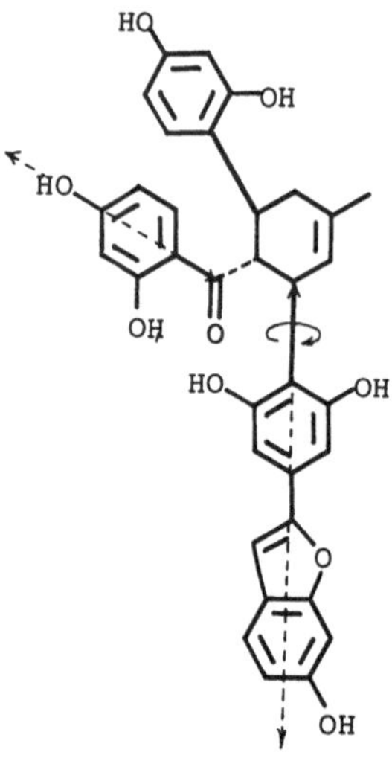   (135)

Fig. 56

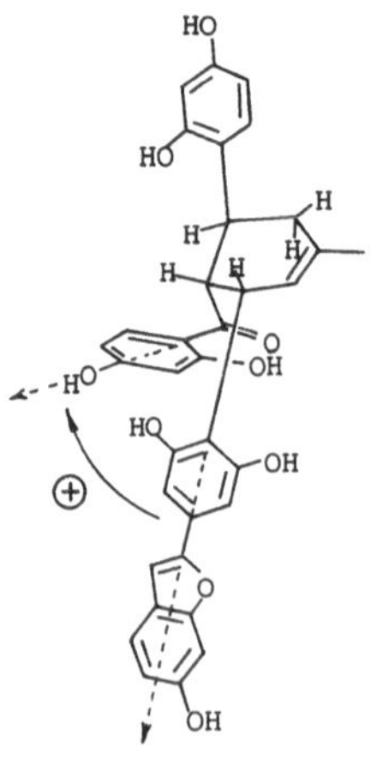

(c) 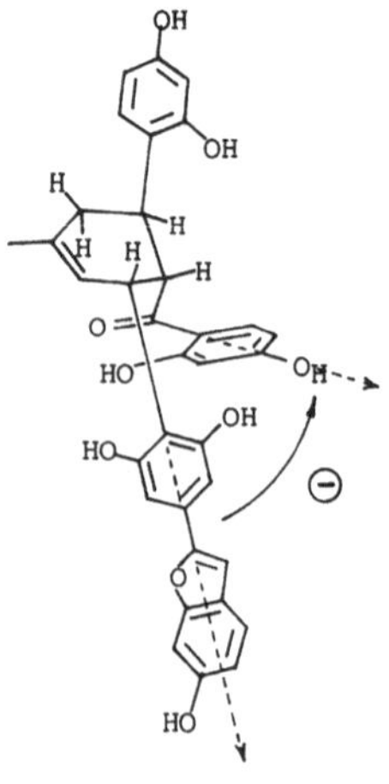   (d)

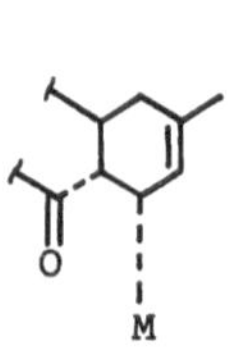

(c)   (130)   (d)

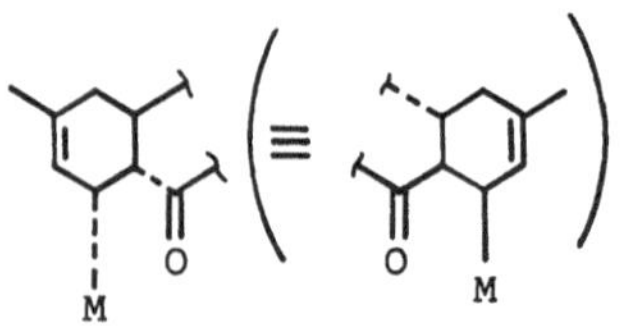

Fig. 57

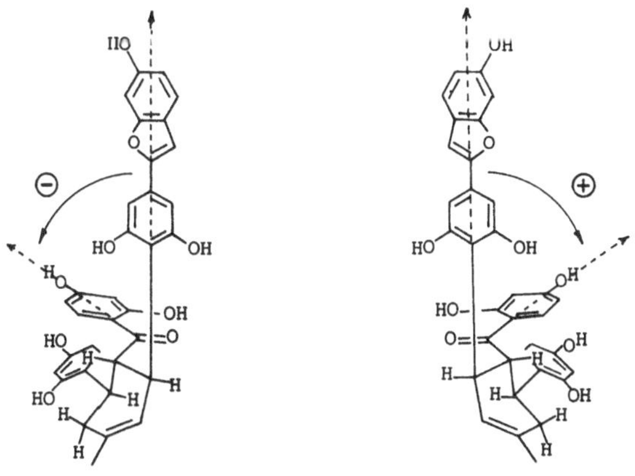

(a)   (b)

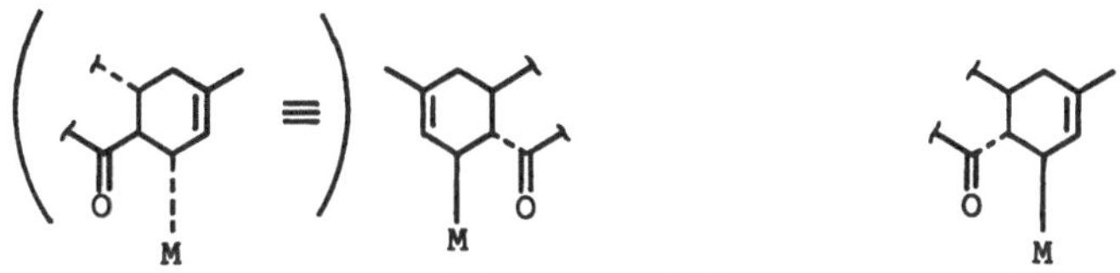

Fig. 58

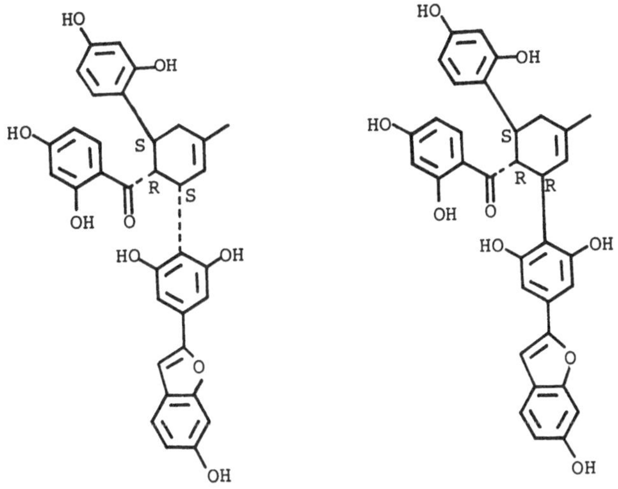

(130)   (135)

Fig. 59

Table 8. *Optical Rotations* $[\alpha]_D$

All-*trans*			Cis-*trans*		
Mulberrofuran J	(135)	−341°	mulberrofuran C	(130)	+153°
Kuwanon X	(145)	−322°	kuwanon Y	(144)	+172°
Sanggenon D	(165)	−145°	sanggenon C	(164)	+304°
Sanggenon E	(175)	− 86°	sanggenon P	(178)	+215°
Kuwanon I	(152)	−454°	kuwanon J	(147)	+85°
Kuwanon G	(25)	−534°	kuwanon Q	(148)	+160°
Kuwanon H	(26)	−536°	kuwanon R	(149)	+56°
Kuwanon K	(153)	−218°	kuwanon V	(150)	+145°
Kuwanon L	(154)	−277°	chalcomoracin	(103)	+193°
Kuwanon N	(155)	−188°			
Kuwanon O	(156)	−243°			
Kuwanon P	(158)	−509°			
Kuwanon W	(157)	−440°			

5. Absolute Configuration of the Chiral Centers on the Cyclohexene Ring of Kuwanon L

The absolute configuration of the chiral centers on the cyclohexene ring of kuwanon L (154) (*133*) was confirmed by the following results (*152*).

Kuwanon L (154) treated with alkali gave a degradation product 185. Comparison of the ^{1}H-NMR and CD spectra of 185 with those of 154, indicated that no epimerization of the chiral centers on the cyclohexene ring of 154 had occurred during treatment with alkali. Oxidation of the pentamethyl ether (185a) with osmium tetraoxide gave a *cis*-diol product (186) which furnished a mono-*p*-bromobenzoate (186a) and a di-*p*-bromobenzoate (186b) (Fig. 60). The CD spectrum of 186b exhibited a positive Cotton effect owing to exciton coupling between the two *p*-bromobenzoyl chromophores (λ_{max} 245 nm) (Fig. 61). The positive exciton coupling is shown clearly in the difference spectrum of 186a and 186b (Fig. 62).

Two possible conformations (A and B in Fig. 63) of the methylcyclohexene ring of 186b can be written on examining the coupling constants involving the ring protons (Fig. 64). Of these two conformations, only the A-conformation can account for the positive Cotton effect owing to exciton coupling between two *p*-bromobenzoyl chromophores. Hence the absolute configuration of the three chiral centers of kuwanon L (154) may be specified as 3″ *R*, 4″ *R*, 5″ *S*. This result is in agree-

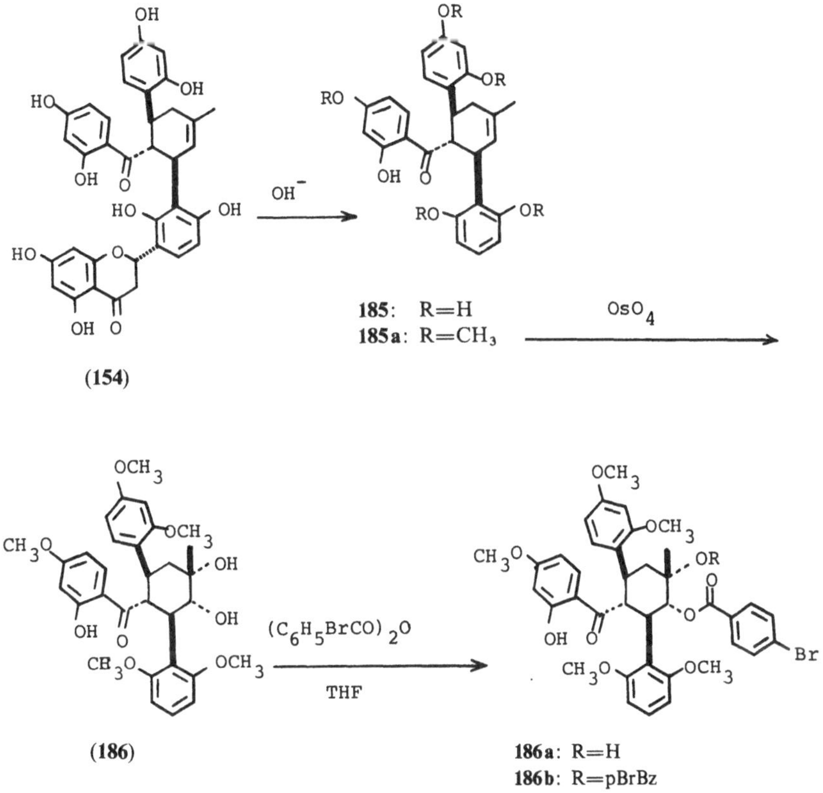

Fig. 60

ment with the result obtained from the CD spectra of mulberrofurans C (130) and J (135) described in the former section.

6. Absolute Configuration of the Ketal Compounds, Mulberrofurans F, G, and K

Structures of mulberrofurans F (131), G (27), and K (179) have been established by correlation with chalcomoracin (103) and with mulberrofuran C (130) using a ketalization reaction (91, 112). Therefore, the configuration of the three chiral centers in the methylcyclohexene ring of the ketal compounds is the same as those of 103 and 130.

The relative configuration of the four chiral centers in 27 was determined by considering the Dreiding model (112) and by X-ray analysis of albanol A pentamethyl ether (115) (Fig. 31). Hence, the absolute

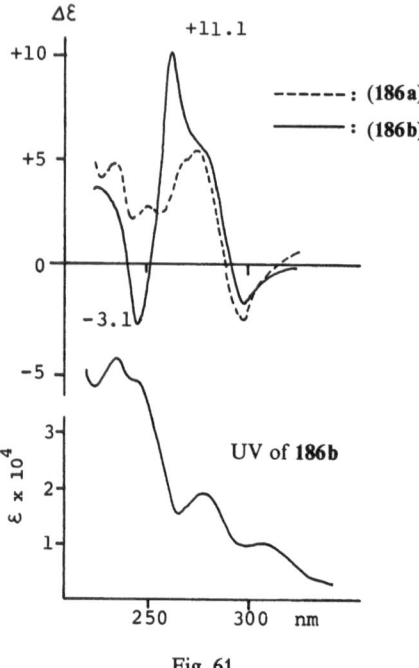

Fig. 61

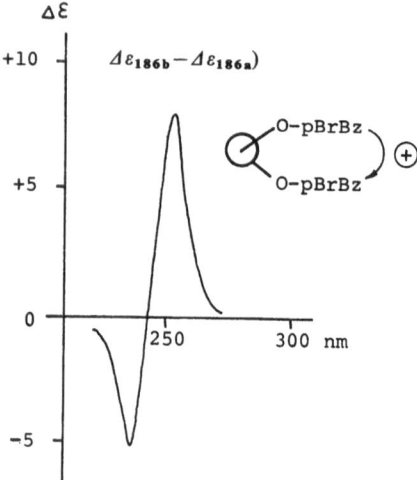

Fig. 62. Differential spectrum of **186a** and **186b** ($\Delta\varepsilon_{186b} - \Delta\varepsilon_{186a}$)

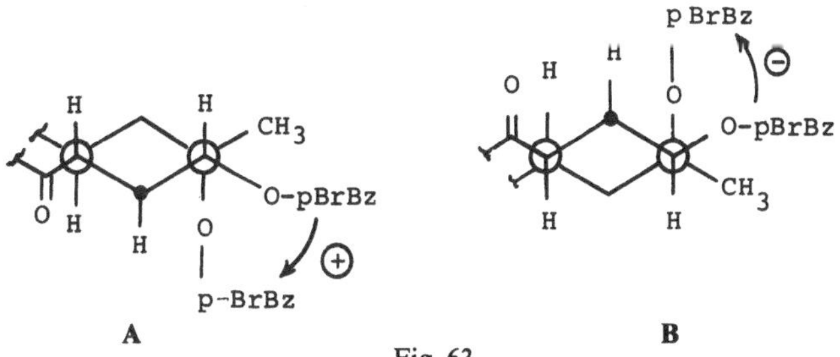

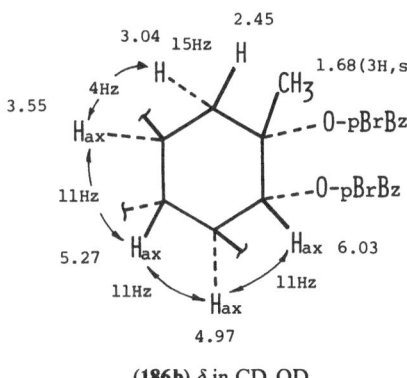

Fig. 63

(186b) δ in CD₃OD

Fig. 64. Chemical shifts (ppm) and coupling constants of proton signals of methylcyclohexene ring of 186b

configuration of these ketal compounds may be specified as 3″ S, 4″ R,
5″ S, 8″ R. This conclusion was confirmed by X-ray crystallographic
analysis. Monobromomulberrofuran G pentamethyl ether (27b) was
prepared from mulberrofuran G (27) and converted to dehydroproduct
187 as shown in Fig. 65. The result of the X-ray analysis of 187 is
depicted in Fig. 66, which shows that the absolute configuration of
the sole chiral center at C-8″ is indeed R and therefore that of mulberrofuran G (27) is indeed 3″ S, 4″ R, 5″ S, 8″ R. Therefore the absolute
configurations of the three chiral centers in the methylcyclohexene ring
of mulberrofuran C (130) is 3″ S, 4″ R, 5″ S, and that of mulberrofuran J (135) 3″ R, 4″ R, 5″ S. The absolute configurations of the Diels-
Alder type adducts between chalcones and dehydroprenylphenols isolated from the mulberry tree are therefore as shown in Fig. 67.

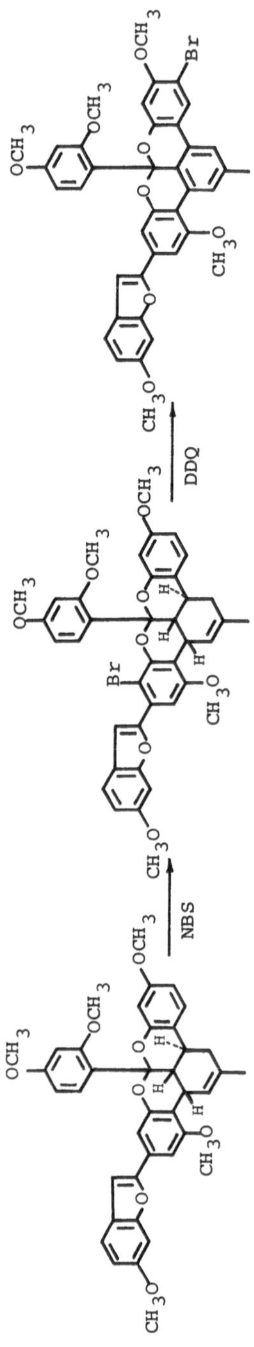

(187)

mp 268–272 °C, [α]$_D$ + 73°

(27b)

Fig. 65

(27a)

DDQ

NBS

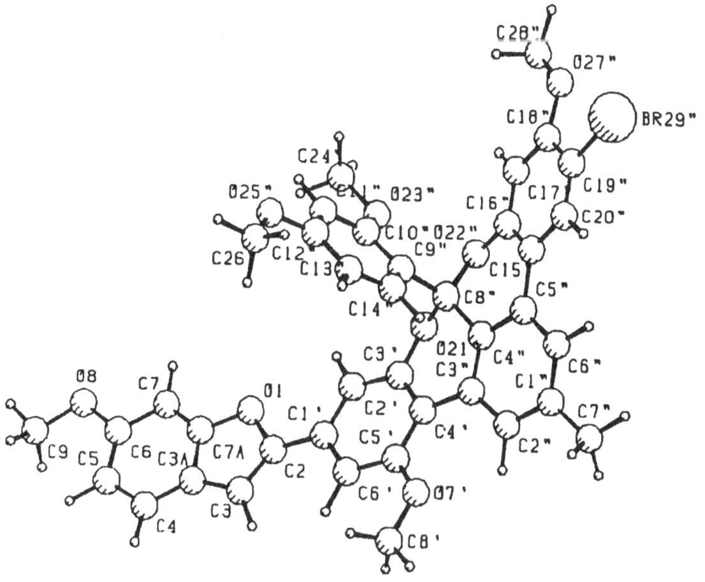

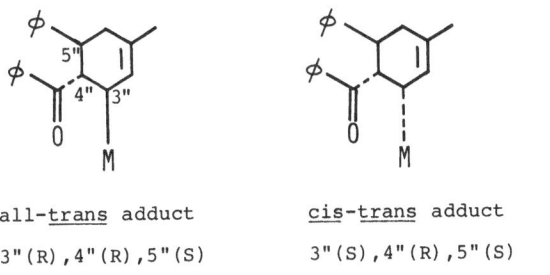

Fig. 66

all-<u>trans</u> adduct

3"(R),4"(R),5"(S)

cis-<u>trans</u> adduct

3"(S),4"(R),5"(S)

Fig. 67

VIII. Phenolic Constituents of *Cudrania tricuspidata* (Carr.) Bur. and *Broussonetia* Sp.

1. Structures of Xanthone and Flavonoid Derivatives from *Cudrania tricuspidata* (Carr.) Bur.

The crude drug "Sang-Bai-Pi" obtained in the market may be adulterated with root bark of *Cudrania tricuspidata* (Carr.) Bur. and *Broussonetia papyrifera* (L.) Vent., both of which belong to the family Moraceae (*155*). *Cudrania tricuspidata* (Carr.) Bur. (Japanese name "Hari-guwa") is a deciduous tree which is distributed throughout China, Korea, and Japan, and whose cortex and root bark have been used as a crude drug (*156*). In the course of our studies on the constituents of *Morus* root bark, we also studied the phenolic constituents of the root bark of *Cudrania tricuspidata* (Carr.) Bur., and characterized four xanthones, cudraxanthones A (**188**) (*157*), mp 212–216 °C, B (**189**) (*157*), mp 163–167 °C, C (**190**) (*157*), amorphous powder, D (**191**) (*157*), mp 127–132 °C, and three new flavonoids, cudraflavones A (**192**) (*158*), mp 265–272 °C, [α]$_D$ +27.3°, B (**193**) (*158*), amorphous powder, and cudraflavanone A (**194**) (*159*), mp 194 °C, [α]$_D$ −109.8° (Fig. 68). The ^{13}C-NMR spectra of xanthone derivatives obtained from *Cudrania tricuspidata* are listed in Table 9.

Table 9. ^{13}C-NMR Chemical Shifts (ppm) of Cudraxanthones A (**188**), B (**189**), C (**190**) and D (**191**)

	CA (**188**)	CB (**189**)	CC (**190**)	CD (**191**)
C-1	160.6	161.8	161.9	162.1
C-2	99.1	100.2	100.2	95.3
C-3	163.4	162.0	161.9	164.8
C-4	100.4	108.2	108.9	112.4
C-4a	154.6	155.2	155.2	154.4
C-4b	149.4	151.0	155.0	153.4
C-5	120.8	101.9	101.2	101.0
C-6	124.2	152.6	154.7	153.4
C-7	151.8	136.9	142.8	139.7
C-8	115.6	119.6	137.0	135.6
C-8a	119.0	109.0	111.8	111.0
C-9	181.5	182.8	182.3	183.3
C-9a	104.4	104.6	104.5	104.2
C-11	115.1	40.9	40.9	41.1
C-12	126.8	28.0	28.1	29.4
C-13	78.1	28.0	28.1	29.4
C-14	28.3	149.3	149.3	150.9
C-15	28.3	113.3	113.2	106.9
C-16	117.6	120.9	26.5	26.0
C-17	132.8	132.4	123.0	121.5
C-18	75.6	[75.8]	132.2	127.2
C-19	27.4	27.3	25.8	25.9
C-20	27.4	27.3	18.2	18.1
			C-7-OCH$_3$ 62.1	C-3-OCH$_3$ 55.5

Measured in CDCl$_3$. [ ] acetone-d$_6$.
CA cudraxanthone A, CB cudraxanthone B, CC cudraxanthone C, CD cudraxanthone D.

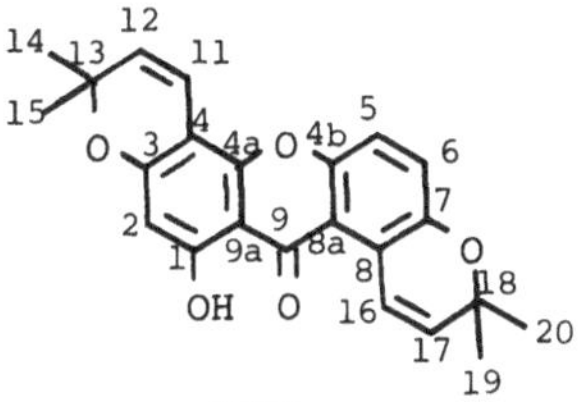

(188)

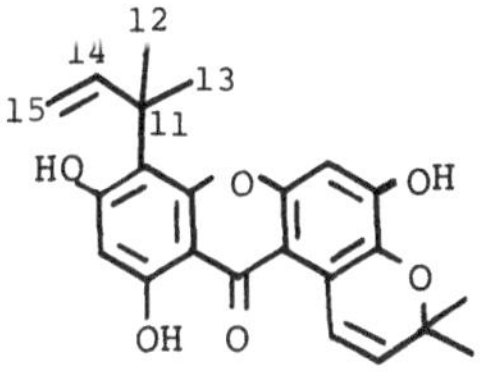

(189)

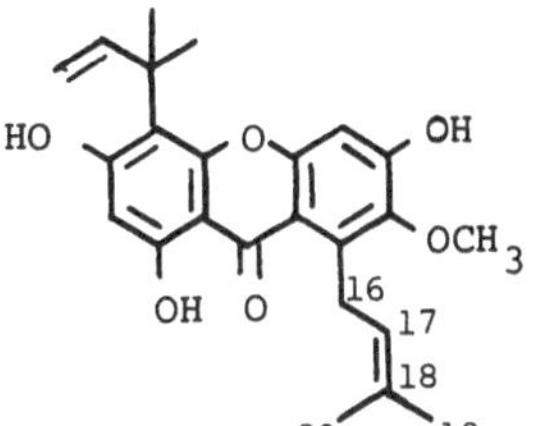

(190)

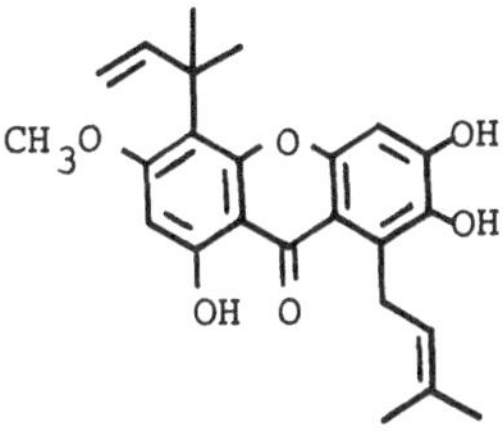

(191)

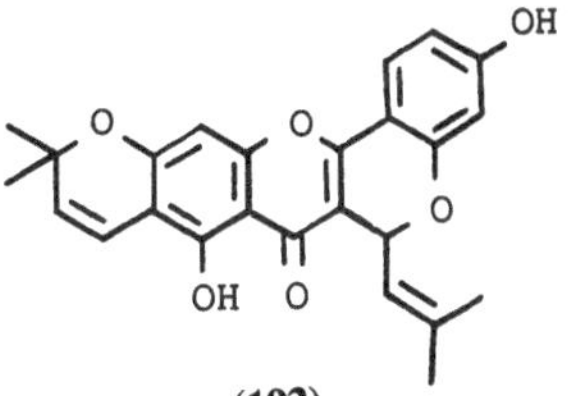

(192)

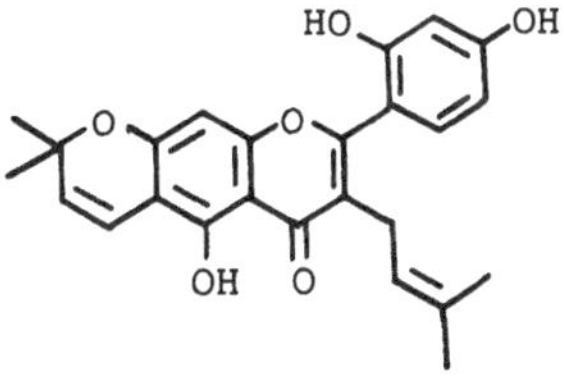

(193)

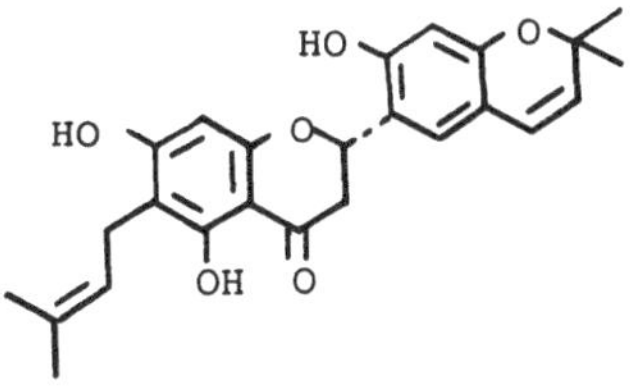

(194)

Fig. 68

2. Structures of 1,3-Diphenylpropane and Flavonoid Derivatives from *Broussonetia* Sp.

Broussonetia papyrifera (L.) Vent. (Japanese name "Kajinoki") is a deciduous tree which is distributed throughout Southeast Asia, China, and Japan. *B. kazinoki* Sieb. (Japanese name "Himekouzo") is also a deciduous tree distributed over Korea, China, Taiwan, and Japan. The cortex of these plants has been used as a raw material for paper and also as a crude drug (*156*). From the cortex of *B. papyrifera* (L.) Vent. the following compounds were isolated: two flavonols, broussoflavonols A (**195**) (*160*), amorphous powder, B (**196**) (*160*), mp 178–179 °C, and two chalcones, broussochalcones A (**197**) (*160*), amorphous

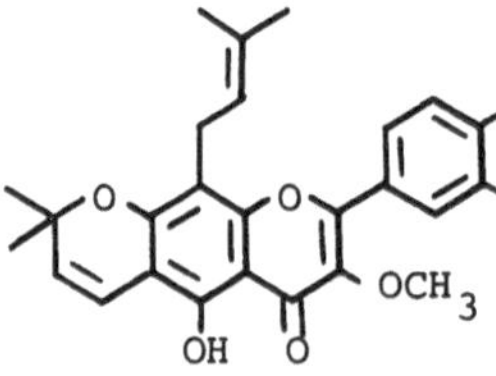

(**195**)

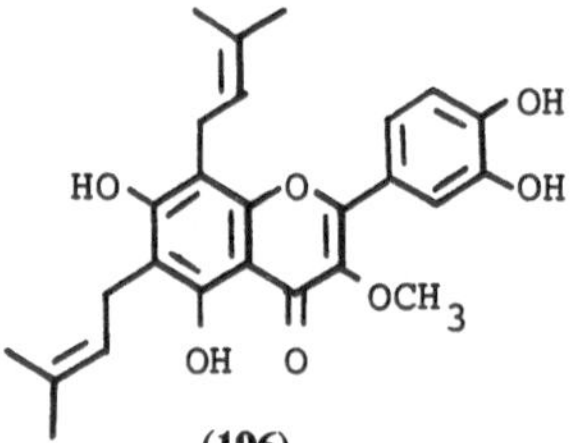

(**196**)

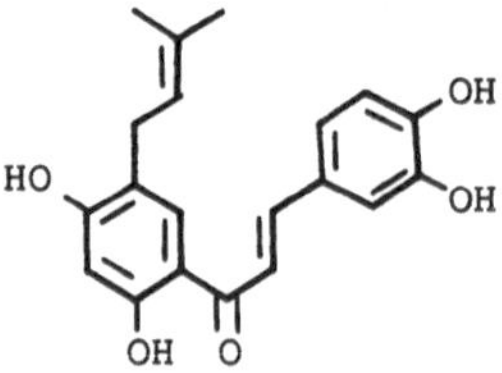

(**197**)

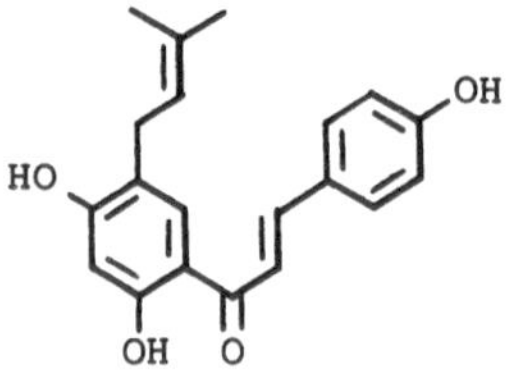

(**198**)

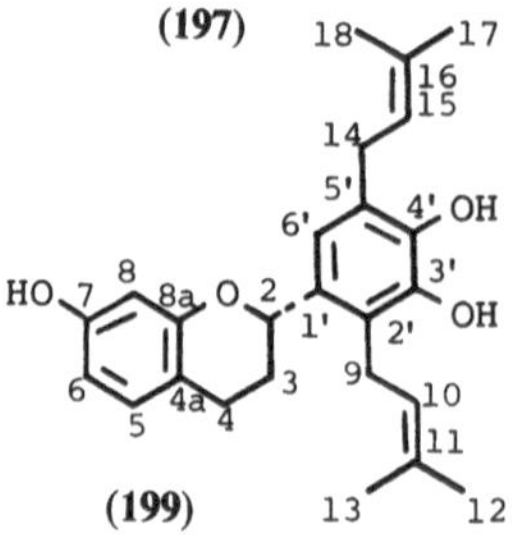

(**199**)

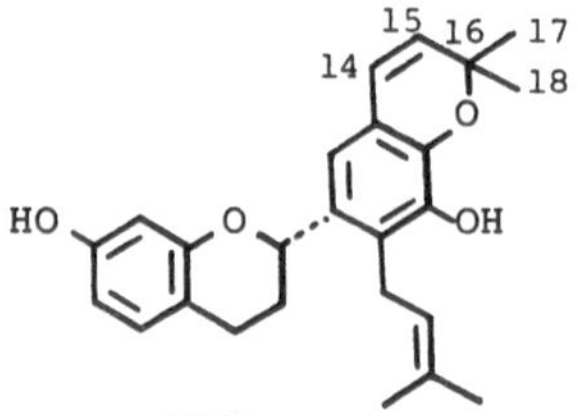

(**200**)

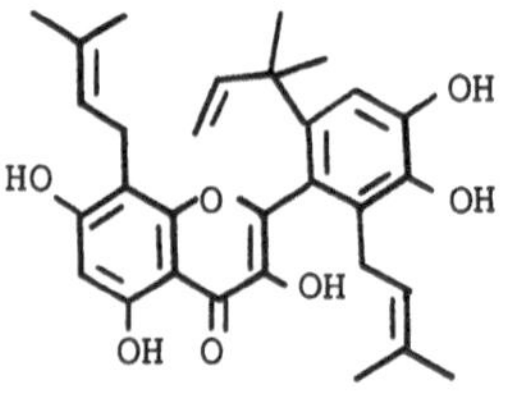

(**201**)

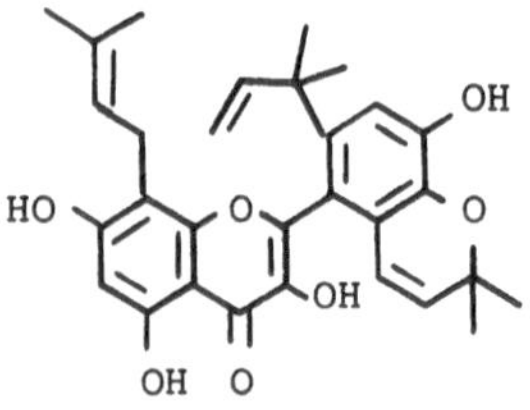

(**202**)

Fig. 69 (No. 1)

powder, B (**198**) (*160*), mp 168–170 °C, and two flavans, kazinols A (**199**) (*161*), oily substance, $[\alpha]_D$ −11°, B (**200**) (*161*), oily substance, $[\alpha]_D$ −20°, while from the root bark of *B. papyrifera* (L.) Vent., two flavonols, broussoflavonols C (**201**) (*162*), 173–176 °C, D (**202**) (*162*), mp 102–110 °C, and two flavans, kazinols I (**203**) (*163*), mp 148–150 °C, $[\alpha]_D$ −2.7°, O (**204**) (*164*), mp 158–160 °C, $[\alpha]_D$ −31.6°.

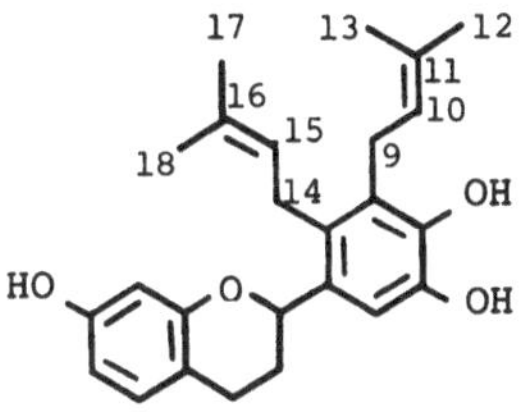

(**203**)

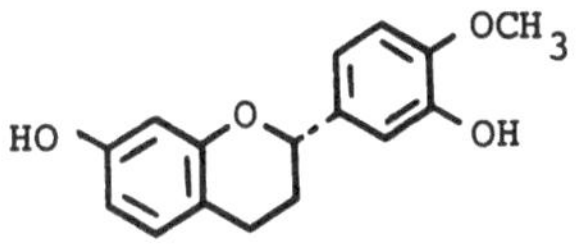

(**204**)

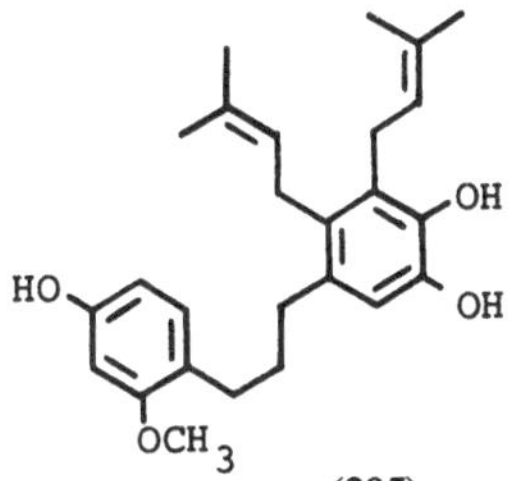

(**205**)

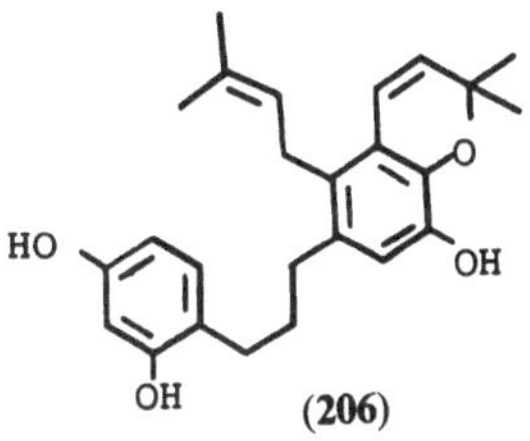

(**206**)

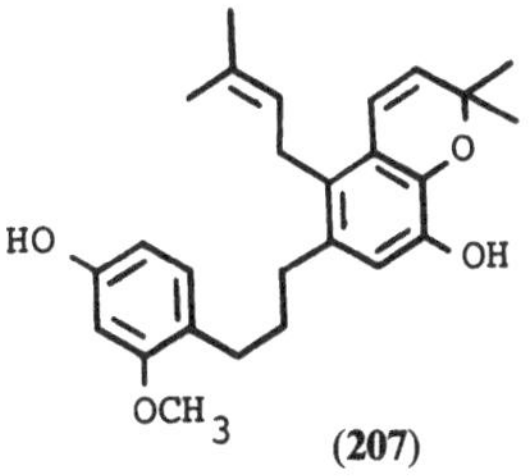

(**207**)

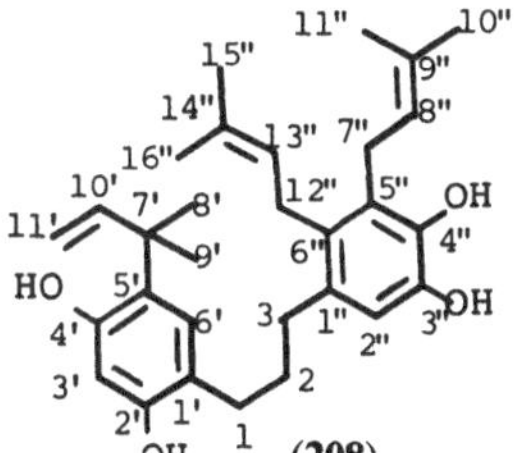

(**208**)

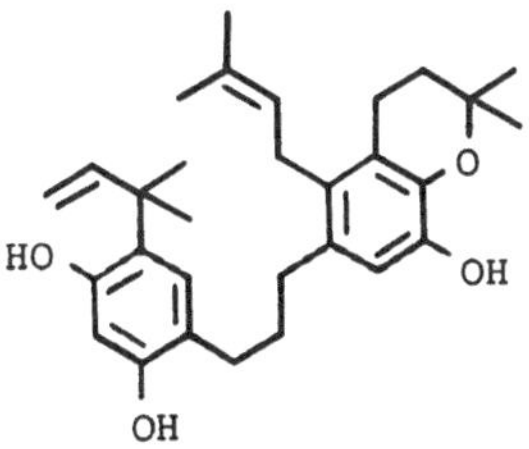

(**209**)

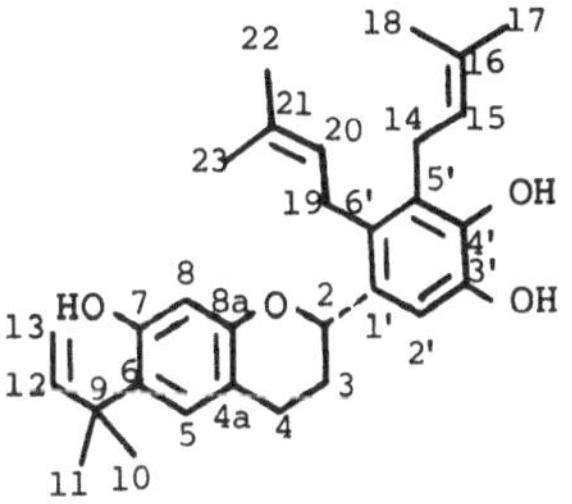

(**210**)

Fig. 69 (No. 2)

From the cortex of *B. kazinoki* Sieb. the following three 1,3-diphen-
ylpropane derivatives with isoprenoid substituents were isolated: kazinols J (**205**) (*165*), mp 116–118 °C; M (**206**) (*165*), oily substance, N (**207**)
(*165*), oily substance. The following ten 1,3-diphenylpropane derivatives were also isolated from the root bark of *B. kazinoki* Sieb: kazinols C (**208**) (*163*), oily substance; D (**209**) (*163*), oily substance; E (**210**)
(*163*), mp 147 °C, [α]$_D$ +0.30°; F (**211**) (*163*), mp 108–109 °C, G (**212**)
(*163*), oily substance; H (**213**) (*163*), oily substance [α]$_D$ +0.53°;
K (**214**) (*163*), oily substance; L (**215**) (*165*), oily substance; P (**216**)
(*166*), mp 163–168 °C, [α]$_D$ 0°. For locating the isoprene units of the
phenolic compounds described in this chapter, nuclear Overhauser effect measurements are very useful. The structure assigned to kazinol P
was confirmed by the following synthesis: kazinol C (**208**) was oxidized

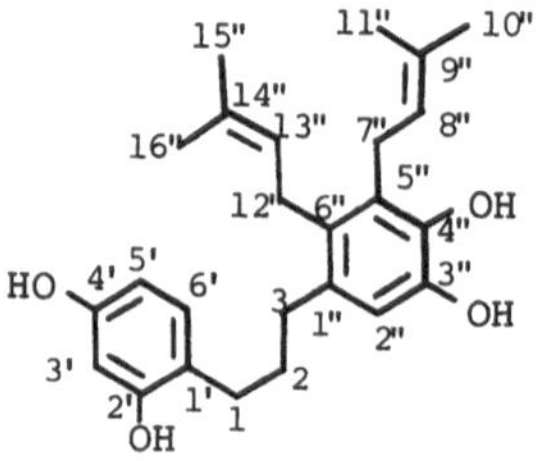

(**211**) 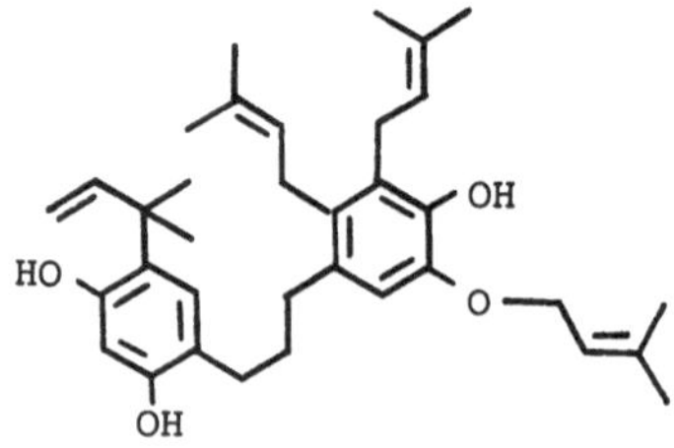   (**212**)

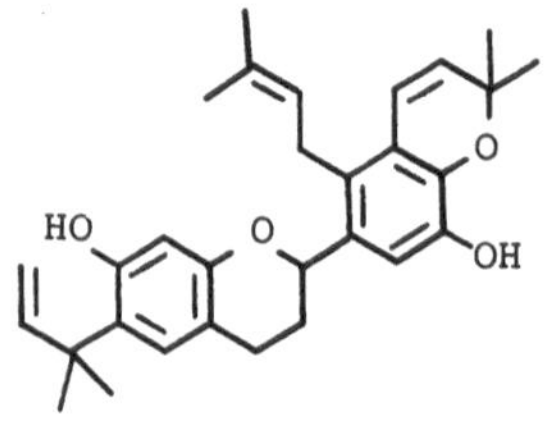

(**213**) 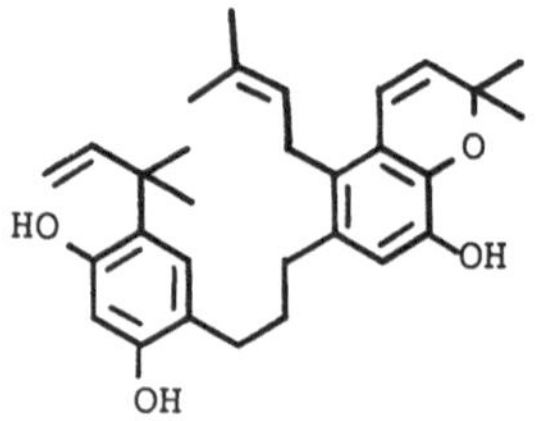   (**214**)

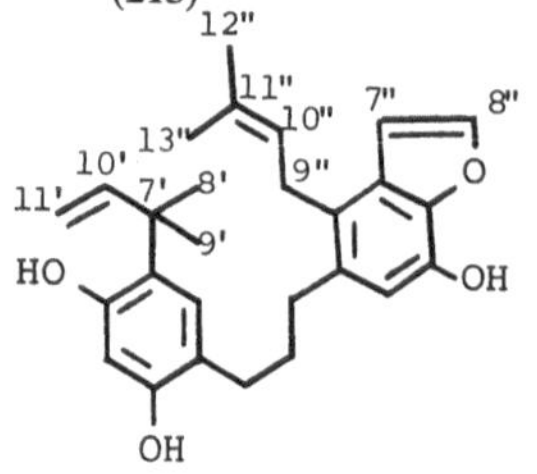

(**215**) 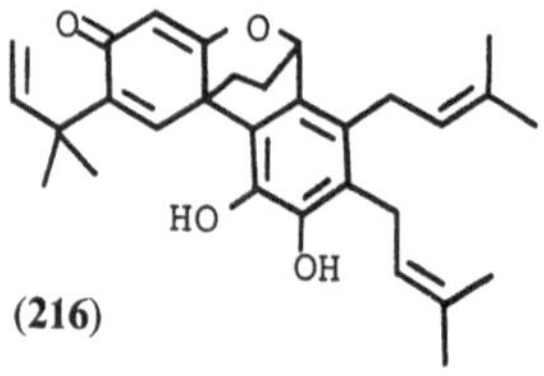   (**216**)

Fig. 69 (No. 3)

with the complex $[Fe(DMF)_3Cl_2][FeCl_4]$ or manganese dioxide to give **216** and kazinol E (**210**), while **210** was recovered unchanged by the same treatment (*167*).

Somewhat earlier Takasugi and his co-workers had reported the isolation and structure determination of a series of antifungal compounds from diseased or wounded paper mulberry trees (*Broussonetia papyrifera* Vent.) (*168–172*) (Fig. 70).

From diseased shoot cortical tissues of the paper mulberry, six 1,3-diphenylpropane derivatives, a catechin, a flavan, two chalcones, and a coumarin, were isolated as phytoalexins. They were brousson-

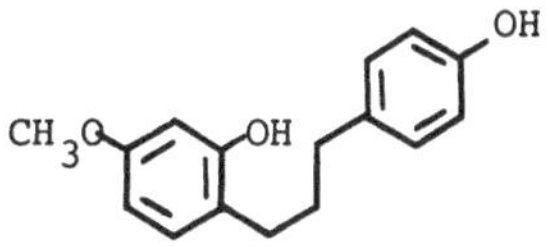

(**217**)

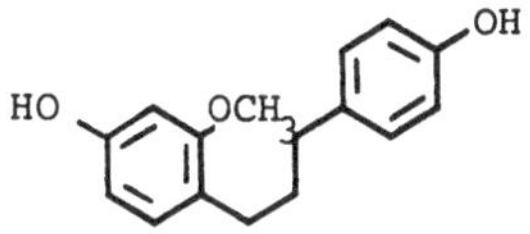

(**218**)

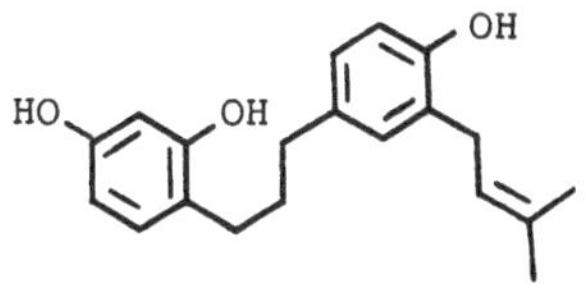

(**219**)

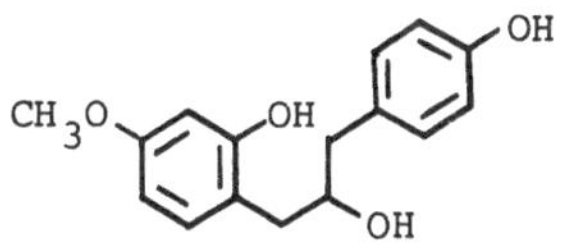

(**220**)

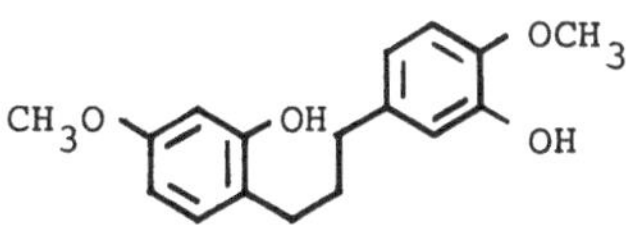

(**221**)

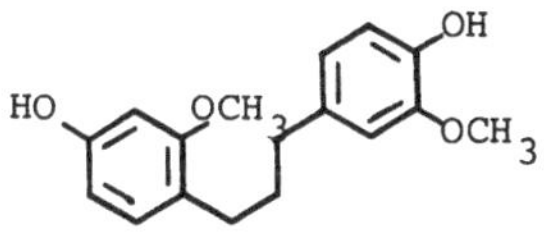

(**222**)

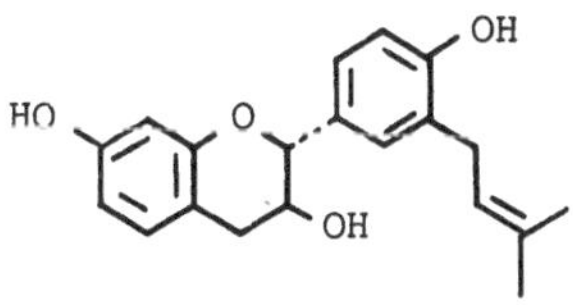

(**223**)

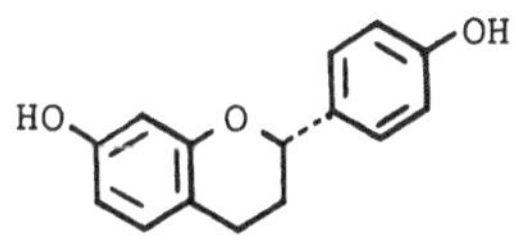

(**224**)

Fig. 70 (No. 1)

ins A (**217**) (*169*), mp 101–101.5 °C; B (**218**) (*169*), mp 99.5–100 °C; C (**219**) (*170, 171*), oily substance; D (**220**) (*171*), mp 73–74 °C, $[\alpha]_D$ −0.8; E (**221**) (*171*), mp 72–74 °C, $[\alpha]_D$ 0°; F (**222**) (*171*), oily substance, $[\alpha]_D$ 0°; broussinol (**223**) (*171*), mp 127–128 °C, $[\alpha]_D$ −21.8°; demethylbroussin (**224**) (*171*), mp 194–195 °C, $[\alpha]_D$ −30.5°; isoliquiritigenin (**225**) (*171*), 4,4′-dihydroxy-2′-methoxychalcone (**226**) (*171*), and marmesin (**227**) (*169*). Two stress metabolites, spirobroussonins A (**228**) (*172*), mp 253–254 °C, $[\alpha]_D$ 0°, and B (**229**) (*172*), mp 232–233 °C, $[\alpha]_D$ 0°, were also isolated from diseased paper mulberry; their structures were determined by synthesis of their methyl ethers from the 1,3-diphenylpropane derivatives.

From wounded xylem tissue of paper mulberry trees, two phytoalexins, broussin (**230**) (*170*), mp 120–122 °C, $[\alpha]_D$ −17.4°, and broussonin C (**219**) (*170*), were isolated. The co-occurrence of the optically active flavans (**210, 230**) and 1,3-diphenylpropanes (**208, 219**) in the same paper mulberry tissue suggests a close biosynthetic relationship between

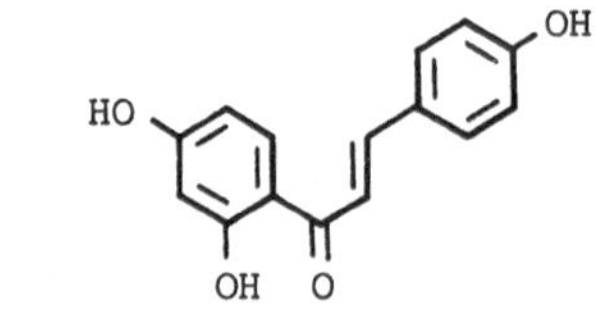

(225)

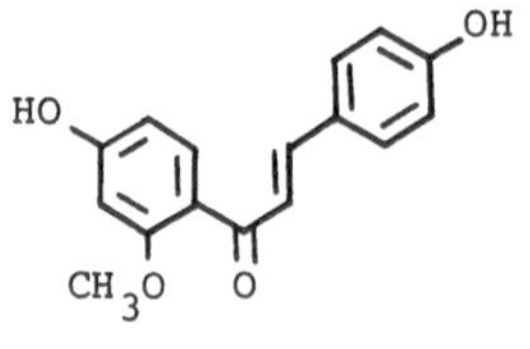

(226)

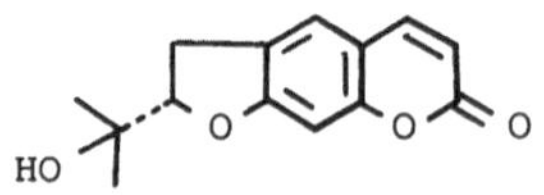

(227)

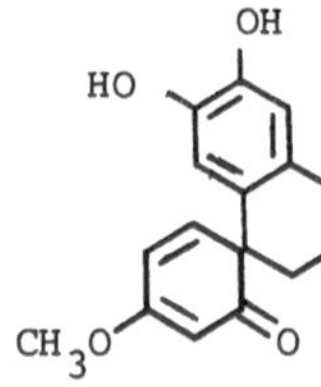

(228)

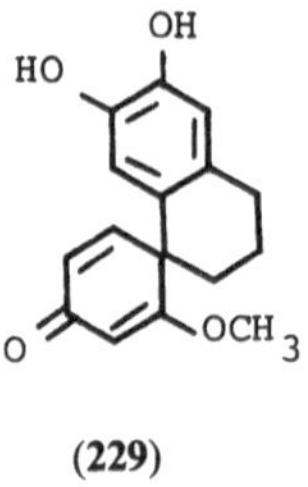

(229)

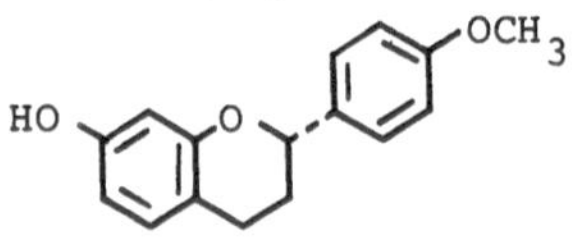

(230)

Fig. 70 (No. 2)

the two types of compounds, and indicates that 1,3-diphenylpropanes can be regarded as flavonoids from the biogenetic view point (*170*). The spiro-compounds such as kazinol P (**216**), spirobroussonins A (**228**) and B (**229**), are regarded biogenetically as variants of 1,3-diphenylpropane derivatives and possess no optical activity (*166, 172*). The ^{13}C-NMR spectra of flavonols and 1,3-diphenylpropane derivatives of *Broussonetia* sp. are listed in Table 10.

Table 10. ^{13}C-*NMR Chemical Shifts (ppm) of Kazinols A* (**199**), *B* (**200**), *C* (**208**), *E* (**210**), *F* (**211**), *I* (**203**), *and L* (**215**)

	KZA (**199**)[1]	KZB (**200**)[1]	KZI (**203**)[1]	KZE (**210**)[1]		KZC (**208**)[1]	KZF (**211**)[2]	KZL (**215**)[1]
C-2	75.6	74.8	74.8	75.0	C-1	32.9[a]	33.6[a]	32.3[a]
C-3	25.0	24.8	25.1	25.6	C-2	29.7	30.4	29.5
C-4	29.4	29.9	29.7	29.9	C-3	31.6[a]	33.1[a]	31.9[a]
C-4a	113.5	113.9	113.7	113.9	C-1'	120.7	120.5	119.4
C-5	129.5	130.0	129.6	127.0	C-2'	153.2	157.3	152.9
C-6	107.6	108.1	107.6	125.1	C-3'	105.0	103.4	104.4
C-7	154.3	155.1	154.1	153.5	C-4'	153.2	156.6	152.4
C-8	103.2	103.7	103.2	105.6	C-5'	124.4	107.3	123.5
C-8a	155.5	156.1	155.5	155.1	C-6'	127.7	131.1	127.0
C-1'	130.7	131.6	130.6	131.0	C-1''	133.0	132.7	134.6
C-2'	124.6	125.8	110.7	111.3	C-2''	114.2	114.6	111.7
C-3'	141.7	142.2	141.7	142.4	C-3''	141.7	143.0	141.1
C-4'	141.3	138.6	141.3	141.7	C-4''	140.5	142.2	138.2
C-5'	123.0	119.0	129.6	130.2	C-5''	130.5	130.4	121.9
C-6'	118.5	122.3	126.4	127.0	C-6''	127.0	127.7	128.4
C-9	25.1[a]	25.3	27.2[a]	39.8	C-7'	39.8		39.0
C-10	122.0[b]	122.0	121.9[b]	27.2	C-8'	27.3		27.1
C-11	132.3[c]	132.3	132.6[c]	27.2	C-9'	27.3		27.1
C-12	25.6[d]	25.7	25.4[d]	148.3	C-10'	148.5		147.7
C-13	17.8	17.9	17.9	113.0	C-11'	113.1		112.6
C-14	29.3[a]	115.1	25.7[a]	27.2[a]	C-7''	27.8[b]	28.1[b]	105.7
C-15	121.5[b]	130.1	123.3[b]	122.8[b]	C-8''	122.7[c]	124.9[c]	143.8
C-16	133.6[c]	77.4	131.1[c]	132.4[c]	C-9''	133.3[d]	130.7[d]	28.4
C-17	25.5[d]	28.0	25.6[d]	25.6	C-10''	25.7	25.8	123.1
C-18	17.8	28.2	17.9	17.9	C-11'	18.0	18.0	130.4
C-19				25.6[a]	C-12''	26.1[b]	26.2[b]	25.5
C-20				124.1[b]	C-13''	124.4[c]	125.8[c]	17.9
C-21				131.4[c]	C-14''	130.9[d]	130.7[d]	
C-22				25.6	C-15''	25.7	25.8	
C-23				17.9	C-16''	18.0	18.0	

Solvent: [1] CDCl$_3$. [2] acetone-d$_6$.
[a-d] Assignments may be reversed in vertical column.
KZA kazinol A, KZB kazinol B, KZC kazinol C, KZE kazinol E, KZF kazinol F, KZI kazinole I, KZL kazinol L.

IX. Biological Activities of Phenolic Constituents of Mulberry Tree and Related Plants

Studies of the phenolic constituents of *Morus* root bark were originally undertaken to characterize the hypotensive components of the root bark. Kuwanons G (**25**) (*33*), H (**26**) (*34*), M (**160**) (*136*), mulberrofurans C (**130**) (*111*), F (**131**) (*112*), and G (**27**) (*112*) were shown to be hypotensive components of the root bark of the cultivated mulberry tree. Compounds **25**, **26**, **130**, **131** and **27** were almost equally effective in causing a transient decrease in arterial blood pressure in doses of 0.1–1 mg/Kg (i.v.) in rabbits. Kuwanon M (**160**) (*136*) showed hypotensive action in hypertensive rats (2 mg/Kg, i.v.). On the other hand, sanggenons C (**164**) (*140*) and D (**165**) (*141*) were characterized as the hypotensive compounds of the crude medicine "Sang-Bai-Pi". Sanggenon C (**164**) showed a marked hypotensive effect (1 mg/Kg, i.v.) in rabbits, and **165** showed the same effect (0.5–2.0 mg/Kg, i.v.) in rats. The mechanism of the hypotensive action of **25** and **26** has been discussed (*95*).

Antimicrobial activities of some flavonoids with isoprenoid substituents and Diels-Alder type adducts obtained from *Morus* root bark were also studied. All of the tested compounds were effective against gram-positive bacteria and against some species of fungi, whereas they were inactive against gram-negative bacteria, as shown in Table 11.

ISHITSUKA *et al.* screened natural products for antirhinoviral activity and reported that a flavone (Ro 09-0179), a potent antipicornavirus agent, was isolated from a Chinese medicinal herb (*Agastache rugosa* Kuntze) (*173*). We studied the antiviral effects on rhinovirus type 2 and found that morusin hydroperoxide (**49**), oxydihydromorusin (**34**), chalcomoracin (**103**), and compound A (**31**) had marginal anti-rhino viral activities (MIC, 1.25–2.5 μg/ml) (*48*).

The phenolic constituents and their derivatives were tested for their inhibitory effects on the beef heart cyclic AMP phosphodiesterase. Almost all of the phenolic compounds isolated from the mulberry tree showed significant inhibitory activities (*174*).

Formation of some products from $[1 - {}^{14}C]$ arachidonic acid was studied in rat platelets in the presence of phenolic compounds isolated from the Morus root bark. Morusin (**24**) was found to inhibit the formation of 12-hydroxy-5,8,10-heptadecatrienoic acid (HHT) and thromboxane B_2 (cyclooxygenase products) more strongly than the formation of 12-hydroxy-5,8,10,14-eicosatetraenoic acid (12-HETE) (12-lipoxygenase product). Oxydihydromorusin (**34**) and kuwanon C (**33**) were also found to inhibit the formation of thromboxane B_2 more

Table 11. Antimicrobial Spectra of Morusin, Kuwanons, and Sanggenons

Test organism	MIC (µg/ml)							
	Morusin (24)	Kuwanons			Sanggenons			
		C (33)	J (147)	L (154)	A (163)	B (174)	C (164)	D (165)
Staphylococcus aureus 209P*	12.5	1.56	3.12	1.56	6.25	6.25	3.12	6.25
Streptococcus faecalis*	50.0	50.0	6.25	3.12	6.25	3.12	25.0	100.0
Bacillus subtilis PCI 219*	1.56	1.56	3.12	3.12	3.12	6.25	1.56	6.25
Mycobacterium smegmatis*	1.56	3.12	50.0	12.5	3.12	25.0	100.0	>100.0
Escherichia coli F$_1$*	>100.0	>100.0	>100.0	>100.0	>100.0	>100.0	>100.0	>100.0
Pseudomonas aeruginosa*	>100.0	>100.0	>100.0	>100.0	>100.0	>100.0	>100.0	>100.0
Candida albicans**	>100.0	100.0	>100.0	>100.0	100.0	>100.0	100.0	>100.0
Saccharomyces cerevisiae**	>100.0	>100.0	>100.0	>100.0	>100.0	>100.0	50.0	>100.0
Aspergillus niger**	>100.0	>100.0	>100.0	>100.0	>100.0	>100.0	>100.0	>100.0
Penicillum glaucum**	>100.0	100.0	50.0	>100.0	100.0	>100.0	100.0	>100.0
Trichophyton mentagrophytes***	12.5	12.5	25.0	50.0	50.0	25.0	12.5	25.0
Microsporum gypseum**	25.0	12.5	25.0	50.0	50.0	50.0	25.0	50.0
Fusarium graminearum****	>100.0	>100.0	>100.0	>100.0	>100.0	>100.0	50.0	100.0
Piricularia oryzae****	3.12	6.25	100.0	50.0	6.25	25.0	12.5	25.0

Agar dilution method. Medium; * Heat infusion agar. ** Yeast morphology agar. *** Sabouraud agar. **** Poteto dextrose agar. Incubation for 24 or 48 h; 1 week depending on microorganism.

strongly than the formation of HHT and 12-HETE. Mulberrofuran A
(**104**) inhibited the formation of HHT and thromboxane B_2, but increased the formation of 12-HETE (*175*). Kuwanons G (**25**), and
H (**26**), sanggenon C (**164**) and mulberrofuran Q (**143**) at concentrations of 10^{-3} to 10^{-4} M inhibited the formation of HHT and thromboxane B_2. However, they increased the formation of 12-HETE. Sanggenon D (**165**) and mulberrofuran J (**135**) at a concentration of 10^{-3} M
inhibited the formation of HHT, thromboxane B_2, and 12-HETE. Mulberrofuran G (**27**) at a concentration of 10^{-3} M inhibited the formation
of HHT, thromboxane B_2, and 12-HETE, while it inhibited the formation of 12-HETE at 10^{-5} M without affecting the formation of HHT
and thromboxane B_2 (*176*).

Fujiki *et al.* reported that flavonoids inhibited the *in vitro* and *vivo*
effects induced by the tumor promoter teleocidin, such as incorporation
of ^{32}Pi into phospholipids, induction of ornithine decarboxylase (ODC)
activity on mouse skin, and so on (*177*). We studied the anti-tumor
promoting activity of the phenolic compounds isolated from *Morus*
root bark and its related plants. Most prenylphenols inhibited the specific ^{3}H-TPA (12-*O*-tetradecanoylphorbol-13-acetate) binding to a
mouse skin particulate. Some prenylphenols also inhibited both the
activation of protein kinase C with teleocidin and the induction of
ODC activity by 11.4 nmol of teleocidin in mouse skin (11.4 nmol of
24; 43% of inhibition) (Table 12). The inhibitory effect of **24** on the
tumor promoting activity of teleocidin was examined in skin carcinogenesis in a mouse initiated with 7,12-dimethylbenz(*a*)anthracene
(DMBA). The percentage of tumor-bearing mice in the group treated
with DMBA (100 µg) and teleocidin (2.5 µg × 2/week) was 100% in
week 20, whereas it was 60% in the group treated with DMBA, teleocidin and morusin (**24**) (1 mg × 2/week). The average number of tumors
per mouse in week 20 was also reduced from 5.3 to 1.1 in the presence

Table 12. *Inhibition of Specific ^{3}H-TPA Binding, Activation of Protein Kinase C, and ODC Induction with Morus flavonoids*

	Inhibition of specific ^{3}H-TPA binding (ED_{50} µM)	Inhibition of activation of protein kinase C (ED_{50} µM)	Inhibition of ODC induction (%)
Morusin (**24**)	57	80	43
Kuwanon G (**25**)	99	40	34
Kuwanon M (**160**)	85	22	25
Mulberrofuran G (**27**)	34	46	10
Sanggenon D (**165**)	60	42	17

Table 13. *Effect of Morusin* (**24**) *on Tumor Promotion with Teleocidin*

	Percentage of tumor-bearing mice	Average number of tumors/mouse
DMBA + teleocidin	100	5.3
DMBA + **24** + teleocidin	60	1.1
DMBA + **24**	0	0 in week 20

of morusin (*178*). It is noteworthy that some of the phenolic components of the *Morus* root bark exhibited positive anti-tumor promoter
activities in all three test systems *in vitro*, and that morusin (**24**) showed
anti-tumor promoting activity in two-stage mouse skin carcinogenesis
experiment (*179*) (Table 13).

Acknowledgements

I am grateful to Drs. T. FUKAI and Y. HANO, and Miss H. TSUBURA for their help in
preparing the manuscript. I also wish to thank Dr. J. UNO, Chiba University for supplying
the unpublished data on antimicrobial spectra.

References

1. KITAMURA, S., and G. MURATA: Coloured Illustrations of Woody Plants of Japan
   (Genshoku Nihon Shokubutsu Zukan, Mokuhonhen), Vol. II, p 231, Osaka: Hoikusha Publishing Co., 1980.
2. TAKAGI, K.: "Saisogaku," pp 39–46, Tokyo: Nihon Gakujutsu Shinkokai, 1952.
3. NANBA, T.: Coloured Illustrations of Wakanyaku (Genshoku Wakanyaku Zukan),
   Vol. II, pp 154–155, Osaka: Hoikusha Publishing Co., 1980.
4. KIMURA, K., and T. KIMURA: Medicinal Plants of Japan in Color (Genshoku Nihon
   Yakuyosyokubutsu Zukan), p 19–20, Osaka: Hoikusha Publishing Co., 1981.
5. FUKUTOME, K.: Hypotensive Action of the Extract of Mulberry Tree. Nihon Seiri-
   gaku Zasshi (J. Physiolg. Soc. Japan) **3**, 172 (1938).
6. OHISHI, T.: On the Hypotensive Action of Mulberry Root Bark. Sanshi Shikenjo
   Iho (Technical Bull. Sericultural Experiment Station) **59**, 1 (1941).
7. SUZUKI, B., and T. SAKUMA: On the Hypotensive Components of the Mulbery Tree.
   Sanshi Shikenjo Iho (Technical Bull. of Sericultural Experiment Station) **59**, 9 (1941).
8. KATAYANAGI, M., H. WAKANA, and T. KIMURA: Studies on the Hypotensive Constituents of *Morus* Root Bark, 1, 12th Annual Meeting of Pharmaceutical Society of
   Japan, Abstract Papers, 289, April, 1959, Osaka, Japan.
9. TANEMURA, I.: Studies on the Hypotensive Constituents of *Morus* Root Bark. Nihon
   Yakurigaku Zasshi (Folia Pharmacol. Japan) **56**, 704 (1960).
10. HSU, C.-S.: Studies on the Hypotensive Effect of Some Extracts of Cortex Mori
    Radicis Produced in China and Japan. Kansai Ika Daigaku Zasshi (The J. of Kansai
    Medical Univ.) **16**, 110 (1964).
11. YAMATAKE, Y., M. SHIBATA, and M. NAGAI: Pharmacological Studies on Root Bark
    of the Mulberry Tree (*Morus alba* L.). Japan J. Pharmacol. **26**, 461 (1976).
12. TSUKAMOTO, T., and T. OHTAKI: Components of a Mulberry Bark. I. Yakugaku
    Zasshi (J. Pharmaceutical Soc. Japan) **68**, 287 (1948).

13. Kondo, Y., and T. Takemoto: A New Diglyceride from Root-Barks of *Morus alba* l. Chem. Pharm. Bull. (Japan) **21**, 2265 (1973).
14. Yagi, M., T. Kouno, Y. Aoyagi, and H. Murai: The Structure of Moranoline, a Piperidine Alkaloid from *Morus* Species. Nippon Nogeikagaku Kaishi (J. Agri. Chem. Soc. Japan) **50**, 571 (1976).
15. Uno, T.: Isolation of Umbelliferone and Scopoletin from the Root Bark of the Mulberry Tree. Sanshi Shikenjo Hokoku (Bull. Sericul. Exp. Sta.) **24**, 437 (1970).
16. Uno, T., A. Isogai, A. Suzuki, and A. Shirata: Isolation and Identification of Ethyl β-Resorcylate (Ethyl 2,4-Dihydroxybenzoate) and 5,7-Dihydroxychromone from the Root Bark of the Mulberry Tree (*Morus alba* l.) and Their Biological Activity. Nihon Sanshigaku Zasshi (J. Sericul. Sci. Japan) **50**, 422 (1981).
17. Shibata, H., I. Mikoshiba, and S. Shimizu: Isolation of β-Tocopherol from the Root Bark of The Mulberry Tree. Agric. Biol. Chem. **38**, 1745 (1974).
18. Venkataraman, K.: Wood Phenolics in the Chemotaxonomy of the Moraceae. Phytochemistry **11**, 1571 (1972).
19. Venkataraman, K.: Recent Work on Some Natural Phenolic Pigments. Recent Dev. Chem. Nat. Carbon Comp. **7**, 39 (1976).
20. Deshpande, V.H., P.C. Parthasarathy, and K. Venkataraman: Four Analogues of Artocarpin and Cycloartocarpin from *Morus alba*, Tetrahedron Letters 1715 (1968).
21. Deshpande, V.H., P.V. Wakharkar, and A.V. Rama Rao: Wood Phenolics of *Morus* Species: Part V – Isolation of a New Flavone, Mulberranol and a Novel Phenol, Alboctalol from *Morus alba*. Indian J. Chem. **14B**, 647 (1976).
22. Deshpande, V.H., A.V. Rama Rao, K. Venkataraman, and P.V. Wakharkar: Wood Phenolics of *Morus* Species: Part III – Phenolic Constituents of *Morus rubra* Bark. Indian J. Chem. **12**, 431 (1974).
23. Dave, K.G., and K. Venkataraman: The Colouring Matters of the Wood of *Artocarpus integrifolia*: Part I – Artocarpin. J. Sci. Industr. Res. **15B**, 183 (1956).
24. Dave, K.G., R. Mani, and K. Venkataraman: The Colouring Matters of the Wood of *Artocarpus integrifolia*: Part III – Constitution of Artocarpin and Synthesis of Tetrahydroartocarpin Dimethyl Ether. J. Sci. Industr. Res. **20B**, 112 (1961).
25. Radhakrishnan, P.V., and A.V. Rama Rao: Colouring Matters of the Wood of *Artocarpus heterophyllus*: Part IV – Constitution of Artocarpesin and Norartocarpetin, and Synthesis of Dihydroartocarpesin Tetramethyl Ether. Indian J. Chem. **4**, 406 (1966).
26. Parthasarathy, P.C., P.V. Radhakrishnan, S.S. Rathi, and K. Venkataraman: Colouring Matters of the Wood of *Artocarpus heterophyllus*: Part V – Cycloartocarpesin and Oxydihydroartocarpesin, Two New Flavones. Indian J. Chem. **7**, 101 (1969).
27. Rama Rao, A.V., M. Varadan, and K. Venkataraman: Colouring Matters of the Wood of *Artocarpus heterophyllus*: Part VI – Cycloheterophyllin, a Flavone Linked to Three Isoprenoid Groups. Indian J. Chem. **9**, 7 (1971).
28. Nair, P.M., A.V. Rama Rao, and K. Venkataraman: Cycloartocarpin, Tetrahedron Letters 125 (1964).
29. Rama Rao, A.V., M. Varadan, and K. Venkataraman: Colouring Matters of the Wood of *Artocarpus heterophyllus*: Part VII – Isocycloheterophyllin, a New Flavone. Indian J. Chem. **11**, 298 (1973).
30. Rama Rao, A.V., S.S. Rathi, and K. Venkataraman: Chaplasin, a Flavone Containing an Oxepine Ring from the Heartwood of *Artocarpus chaplasha* Roxb. Indian J. Chem. **10**, 905 (1972).
31. Pendse, A.D., R. Pendse, A.V. Rama Rao, and K. Venkataraman: Integrin, Cyclointegrin and Oxyisocyclointegrin, Three New Flavones from the Heartwood of *Artocarpus integer*. Indian J. Chem. **14B**, 69 (1976).

32. NOMURA, T., T. FUKAI, S. YAMADA, and M. KATAYANAGI: Studies on the Constituents of the Cultivated Mulberry Tree I. Three New Prenylflavones from the Root Bark of *Morus alba* L. Chem. Pharm. Bull. (Japan) **26**, 1394 (1978).

33. NOMURA, T., and T. FUKAI: Kuwanon G, a New Flavone Derivative from the Root Barks of the Cultivated Mulberry Tree (*Morus alba* L.). Chem. Pharm. Bull. (Japan) **28**, 2548 (1980).

34. NOMURA, T., T. FUKAI, and T. NARITA: Hypotensive Constituent, Kuwanon H, a New Flavone Derivative from the Root Bark of the Cultivated Mulberry Tree (*Morus alba* L.). Heterocycles **14**, 1943 (1980).

35. FUKAI, T., Y. HANO, K. HIRAKURA, T. NOMURA, J. UZAWA, and K. FUKUSHIMA: Structures of Mulberrofurans F and G, Two Natural Hypotensive Diels-Alder Type Adducts from the Cultivated Mulberry Tree (*Morus lhou* (Ser.) Koidz.). Heterocycles **22**, 473 (1984).

36. MABRY, T.J., K.R. MARKHAM, and M.B. THOMAS: The Systematic Identification of Flavonoids, Chapter V. New York: Springer 1970.

37. ARNONE, A., G. CARDILLO, L. MERLINI, and R. MONDELLI: Natural Chromens V. NMR Effects of Acetylation and Long-Range Coupling as a Tool for Structural Elucidation of Hydroxychromenes, Tetrahedron Letters 4201 (1967).

38. KONNO, C., Y. OSHIMA, and H. HIKINO: Morusinol, Isoprenoid Flavone from *Morus* Root Barks. Planta medica **32**, 118 (1977).

39. MARKHAM, K.R., and V.M. CHARI (with T.J. Mabry, Section 2.5): Carbon-13 NMR Spectroscopy of Flavonoids In: The Flavonoids, Advances in Research (HARBORNE, J.B., and T.J. MABRY eds.), p 19. New York: Chapman and Hall. 1982.

40. NOMURA, T., T. FUKAI, and M. KATAYANAGI: Studies on the Constituents of the Cultivated Mulberry Tree III. Isolation of Four New Flavones, Kuwanon A, B, C, and Oxydihydromorusin from the Root Bark of *Morus alba* L. Chem. Pharm. Bull. (Japan) **26**, 1453 (1978).

41. NOMURA, T., T. FUKAI, S. YAMADA, and M. KATAYANAGI: Studies on the Constituents of the Cultivated Mulberry Tree II. Photooxidative Cyclization of Morusin. Chem. Pharm. Bull. (Japan) **26**, 1431 (1978).

42. NOMURA, T., and T. FUKAI: Constituents of the Cultivated Mulberry Tree. VII. Isolation of Three New Isoprenoid Flavanones, Kuwanon D, E, and F from Root Bark of *Morus alba* L. Planta medica **42**, 79 (1981).

43. HOROWITZ, R.M.: Detection of Flavanones by Reduction with Sodium Borohydride. J. Org. Chem. **22**, 1733 (1957).

44. JEFFERIES, P.R., and G.K. WORTH: The Chemistry of the Western Australian Rutaceae – VI. Two Novel Coumarins from *Eriostemon brucei*. Tetrahedron **29**, 903 (1973).

45. BEGLEY, M.J., L. CROMBIE, R.W. KING, D.A. SLACK, and D.A. WHITING: Chromens and Citrans Derived from Phloroacetophenone and Phloroglucinaldehyde by Citral Condensation: Regioselectivity, Mechanism, and X-ray Crystal Structures. J. Chem. Soc. Perkin I 2393 (1977).

46. CROMBIE, L., and R. PONSFORD: Synthesis of Cannabinoids by Pyridine-Catalysed Citral-Olivetol Condensation: Synthesis and Structure of Cannabicyclol, Cannabichromen, (Hashish Extractives), Citrylidene-Cannabis, and Related Compounds. J. Chem. Soc. (C) 796 (1971).

47. FUKAI, T., Y. HANO, K. HIRAKURA, T. NOMURA, and J. UZAWA: Constituents of the Cultivated Mulberry Tree XXVII. Structures of a Novel 2-Arylbenzofuran Derivative and Two Flavone Derivatives from the Cultivated Mulberry Tree (*Morus lhou* Koidz.). Chem. Pharm. Bull. (Japan) **33**, 4288 (1985).

48. NOMURA, T.: Phenolic Constituents of the Root Barks of the Mulberry Tree, 20th Symposium on Phytochemistry, Abstract Papers, p 1, Jan. 1984, Tokyo, Japan.

49. NOMURA, T., and T. FUKAI: Prenylflavonoids from the Root Bark of the Cultivated Mulberry Tree. Heterocycles 15, 1531 (1981).
50. NOMURA, T., Y. SAWAURA, T. FUKAI, S. YAMADA, and S. TAMURA: Studies on the Constituents of the Cultivated Mulberry Tree V. The Synthesis of Tetrahydro-kuwanon C Tetramethyl Ether. Heterocycles 9, 1355 (1978).
51. FINNEGAN, R.A., B. GILBERT, E.J. EISENBRAUN, and C. DJERASSI: Naturally Occurring Oxygen Heterocycles VIII. Synthesis of Some Coumarins Related to Mammein. J. Org. Chem. 25, 2169 (1960).
52. a) MAHAL, H.S., and K. VENKATARAMAN: Synthetical Experiments in the Chromone Group. Part XIV. The Action of Sodamide on 1-Acyloxy-2-acetonaphthones. J. Chem. Soc. 1767 (1934);
    b) NAKAZAWA, K., and T. MIYATA: Synthetic Methods of Organic Compounds (The Society of Synthetic Organic Chemistry Japan, ed.), 13, 111, Tokyo: Giho-do. 1966.
53. CHARI, V.M., S. AHMAD, and B.-G. ÖSTERDAHL: ^{13}C NMR Spectra of Chromeno- and Prenylated Flavones. Structure Revision of Mulberrin, Mulberrochromene, Cyclomulberrin and Cyclomulberrochromene. Z. Naturforsch. 33b, 1547 (1978).
54. WENKERT, E., and H.E. GOTTLIEB: Carbon-13 Nuclear Magnetic Resonance Spectroscopy of Naturally Occurring Substances 49. Carbon-13 Nuclear Magnetic Resonance Spectroscopy of Flavonoid and Isoflavonoid Compounds. Phytochemistry 16, 1811 (1977).
55. NOMURA, T., and T. FUKAI: On the Structures of Mulberrin, Mulberrochromene, Cyclomulberrin, and Cyclomulberrochromene. Heterocycles 12, 1289 (1979).
56. BARTLETT, P.D., and R.R. HIATT: A Series of Tertiary Butyl Peresters Showing Concerted Decomposition. J. Am. Chem. Soc. 80, 1398 (1958).
57. CROMBIE, L., D.E. GAMES, N.J. HASKINS, and G.F. REED: Extractives of Mammea americana L. Part III. Identification of New Coumarin Relatives of Mammea B/BA, B/BB, and B/BC Having 5,6-Annulation and Higher Oxidation Levels. J. Chem. Soc. Perkin I 2241 (1972).
58. MATSUURA, T., and H. MATSUSHIMA: Photoinduced Reactions-XXII Photooxidative Cyclozation of 3-Methoxyflavones. Tetrahedron 24, 6615 (1968).
59. NAKASHIMA, R., K. OKAMOTO, and T. MATSUURA: Photoinduced Reactions. XCIV. Photoreactions of Flavanones. Bull. Chem. Soc. Jpn. 49, 3355 (1976).
60. NOMURA, T., T. FUKAI, and M. AMAGAI: The Photoreaction of Morusin Trimethyl Ether. Heterocycles 12, 1529 (1979).
61. NOMURA, T., and T. FUKAI: Studies on the Constituents of the Cultivated Mulberry Tree IV. On the Reaction Mechanism of Photo-oxidative Cyclization of Morusin. Heterocycles 9, 635 (1978).
62. MATSUURA, T., and I. SAITO: Photooxidation of Heterocyclic Compounds In: Photochemistry of Heterocyclic Compounds (BUCHARDT, O., ed.), p. 456, New York: John Wiley and Sons, 1976.
63. a) STENBERG, V.I., R.D. OLSON, C.T. WANG, and N. KULEVSKY: The Role of Charge-Transfer Complexes in the Photooxidation of Ethers with Oxygen. J. Org. Chem. 32, 3227 (1967);
    b) MAEDA, K., A. NAKANE, and H. TSUBOMURA: The Photo-oxygenation Reactions of Diethyl Ether, p-Phenylenediamine, and N,N-Dimethylaniline. Bull. Chem. Soc. Jpn. 48, 2448 (1975);
    c) TSUBOMURA, H., and M. HORI: Electronic Structures and Reactivities of Excited Oxygen Molecules, Yuki Gosei Kagaku Kyokai Shi (J. Syn. Org. Chem. Jpn.) 26, 929 (1968).
64. SHANI, A., and R. MECHOULAM: Cannabielsoic acids, Isolation and Synthesis by a Novel Oxidative Cyclization, Tetrahedron 30, 2437 (1974).
65. NOMURA, T., T. FUKAI, and M. KATAYANAGI: Oxidative Cyclization of Morusin with Manganese Dioxide. Heterocycles 6, 1847 (1977).

66. NOMURA, T., and T. FUKAI: The Photo-sensitized Oxidation of Morusin. Heterocycles **8**, 443 (1977).
67. NOMURA, T., T. FUKAI, and J. MATSUMOTO: Oxidative Cyclization of Morusin, J. Heterocyclic Chem. **17**, 641 (1980).
68. OUANNÈS, C., and T. WILSON: Quenching of Singlet Oxygen by Tertiary Aliphatic Amines. Effect of DABCO. J. Am. Chem. Soc. **90**, 6527 (1968).
69. TAKASUGI, M.: Phytoalexins Produced by Mulberry Tree. Kagaku to Seibutsu **19**, 161 (1981).
70. MASAMUNE, T., M. TAKASUGI, and A. MURAI: On the Chemistry of Phytoalexins, Yuki Gosei Kagaku Kyokai Shi (J. Syn. Org. Chem. Jpn.) **43**, 217 (1985).
71. SHIRATA, A., K. TAKAHASHI, M. TAKASUGI, S. NAGAO, S. ISHIKAWA, S. UENO, L. MUÑOZ, and T. MASAMUNE: Antimicrobial Spectra of the Compounds from Mulberry. Sanshi Shikenjo Hokoku (Bull. Sericul. Exp. Sta.) **28**, 793 (1983).
72. TAKASUGI, M., S. NAGAO, L. MUÑOZ, S. ISHIKAWA, T. MASAMUNE, A. SHIRATA, and K. TAKAHASHI: The Structure of Phytoalexins Produced in Diseased Mulberry. 22nd Symposium on the Chemistry of Natural Products, Symposium Paper, p 275, Oct., 1979, Fukuoka, Japan.
73. TAKASUGI, M., S. ISHIKAWA, T. MASAMUNE, A. SHIRATA, and K. TAKAHASHI: Antifungal Substances Produced in the Epidermis of Mulberry Shoots, 42nd Annual Meeting of the Chemical Society of Japan, Abstract Papers, p 352, Sept., 1980, Sendai, Japan.
74. TAKASUGI, M., S. ISHIKAWA, S. NAGAO, T. MASAMUNE, A. SHIRATA, and K. TAKAHASHI: Studies on Phytoalexins of the Moraceae 8. Albanins F and G, Natural Diels-Alder Adducts from Mulberry. Chem. Letters 1577 (1980).
75. OSHIMA, Y., C. KONNO, H. HIKINO, and K. MATSUSHITA: Structure of Moracenin B, a Hypotensive Principle of *Morus* Root Barks. Tetrahedron Letters **21**, 3381 (1980).
76. OSHIMA, Y., C. KONNO, H. HIKINO, and K. MATSUSHITA: Structure of Moracenin A, a Hypotensive Principle of *Morus* Root Barks. Heterocycles **14**, 1287 (1980).
77. TAKASUGI, M., S. ISHIKAWA, T. MASAMUNE, A. SHIRATA, and K. TAKAHASHI: Structure of Antifungal Compounds, Albanin H and Albafuran C, from the Epidermis of Mulberry Shoots, 43rd Annual Meeting of the Chemical Society of Japan, Abstract Papers, p 718, Apr., 1981.
78. TAKASUGI, M., S. ISHIKAWA, and T. MASAMUNE: Studies on Phytoalexins of the Moraceae 11. Albafurans A and B, Geranyl 2-Phenylbenzofurans from Mulberry. Chem. Letters 1221 (1982).
79. TAKASUGI, M., S. ISHIKAWA, S. NAGAO, and T. MASAMUNE: Studies on Phytoalexins of the Moraceae 12. Albafuran C, a Natural Diels-Alder Adduct of a Dehydroprenyl-2-phenylbenzofuran with a Chalcone from Mulberry. Chem. Letters 1223 (1982).
80. TAKASUGI, M., L. MUÑOZ, T. MASAMUNE, A. SHIRATA, and K. TAKAHASHI: Studies on Phytoalexins of the Moraceae 3. Stilbene Phytoalexins from Diseased Mulberry. Chem. Letters 1241 (1978).
81. TAKASUGI, M., S. NAGAO, T. MASAMUNE, A. SHIRATA, and K. TAKAHASHI: Structure of Moracin A and B, New Phytoalexins from Diseased Mulberry. Tetrahedron Letters 797 (1978).
82. TAKASUGI, M., S. NAGAO, S. UENO, T. MASAMUNE, A. SHIRATA, and K. TAKAHASHI: Studies on Phytoalexins of Moraceae 2. Moracin C and D, New Phytoalexins from Diseased Mulberry. Chem. Letters 1239 (1978).
83. TAKASUGI, M., S. NAGAO, T. MASAMUNE, A. SHIRATA, and K. TAKAHASHI: Studies on Phytoalexins of Moraceae 4. Structures of Moracins E, F, G, and H, New Phytoalexins from Diseased Mulberry. Tetrahedron Letters 4675 (1979).
84. TAKASUGI, M., S. NAGAO, and T. MASAMUNE: Studies on Phytoalexins of Mora-

ceae 10. Structure of Dimoracin, a New Natural Diels-Alder Adduct from Diseased Mulberry. Chem. Letters 1217 (1982).

85. BURKE, J.M., and R. STEVENSON: Natural Benzofurans. Synthesis of Moracin A and B. J. Chem. Research (S) 34 (1985).

86. CLOUGH, J.M., I.S. MANN, and D.A. WIDDOWSON: Transition Metal Mediated Organic Synthesis: The Synthesis of Moracin M. Tetrahedron Letters 28, 2645 (1987).

87. TAKASUGI, M., S. NAGAO, T. MASAMUNE, A. SHIRATA, and K. TAKAHASHI: Studies on Phytoalexins of the Moraceae 7. Chalcomoracin, a Natural Diels-Alder Adduct from Diseased Mulberry. Chem. Letters 1573 (1980).

88. NOMURA, T., T. FUKAI, J. UNO, and T. ARAI: Mulberrofuran A, a New Isoprenoid 2-Arylbenzofuran from the Root Bark of the Cultivated Mulberry Tree (*Morus alba* L.). Heterocycles 9, 1593 (1978).

89. FUKAI, T., T. FUJIMOTO, Y. HANO, T. NOMURA, and J. UZAWA: Constituents of the Cultivated Mulberry Tree XXII. Structures of Mulberrofurans B and L, 2-Arylbenzofuran Derivatives from the Root Bark of the Cultivated Mulberry Tree (*Morus lhou* Koidz.). Heterocycles 22, 2805 (1984).

90. NOMURA, T., T. FUKAI, T. SHIMADA, and I.-S. CHEN: Constituents of the Cultivated Mulberry Tree XIII. Components of Root Bark of *Morus australis*. 1. Structure of a New 2-Arylbenzofuran Derivative, Mulberrofuran D. Planta medica 49, 90 (1983).

91. HANO, Y., H. KOHNO, M. ITOH, and T. NOMURA: Constituents of the Cultivated Mulberry Tree XXIX. Constituents of the Chinese Crude Drug "Sang-Bai-Pi" (*Morus* Root Bark) VII. Structures of Three New 2-Arylbenzofuran Derivatives from the Chinese Crude Drug "Sang-Bai-Pi" (*Morus* Root Bark). Chem. Pharm. Bull. (Japan) 33, 5294 (1985).

92. HIRAKURA, K., T. FUJIMOTO, T. FUKAI, and T. NOMURA: Two Phenolic Glycosides from the Root Bark of the Cultivated Mulberry Tree (*Morus lhou*). J. Nat. Prod. 49, 218 (1986).

93. NOMURA, T.: Constituents of the Root Bark of the Mulberry Tree. Kagaku no Ryoiki 36, 596 (1982).

94. MOMOSE, Y., and T. NOMURA: Effect of the Components of *Morus* Root Bark on Blood Pressure, 99th Annual Meeting of Pharmaceutical Socity of Japan, Abstract Paper, p 162, Aug., 1979, Sapporo, Japan.

95. NOMURA, T., T. FUKAI, Y. MOMOSE, and R. TAKEDA: Hypotensive Constituents of the Root Bark of the Mulberry Tree (*Morus alba* L.) and the Mechanism of Their Actions, Third Symposium on the Development and Application of Naturally Occurring Drug Materials, Symposium Paper, p 13, Aug., 1980, Tokyo, Japan.

96. NOMURA, T., T. FUKAI, T. NARITA, S. TERADA, J. UZAWA, Y. IITAKA, M. TAKASUGI, S. ISHIKAWA, S. NAGAO, and T. MASAMUNE: Confirmation of the Structures of Kuwanons G and H (Albanins F and G) by Partial Synthesis. Tetrahedron Letters 22, 2195 (1981).

97. NOMURA, T., T. FUKAI, Y. MOMOSE, and T. NARITA: Structures of the Hypotensive Constituents of the Root Bark of the Mulberry Tree (*Morus alba* L.), 23rd Symposium on the Chemistry of Natural Products, Symposium Paper, p 552, Oct., 1980, Nagoya, Japan.

98. IMASHIMIZU, A.: Studies on the Components of *Morus* Root Bark, Master's thesis, Toho University, p 23 (1982).

99. KUTNEY, J.P., T. INABA, and D.L. DREYER: The Structure of Thamnosin. A Novel Dimeric Coumarin System. J. Am. Chem. Soc. 90, 813 (1968).

100. BELL, A.A., R.D. STIPANOVIC, D.H. O'BRIEN, and P.A. FRYXELL: Sesquiterpenoid Aldehyde Quinones and Derivatives in Pigment Glands of *Gossypium*. Phytochemistry 17, 1297 (1978).

101. MORI, I., Y. NAKACHI, K. UEDA, D. UEMURA, and Y. HIRATA: Isolation and Structure of Alflabene from *Alpinia flabellata* Ridl. Tetrahedron Letters 2291 (1978).

102. JURD, L., R.Y. WONG, and M. BENSON: The Structures of Paraensidimerin A and C, Two Bisquinolinone Alkaloids from *Euxylophora paraensis*. Aust. J. Chem. **35**, 2505 (1982).

103. OSHIMA, Y., C. KONNO, H. HIKINO, and K. MATSUSHITA: Structure of Moracenin C, a Hypotensive Principle of *Morus* Root Barks. Heterocycles **14**, 1461 (1980).

104. OSHIMA, Y., C. KONNO, and H. HIKINO: Structure of Moracenin D, a Hypotensive Principle of *Morus* Root Barks. Heterocycles **16**, 979 (1981).

105. NOMURA, T., T. FUKAI, E. SATO, and K. FUKUSHIMA: The Formation of Moracenin D from Kuwanon G. Heterocycles **16**, 983 (1981).

106. SAUER, J.: Diels-Alder-Reaktionen I: Präparative Aspekte. Angew. Chem. **78**, 233 (1966).

107. SAUER, J.: Diels-Alder-Reaktionen II: Zum Reaktionsmechanismus. Angew. Chem. **79**, 76 (1967).

108. TAKEUCHI, S., J. UZAWA, H. SETO, and H. YONEHARA: New ^{13}C-NMR Techniques Applied to the Pentalenolactone Structure. Tetrahedron Letters 2943 (1977).

109. SETO, H., T. SASAKI, H. YONEHARA, and J. UZAWA: Studies on the Biosynthesis of Pentalenolactone. Part I. Application of Longrange Selective Proton Decoupling (LSPD) and Selective ^{13}C-{^{1}H} NOE in the Structure Elucidation of Pentalenolactone G. Tetrahedron Letters 923 (1978).

110. UZAWA, J., T. NOMURA, and T. FUKAI: LSPD Technique on Kuwanon G, H, and Related Compounds, 101st Annual Meeting of Pharmaceutical Society of Japan, Abstract Paper, p 509, Apr., 1981, Kumamoto, Japan.

111. NOMURA, T., T. FUKAI, J. MATSUMOTO, and T. OHMORI: Constituents of the Cultivated Mulberry Tree. VIII. Components of Root Barks of *Morus bombycis*. Planta medica **46**, 28 (1982).

112. FUKAI, T., Y. HANO, K. HIRAKURA, T. NOMURA, J. UZAWA, and K. FUKUSHIMA: Constituents of the Cultivated Mulberry Tree XXV. Structures of Two Natural Hypotensive Diels-Alder Type Adducts, Mulberrofurans F and G, from the Cultivated Mulberry Tree (*Morus lhou* Koidz.). Chem. Pharm. Bull. (Japan) **33**, 3195 (1985).

113. SHERIF, E.A., R.K. GUPTA, and M. KRISHNAMURTI: Anomalous AlCl$_3$ Induced U.V. Shift of C-Alkylated Polyphenols. Tetrahedron Letters **21**, 641 (1980).

114. HIRAKURA, K., Y. HANO, T. FUKAI, T. NOMURA, J. UZAWA, and K. FUKUSHIMA: Constituents of the Cultivated Mulberry Tree XXI. Structure of Three New Natural Diels-Alder Type Adducts, Kuwanons P and X, and Mulberrofuran J, from the Cultivated Mulberry Tree (*Morus lhou* Koidz.). Chem. Pharm. Bull. (Japan) **33**, 1088 (1985).

115. RAMA RAO, A.V., V.H. DESHPANDE, R.K. SHASTRI, S.S. TAVALE, and N.N. DHANESHWAR: Structures of Albanols A and B, Two Novel Phenols from *Morus alba* Bark, Tetrahedron Letters **24**, 3013 (1983).

116. HANO, Y., T. FUKAI, T. NOMURA, J. UZAWA, and K. FUKUSHIMA: Structure of Mulberrofuran I, a Novel 2-Arylbenzofuran Derivative from the Cultivated Mulberry Tree (*Morus bombycis* Koidz.). Chem. Pharm. Bull. (Japan) **32**, 1260 (1984).

117. HANO, Y., T. FUKAI, H. TSUBURA, and T. NOMURA: Some Pigments and Related Compounds of *Morus* Root Bark, 27th Symposium on the Chemistry of Natural Products, Abstract Paper, p 710, Oct. Hiroshima, Japan, 1985.

118. HANO, Y., M. ITOH, and T. NOMURA: Structures of Kuwanols A and B, Two Novel Stilbene Derivatives from the Cultivated Mulberry Tree (*Morus bombycis* Koidz.). Heterocycles **23**, 819 (1985).

119. HANO, Y., and T. NOMURA: Constituents of the Cultivated Mulberry Tree XXXVI.

Structure of Mulberrofuran P, a Novel 2-Arylbenzofuran Derivative from the Cultivated Mulberry Tree (*Morus alba* L.). Heterocycles **24**, 1381 (1986).

120. Hano, Y., K. Hirakura, T. Someya, and T. Nomura: Structure of Mulberrofuran
M, a Novel 2-Arylbenzofuran Derivative from the Cultivated Mulberry Tree (*Morus
alba* L.). Heterocycles **24**, 1251 (1986).

121. Hano, Y., H. Tsubura, and T. Nomura: Structure of Mulberrofuran Q, a Novel
2-Arylbenzofuran Derivative from the Cultivated Mulberry Tree (*Morus alba* L.),
Heterocycles **24**, 1807 (1986).

122. Uzawa, J., and S. Takeuchi: Application of Selective ^{13}C-$\{^{1}$H$\}$ Nuclear Overhauser
Effects with Low-power ^{1}H-Irradiation in Carbon-13 NMR Spectroscopy. Org. Mag.
Reson. **11**, 502 (1978).

123. Hano, Y., H. Tsubura, and T. Nomura: Constituents of the Cultivated Mulberry
Tree XXXVIII. Structures of Kuwanons Y and Z, Two New Stilbene Derivatives
from the Cultivated Mulberry Tree (*Morus alba* L.). Heterocycles **24**, 2603 (1986).

124. Afzal, M., and G. Al-Oriquat: Biosynthesis of Isoflavonoid and Related Phytoalexins. Heterocycles **19**, 1295 (1982).

125. Narasimhan, R., B. Dhruva, S.V. Paranjpe, D.D. Kulkarni, A.F. Mascarenhas,
and S.B. Davis: Tissue Culture of Some Woody Species. Proc. Indian Acad. Sci.
71 B, 204 (1970).

126. Kulkarni, D.D., D.D. Ghugale, and R. Narasimhan: Chemical Investigations
of Plant Tissues Grown *in vitro*. Isolation of β-Sitosterol from *Morus alba* Callus
Tissue. Indian J. Exp. Biol. **8**, 347 (1970).

127. Seki, H., M. Takeda, K. Tsutsumi, and Y. Ushiki: Callus Culture of the Mulberry
Tree. I. Effect of Concentrations of Auxin and Kinetin on the Callus Culture of
the Mulberry Stem. Nihon Sanshigaku Zasshi (J. Sericul. Sci. Japan) **40**, 81 (1971).

128. Ueda, S., K. Inoue, Y. Shiobara, I. Kimura, and H. Inouye: Quinones and Related
Compounds in Higher Plants. X. Naphtoquinone Derivatives of the Callus Culture
of *Catalpa ovata*. Planta medica **40**, 168 (1980).

129. Ueda, S., T. Nomura, T. Fukai, and J. Matsumoto: Kuwanon J, a New Diels-Alder
Adduct and Chalcomoracin from Callus Culture of *Morus alba* L. Chem. Pharm.
Bull. (Japan) **30**, 3042 (1982).

130. Ueda, S., J. Matsumoto, and T. Nomura: Four New Natural Diels-Alder Type
Adducts. Mulberrofuran E, Kuwanon Q, R, and V from Callus Culture of *Morus
alba* L. Chem. Pharm. Bull. (Japan) **32**, 350 (1984).

131. Ikuta (nee Matsumoto), J., T. Fukai, T. Nomura, and S. Ueda: Constituents
of the Cultivated Mulberry Tree XXXV. Constituents of *Morus alba* L. Cell Cultures.
(1). Structures of Four New Natural Diels-Alder Type Adducts, Kuwanons J, Q,
R, and V. Chem. Pharm. Bull. (Japan) **34**, 2471 (1986).

132. Nomura, T., T. Fukai, J. Matsumoto, A. Imashimizu, S. Terada, and M. Hama:
Constituents of the Cultivated Mulberry Tree X. Structure of Kuwanon I, a New
Natural Diels-Alder Adduct from the Root Bark of *Morus alba*. Planta medica
46, 167 (1982).

133. Nomura,T., T. Fukai, Y. Hano, K. Nemoto, S. Terada, and T. Kuramochi: Constituents of the Cultivated Mulberry Tree XII. Isolation of Two New Natural Diels-
Alder Adducts from Root Bark of *Morus alba*. Planta medica **47**, 151 (1983).

134. Hano, Y., K. Hirakura, T. Nomura, S. Terada, and K. Fukushima: Constituents
of the Cultivated Mulberry Tree XVI. Components of Root Bark of *Morus lhou*.
1. Structures of Two New Natural Diels-Alder Adducts, Kuwanons N and O. Planta
medica **50**, 127 (1984).

135. Hirakura, K., T. Fukai, Y. Hano, and T. Nomura: Constituents of the Cultivated
Mulberry Tree 20. Constituents of the Root Bark of *Morus lhou*. 2. Kuwanon W,
a Natural Diels-Alder Type Adduct from the Root Bark of *Morus lhou*. Phytochemistry **24**, 159 (1985).

136. NOMURA, T., T. FUKAI, Y. HANO, and H. IKUTA: Kuwanon M, a New Diels-Alder Adduct from the Root Barks of the Cultivated Mulberry Tree (*Morus lhou* (Ser) Koidz.). Heterocycles **20**, 585 (1983).

137. KOHNO, H., T. TAKABA, T. FUKAI, and T. NOMURA: Constituents of the Cultivated Mulberry Tree XXXIX. Structure of Mulberrofuran R, a Novel 2-Arylbenzofuran Derivative from the Cultivated Mulberry Tree (*Morus lhou* Koidz.). Heterocycles **26**, 759 (1987).

138. HIRAKURA, K., I. SAIDA, T. FUKAI, and T. NOMURA: Mulberroside B, a New C-Glucosylcoumarin from the Cultivated Mulberry Tree (*Morus lhou* Koidz.). Heterocycles **23**, 2239 (1985).

139. NOMURA, T., T. FUKAI, and Y. HANO: Constituents of the Cultivated Mulberry Tree IX. Constituents of the Chinese Crude Drug "Sang-Bai-Pi" (*Morus* Root Bark) I. Structure of a New Flavanone Derivative, Sanggenon A, Planta medica **47**, 30 (1983).

140. NOMURA, T., T. FUKAI, Y. HANO, and J. UZAWA: Structure of Sanggenon C, a Natural Hypotensive Diels-Alder Adduct from Chinese Crude Drug "Sang-Bai-Pi" (*Morus* Root Barks). Heterocycles **16**, 2141 (1981).

141. NOMURA, T., T. FUKAI, Y. HANO, and J. UZAWA: Structure of Sanggenon D, a Natural Hypotensive Diels-Alder Adduct from Chinese Crude Drug "Sang-Bai-Pi" (*Morus* Root Barks). Heterocycles **17**, 381 (1982).

142. SUN, J.-Y., J.-Y. LOU, S. SUZUKI, Y. HANO, and T. NOMURA: On the Components of the Chinese *Morus* Root Bark. 2., 34th Annual Meeting of the Japanese Society of Pharmacognosy, Abstract Papers, p 155, Oct., 1987, Osaka, Japan.

143. HANO, Y., M. ITOH, N. KOYAMA, and T. NOMURA: Constituents of the Cultivated Mulberry Tree XIX. Constituents of the Chinese Crude Drug "Sang-Bai-Pi" (*Morus* Boot Bark) V. Structures of Three New Flavanones, Sanggenons L, M, and N. Heterocycles **22**, 1791 (1984).

144. NOMURA, T., T. FUKAI, Y. HANO, and K. TSUKAMOTO: Constituents of the Cultivated Mulberry Tree XIV. Constituents of the Chinese Crude Drug "Sang-Bai-Pi" (*Morus* Root Barks) III. Structure of a New Flavanone Derivative, Sanggenon F. Heterocycles **20**, 661 (1983).

145. HANO, Y., and T. NOMURA: Constituents of the Cultivated Mulberry Tree. XV. Constituents of the Cultivated Mulberry Tree. XV. Constituents of the Chinese Crude Drug "Sang-Bai-Pi" (*Morus* Root Barks) IV. Structures of Four New Flavonoids, Sanggenon H, I, J and K. Heterocycles **20**, 1071 (1983).

146. HANO, Y., M. ITOH, T. FUKAI, T. NOMURA, and S. URANO: Constituents of the Cultivated Mulberry Tree XXVIII. Constituents of the Chinese Crude Drug "Sang-Bai-Pi" (*Morus* Root Bark) VI. Revised Structure of Sanggenon B. Heterocycles **23**, 1691 (1985).

147. HANO, Y., H. KOHNO, S. SUZUKI, and T. NOMURA: Constituents of the Cultivated Mulberry Tree XXXVII. Constituents of the Chinese Crude Drug "Sang-Bai-Pi" (*Morus* Root Bark) VIII. Structures of Sanggenons E and P, Two New Diels-Alder Type Adducts from the Chinese Crude Drug "Sang-Bai-Pi" (*Morus* Root Bark). Heterocycles **24**, 2285 (1986).

148. FUKAI, T., Y. HANO, T. FUJIMOTO, and T. NOMURA: Structure of Sanggenon G, a New Diels-Alder Adduct from the Chinese Crude Drug "Sang-Bai-Pi" (*Morus* Root Barks). Heterocycles **20**, 611 (1983).

149. HANO, Y., and T. NOMURA: Structure of Sanggenon O, a Natural Diels-Alder Type Adduct from Chinese Crude Drug "Sang-Bai-Pi" (*Morus* Root Bark). Heterocycles **23**, 2499 (1985).

150. NOMURA, T., T. FUKAI, Y. HANO, and S. URANO: Constituents of the Cultivated Mulberry Tree XI. Constituents of the Chinese Crude Drug "Sang-Bai-Pi" (*Morus*

Root Bark). II. Structure of a New Flavanone Derivative, Sanggenon B. Planta medica **47**, 95 (1983).

151. HANO, Y., H. KOHNO, T. FUKAI, and T. NOMURA: Absolute Configuration of Diels-Alder Type Adducts from the *Morus* Root Bark, 28th Symposium on the Chemistry of Natural Products, Abstract Papers, p 1, Oct., 1986, Sendai, Japan.

152. HANO, Y., S. SUZUKI, H. KOHNO, and T. NOMURA: Absolute Configuration of Kuwanon L, a Natural Diels-Alder Type Adduct from the *Morus* Root Bark, Heterocycles **27**, 75 (1988).

153. HANO, Y., S. SUZUKI, T. NOMURA, and Y. IITAKA: Absolute Configuration of Natural Diels-Alder Type Adducts from the *Morus* Root Bark, Heterocycles, in press.

154. HARADA, N., and K. NAKANISHI: "Circular Dichroism Spectroscopy – Exciton – Coupling in Organic Stereochemistry". Tokyo: Tokyokagakudojin. 1982.

155. YANG, Z.-Y.: Identification of Cortex Mori Radicis and its Adulterants Root Barks of *Cudrania tricuspidata* and *Broussonetia papyrifera*. Chinese J. Pharmaceutical Analysis **1**, 94 (1981).

156. AKAMATSU, K.: "Wakanyaku", p 511. Tokyo: Ishiyaku Shuppan. 1970.

157. FUJIMOTO, T., Y. HANO, and T. NOMURA: Constituents of the Cultivated Mulberry Tree. XVII. Components of Root Bark of *Cudrania tricuspidata* 1. Structures of Four New Isoprenylated Xanthones, Cudraxanthones A, B, C and D. Planta medica **50**, 218 (1984).

158. FUJIMOTO, T., Y. HANO, T. NOMURA, and J. UZAWA: Constituents of the Cultivated Mulberry Tree. XVIII. Components of Root Bark of *Cudrania tricuspidata* 2. Structures of Two New Isoprenylated Flavones, Cudraflavones A and B. Planta medica **50**, 161 (1984).

159. FUJIMOTO, T., and T. NOMURA: Constituents of the Cultivated Mulberry Tree XXIV. Components of Root Bark of *Cudrania tricuspidata* 3. Isolation and Structure Studies on the Flavonoids. Planta medica 190 (1985).

160. MATSUMOTO, J., T. FUJIMOTO, C. TAKINO, M. SAITOH, Y. HANO, T. FUKAI, and T. NOMURA: Constituents of the Cultivated Mulberry Tree XXVI. Components of *Broussonetia papyrifera* (L.) Vent. 1. Structures of Two New Isoprenylated Flavonols and Two Chalcone Derivatives. Chem. Pharm. Bull. (Japan) **33**, 3250 (1985).

161. IKUTA (nee MATSUMOTO), J., Y. HANO, and T. NOMURA: Constituents of the Cultivated Mulberry Tree XXXI. Components of *Broussonetia papyrifera* (L.) Vent. 2. Structures of Two New Isoprenylated Flavans, Kazinols A and B. Heterocycles **23**, 2835 (1985).

162. FUKAI, T., J. IKUTA (nee MATSUMOTO), and T. NOMURA: Constituents of the Cultivated Mulberry Tree XXXIII. Components of *Broussonetia papyrifera* (L.) Vent. III. Structures of Two New Isoprenylated Flavonols, Broussoflavonols C and D. Chem. Pharm. Bull. (Japan) **34**, 1987 (1986).

163. IKUTA (nee MATSUMOTO), J., Y. HANO, T. NOMURA, Y. KAWAKAMI, and T. SATO: Constituents of the Cultivated Mulberry Tree XXXII. Components of *Broussonetia kazinoki* Sieb. 1. Structures of Two New Isoprenylated Flavans and Five New Isoprenylated 1,3-Diphenylpropane Derivatives. Chem. Pharm. Bull. (Japan) **34**, 1968 (1986).

164. KATO, S., Y. HANO, T. FUKAI, and T. NOMURA: Studies on the Components of *Broussonetia* sp. 3., 106th Annual Meeting of the Pharmaceutical Society of Japan, Abstract Papers, p 162, Apr., 1986, Chiba, Japan.

165. KATO, S., T. FUKAI, J. IKUTA (nee MATSUMOTO), and T. NOMURA: Constituents of the Cultivated Mulberry Tree XXXIV. Components of *Broussonetia kazinoki* Sieb. (2). Structures of Four New Isoprenylated 1,3-Diphenylpropane Derivatives, Kazinols J, L, M, and N. Chem. Pharm. Bull. (Japan) **34**, 2448 (1986).

166. KATO, S., Y. HANO, T. FUKAI, Y. KOSUGE, and T. NOMURA: Kazinol P, a Novel

Isoprenylated Spiro-compound from *Broussonetia kazinoki* Sieb. Heterocycles **24**, 2141 (1986).

167. KATO, S., Y. HANO, T. FUKAI, T. NOMURA, and J.-Y. SUN: Studies on the Components of *Broussonetia* sp. (4), 107th Annual Meeting of Pharmaceutical Society of Japan, Abstract Paper, p 348, Apr., 1987, Kyoto, Japan.

168. SHIRATA, A., K. TAKAHASHI, M. TAKASUGI, M. ANETAI, and T. MASAMUNE: Production of Phytoalexins in Shoot Cortex of Paper Mulberry and Their Antimicrobial Spectra. Sanshi Shikenjo Hokoku (Bull. Sericul. Exp. Stat.) **28**, 781 (1983).

169. TAKASUGI, M., M. ANETAI, T. MASAMUNE, A. SHIRATA, and K. TAKAHASHI: Studies on Phytoalexins of the Moraceae 5. Broussonins A and B, New Phytoalexins from Diseased Paper Mulberry. Chem. Letters 339 (1980).

170. TAKASUGI, M., Y. KUMAGAI, S. NAGAO, T. MASAMUNE, A. SHIRATA, and K. TAKAHASHI: Studies on Phytoalexins of the Moraceae. 6. The Co-occurrence of Flavan and 1,3-Diphenylpropane Derivatives in Wounded Paper Mulberry. Chem. Letters 1459 (1980).

171. TAKASUGI, M., N. NIINO, S. NAGAO, M. ANETAI, T. MASAMUNE, A. SHIRATA, and K. TAKAHASHI: Studies on Phytoalexins of the Moraceae 13. Eight Minor Phytoalexins from Diseased Paper Mulberry. Chem. Letters 689 (1984).

172. TAKASUGI, M., N. NIINO, M. ANETAI, T. MASAMUNE, A. SHIRATA, and K. TAKAHASHI: Studies on Phytoalexins of the Moraceae 14. Structure of Two Stress Metabolites, Spirobroussonin A and B, from Diseased Paper Mulberry. Chem. Letters 693 (1984).

173. ISHITSUKA, H., C. OHSAWA, T. OHIWA, I. UMEDA, and Y. SUHARA: Antipicornavirus Flavone Ro 09-0179, Antimicrob. Agents Chemother. **22**, 611 (1982).

174. NIKAIDO, T., T. OHMOTO, T. NOMURA, T. FUKAI, and U. SANKAWA: Inhibition of Adenosine 3′,5′-Cyclic Monophosphate Phosphodiesterase by Phenolic Constituents of Mulberry Tree. Chem. Pharm. Bull. (Japan) **32**, 4929 (1984).

175. KIMURA, Y., H. OKUDA, T. NOMURA, T. FUKAI, and S. ARICHI: Effects of Flavonoids and Related Compounds from Mulberry Tree on Arachidonate Metabolism in Rat Platelet Homogenates. Chem. Pharm. Bull. (Japan) **34**, 1223 (1986).

176. KIMURA, Y., H. OKUDA, T. NOMURA, T. FUKAI, and S. ARICHI: Effects of Phenolic Constituents from the Mulberry Tree on Arachidonate Metabolism in Rat Platelets. J. Nat. Prod. **49**, 639 (1986).

177. FUJIKI, H., T. HORIUCHI, K. YAMASHITA, H. HAKII, M. SUGANUMA, H. NISHINO, A. IWASHIMA, Y. HIRATA, and T. SUGIMURA: Inhibition of Tumor Promotion by Flavonoids In: Progress in Clinical and Biological Research, Vol. 213. Plant Flavonoids in Biology and Medicine. Biochemical Pharmacological, and Structure – Activity Relationships (V. CODY, E. MIDDLETON, JR., J.B. HARBORNE, eds.), p. 429. New York: Alan R. Liss, Inc. 1986.

178. NOMURA, T., T. FUKAI, and H. FUJIKI: Chemistry and Biological Activity of *Morus* Flavonoids, 2nd International Symposium of Plant Flavonoids in Biology and Medicine, Abstract Papers, p 29, Aug., 1987, Strasbourg, France.

179. NOMURA, T., T. FUKAI, Y. HANO, S. YOSHIZAWA, M. SUGANUMA, and H. FUJIKI: Chemistry and Anti-tumor Promoting Activity of *Morus* Flavonoids In: Plant Flavonoids in Biology and Medicine II. Biochemical, Cellular and Medicinal Properties. (V. CODY, E. MIDDLETON, JR. and J.B. HARBORNE, eds.). New York: Alan R. Liss, Inc. 1988, in press.

(Received October 10, 1987)

N-Hydroxyamino Acids and Their Derivatives

By Andrzej Chimiak and Maria J. Milewska
Department of Organic Chemistry
Technical University, Gdańsk, Poland

Contents

DCCI = dicyclohexylcarbodiimide, EEDQ = 1-ethoxycarbonyl-2-ethoxy-1,2-dihydroquin-oline, DEAD = diethyl azodicarboxylate, MPP = monoperphtalic acid, TFA = trifluoro-acetic acid, TEA = triethylamine, Py = pyridine, Z = benzyloxycarbonyl, trOC = trichloro-ethoxycarbonyl, Su = succinimidyl, PNP = p-nitrophenyl, Pr = propyl, Bu = butyl, He = hexyl, Ph = phenyl, Bzl = benzyl, p-HO-Bzl = p-hydroxybenzyl and other acc. IUPAC rules.

I. Introduction, Scope and Nomenclature

N-Hydroxyamino acids should, at present, be treated as a new, separate, characteristic group of amino acids. This is necessitated by the particular biological action of these compounds and their deriva-tives. Although α-N-hydroxyamino acids were first discovered by MILL-ER and PLÖCHL (*1*) as early as 1893, they have aroused particular inter-est mainly in the last twenty years.

N-Hydroxyamino acids are similar in structure to amino acids but with the hydroxyamine group replacing the amine group. Therefore α-N-hydroxyamino acids can be represented by formula (**1**) and ω-N-hydroxyamino acids by formula (**2**).

HO—NH—CR$_1$R$_2$—COOH   HO—NH—(CH$_2$)$_n$—CR$_1$(X)—COOH

(**1**)   (**2**)

R—C—N—CR$_1$R$_2$—COOH   R—C—N—(CH$_2$)$_n$—CR$_1$(X)—COOH
$\quad$‖ $\quad$| $\quad$‖ $\quad$|
$\quad$O $\quad$OH $\quad$O $\quad$OH

(**3**) R, R$_1$, R$_2$ = H, alkyl or aryl   (**4**) X = H or NH$_2$

Scheme 1

It is necessary to distinguish these compounds nomenclaturally from hydroxyamino acids which are amino acids with a hydroxyl group in the side chain, such as serine or threonine. According to IUPAC rules (*2*) α- and ω-N-hydroxyamino acids can be symbolized also by shortened formulas indicating which amine group is substituted by an hydroxyamine group. For example, formulas HO-Ala and Orn(OH) denote α-N-hydroxyalanine and δ-N-hydroxyornithine, respectively. Names of the title compounds (**1, 2**) are formed from the names of the corresponding amino acids or, at times, from less popular systemat-ic names (e.g. N-hydroxyalanine or α-N-hydroxyaminopropionic acid). Obviously, regardless of the position of the hydroxyamine function (α or ω), the chemistry of these compounds, as is the case with α- and ω-amino acids, has so many common elements that it is advisable to discuss them together.

The presence of the hydroxyamine group causes their acyl derivatives (3, 4) to be N-hydroxyamides (hydroxamic acids). Two oxygen atoms of such N-hydroxyamides can form bidentate ligand structures capable of coordinating a metal ion. This part of the structure, while retaining other important amino acid features, accounts for the fact that natural biopolymers such as peptides which incorporate α- and ω-N-hydroxyamino acid residues may act as bioligands. This aspect of N-hydroxyamino acid residues became clear as a result of the initial basic investigations on the highly specific iron transport system in microorganism cells in which siderophores (ionophores of ferric ions) participate. Many siderophores are N-acyl peptide derivatives of N-hydroxyamino acids, especially N^5-hydroxyornithine and N^6-hydroxylysine.

The significance of N-hydroxyamino acids does not end here. Their derivatives also form complexes with a series of still insufficiently investigated but biologically important metal ions such as molybdenum or vanadium, thus possibly effecting their metabolism. Finally, in investigations on the action of metalloenzymes and organic ligands as their inhibitors, the interaction of derivatives of (3) and (4) with the active centre containing, for example, a zinc atom, may, as is now presumed, determine the biological role of the title compounds and their derivatives.

Taking into account the above generalizations, it is not surprising that many naturally occurring bioactive compounds which have been isolated have been found to be N-hydroxyamino acid derivatives. Such compounds may act as herbicides, antibiotics, growth agents, enzyme inhibitors, antitumor agents, antifungal agents and siderophores. This accounts for a significant number of articles which attempt to explain the insufficiently understood biogenesis of N-hydroxyamino acid residues and the various types of their biological action which are not only restricted to the ligand-metal interactions mentioned in the previous paragraph.

A great number of important publications on the synthesis of the title compounds, their substrates and derivatives, methods of incorporating such residues into complex organic molecules as a part of the total synthesis of the latter, as well as papers on the synthesis of N-hydroxydiketopiperazine and N-hydroxypeptides have appeared in recent years. Several syntheses of analogues of naturally occurring compounds have also been described.

The present review will be devoted mainly to these synthetic aspects.

A description of physical and chemical properties of N-hydroxyamino acids and methods of their analysis will be included and brief mention will be made of the more important natural products containing N-hydroxyamino acid residues as well as the main aspects of their biogenesis.

The first brief review devoted strictly to N-hydroxyamino acids was written by CHIMIAK (3) in 1965. MØLLER (4) in "Cyanide Biology" briefly reviewed the title compounds (1) as intermediates in metabolic transformations of cyanide. A recent review in Japanese by AKIYAMA (5) devoted only two pages to this subject. The small number of articles describing the chemistry of N-hydroxyamino acids was the reason for writing this review.

II. N-Hydroxyamino Acid Residues as Fragments of Natural Products

Free N-hydroxyamino acids (1) or (2) have never been isolated from living organisms, probably because of their instability. As already mentioned, N-hydroxyamino acid residues occur only as fragments of many natural products containing the N-hydroxyamide bond. The isolation of free N^5-hydroxy-L-arginine from the fermentation medium of *Bacillus cereus* (6, 7) may be considered as an exception; however, this compound contains an "N-hydroxyguanidine" rather than an N-hydroxyamine group.

The most important classes and compounds are as follows (Scheme 2):

a) siderophores (the most numerous group): the family of ferrichromes (5), the family of rhodotorulic acid (6), mycobactines (7), derivatives of citric acid (8), fusarinines (9), pyoverdines (10) and exochelines (11),

b) antibiotics: albomycins (12), alanosine (18),

c) tumor inhibitors: hadacidin (13),

d) growth inhibitors: mycelianamide (14) and vanoxonin (15),

e) natural pigments: amavadine (16) (Scheme 4) and pulcherrimic acid (17),

f) miscellaneous: fusigen (19), connatin (20) and aspergillic acid (21).

Derivatives of phosphonic N-hydroxyamino acid analogues (22–25) (Scheme 38, page 236) are also known as naturally occurring compounds.

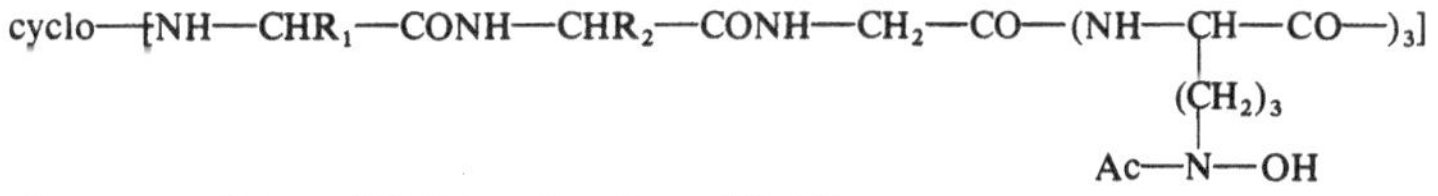

cyclo—[NH—CHR$_1$—CONH—CHR$_2$—CONH—CH$_2$—CO—(NH—CH—CO—)$_3$]
                                                                                                 |
                                                                                            (CH$_2$)$_3$
                                                                                                 |
                                                                                        Ac—N—OH

(5) R$_1$=H, CH$_3$ or CH$_2$OH   R$_2$=H or CH$_2$OH
Ac=CH$_3$CO, HO—(CH$_2$)$_2$—C(CH$_3$)=CHCO,
HOOC—CH$_2$—C(CH$_3$)=CHCO or HOOC—CH$_2$—CO (see also (56))

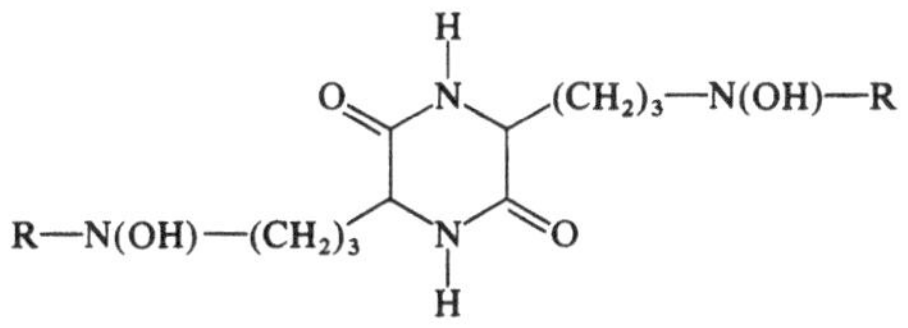

(6) R=CH$_3$CO or HO—(CH$_2$)$_2$—C(CH$_3$)=CHCO

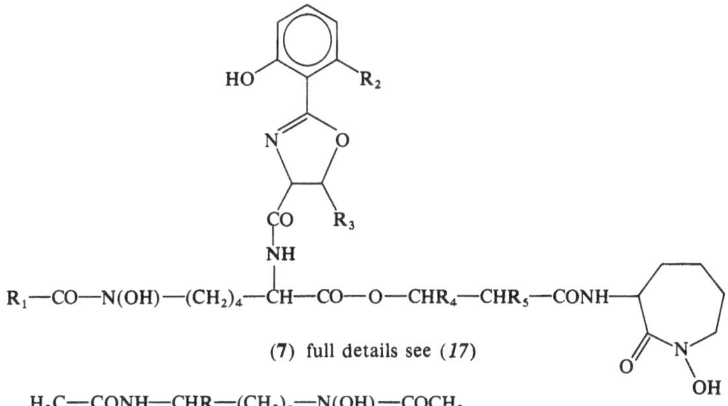

(7) full details see (17)

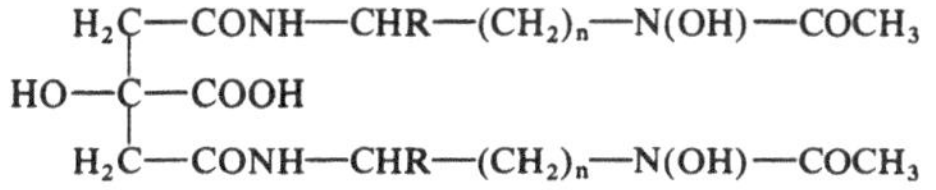

H$_2$C—CONH—CHR—(CH$_2$)$_n$—N(OH)—COCH$_3$
|
HO—C—COOH
|
H$_2$C—CONH—CHR—(CH$_2$)$_n$—N(OH)—COCH$_3$

(8) R=COOH or H   n=2 or 4

HO—[CO—CH(NH$_2$)—(CH$_2$)$_3$—N(OH)—CO—CH=C(CH$_3$)(CH$_2$)$_2$—O]$_n$—H

(9) n=2 or 3

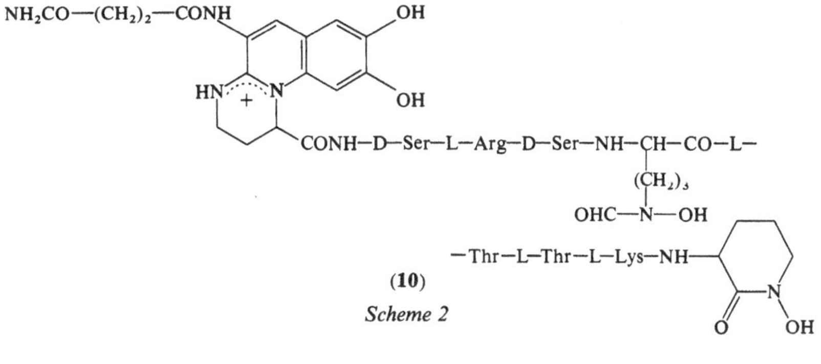

NH$_2$CO—(CH$_2$)$_2$—CONH   OH

HN   N   OH

CONH—D—Ser—L—Arg—D—Ser—NH—CH—CO—L—
                                                        |
                                                   (CH$_2$)$_3$
                                                        |
                                                 OHC—N—OH

—Thr—L—Thr—L—Lys—NH—

(10)

Scheme 2

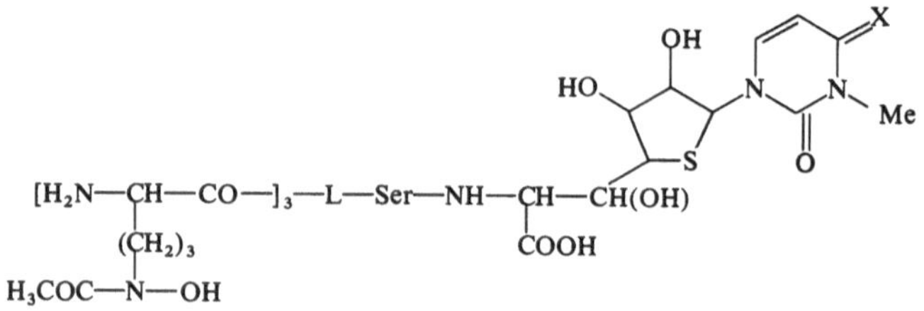

$$[H_2N-\underset{\underset{\underset{H_3COC-N-OH}{|}}{(CH_2)_3}}{CH}-CO-]_3-L-Ser-NH-\underset{\underset{COOH}{|}}{CH}-CH(OH)$$

(12) X = O, NH or N—CONH₂

$$HCO-N(OH)-CH_2-COO^-Na^+$$

(13)

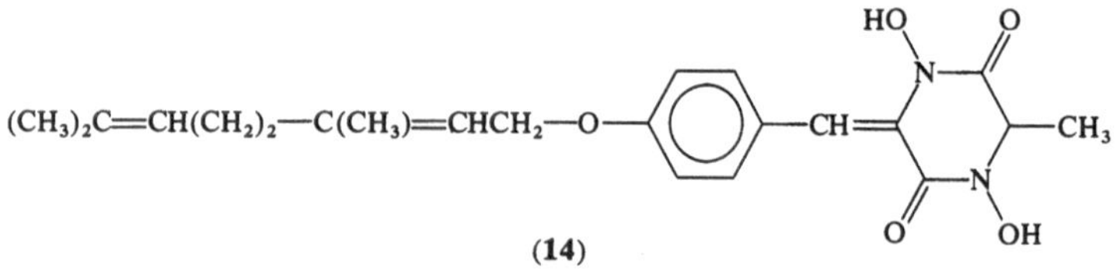

$$(CH_3)_2C=CH(CH_2)_2-C(CH_3)=CHCH_2-O-\text{(phenyl)}-CH=$$

(14)

$$\text{(benzene, HO, OH)}-CO-L-Thr-NH-\underset{\underset{\underset{H_3COC-N-OH}{|}}{(CH_2)_3}}{CH}-COOH$$

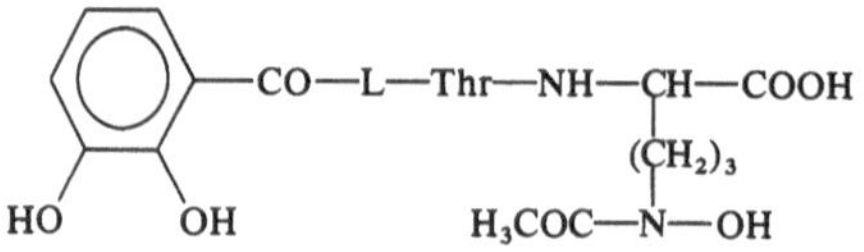

(15)

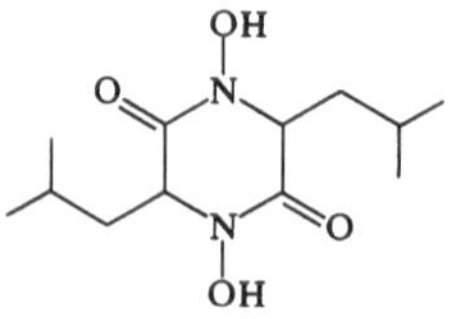

(17)

$$HO-N(NO)-CH_2-CH(NH_2)-COOH$$

L-**(18)**

$$cyclo-[CO-CH(NH_2)-(CH_2)_3-N(OH)-CO-CH=C(CH_3)-(CH_2)_3-O-]_3$$

(19)

$$(CH_3)_2N-CO-N(OH)-(CH_2)_3-CH(NH_2)-COOH$$

(20)

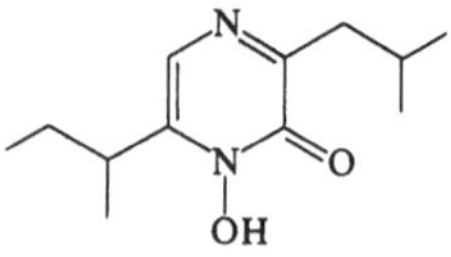

(21)

Scheme 2 (continued)

However it is becoming increasingly difficult to assign naturally-occurring hydroxamic acids unequivocally to such classes as antibiotics, growth factors, tumor inhibitors, cell division factors, pigments and siderophores as the distinctions between there activity classes are becoming less clear.

Several of the above classes of compounds have previously been the subject of review articles. MAEHR's review of antibiotics and natural hydroxamic acids (8) is of special interest, while N-hydroxyamide siderophores have been the subject of many reviews by NEILANDS (9, 10, 11, 12, 13) and others (14, 15, 16). Some of the individual compounds or groups such as the mycobactins (7) (17) or hadacidin (13) (117) have been reviewed separately. Hence detailed consideration of these compounds, aside from the presentation of their formulas in Scheme 2, will be omitted. However Table 1 contains data on the occurrence of α- and ω-N-hydroxyamino acid residues, names of the natural products as well as names of the microorganism strains which produce them. Literature references to the original articles are also included. As may be seen from Table 1 ω-N-hydroxyamino acid residues predominate in natural products. They are capable of forming peptides due to the presence of the α-amino and carboxylic groups. Depending on the structure acylated N-hydroxyamino groups in the ω-position may then participate in the eventual formation of the complex of the ligand with a metal ion. Simpler derivatives of citric acid and N^6-hydroxylysine are found among procaryotes while among higher organisms such as yeasts and fungi, cyclo-hexapeptides of N^5-hydroxyornithine as, for example, in the ferrichromes predominate.

Synthetic methods and syntheses of individual compounds are described in parts VIII and IX.

Table 1. *Occurrence of N-Hydroxyamino Acids in Natural Products*

N-Hydroxy-amino acid	Natural product	Isolated from	Ref.
N-Hydroxyalanine	mycelianamide	*Penicillium griseofulvum*	(19), (20)
N⁵-Hydroxy-L-arginine	—	*Nannizzia gypsea*	(6)
		Bacillus cereus	(7), (21)
N-Hydroxy-L-asparagine	L-asparaginyl-L-N-hydroxy-asparagine	*Mycobacterium avium*	(22)
N-Hydroxy-L-aspartic	L-aspartyl-L-N-hydroxy-aspartyl-D-cycloserine	*Cornebacterium kutscheri*	(23)
N⁵-Hydroxy-N,N-dimethyl-L-citrullin	connatin	*Lyophyllum connatum*	(24)
N-Hydroxy-glycine	hadacidin	*Penicillium frequentans*	(25)
		Penicillium aurantio-violaceum	(26)—(29)
N-Hydroxy-L-leucine	aspergillic acid	*Aspergillus flavus*	(30)
	pulcherriminic acid	*Candida pulcherrima*	(31)—(33)
N⁶-Hydroxy-L-lysine	aerobactin	*Aerobacter aerogenes*	(34)
	exochelins	*Mycobacteria sp.*	(35)
	mycobactins A, F, H, M, N, P, R, S, T,	*Mycobacteria sp.*	(36)
	nocobactin	*Nocardia asteroides*	(37)
N⁵-Hydroxy-D-ornithine	neurosporin	*Neurospora crassa*	(38)
	pseudobactin	*Pseudomonas fluorescens-putida*	(39), (40)
	ferribactin	*Pseudomonas fluorescens-migula*	(41)
	pyoverdine	*Pseudomonas fluorescens*	(42)
N⁵-Hydroxy-L-ornithine	ferrichrome	*Ustilago sphaerogena*	(43)
		Aspergillus niger	(33)
		Penicillium resticulosum	(33)

Compound	Organism	Ref.
N⁵-Hydroxy-L-ornithine	Ustilago sphaerogena	(44)
ferrichrome A	Cryptococcus melibiosus	(45)
ferrichrome C		
ferrichrysin	Aspergillus mellasus	(46)
	Aspergillus ochraceous	(47)
ferricrocin	Aspergillus versicolor	(46)
	Epicoccum purpurascens	(48)
	Aspergillus nidulans	(49)
ferrirubin	Penicillium variabile	(50)
	Paecilomyces varioli, Spicaria sp.	(49)
	Penicillium chrysogenum	(47)
	Aspergillus ochraceous	(50)
ferrirhodin	Penicillium versicolor	(51)
tetraglycylferrichrome	Neovossia indica	(48)
triornicin	Epicoccum purpurascens	(52)
sake colorant A	Aspergillus oryzae	(53), (54)
albomycins (grisein) δ_2, ϵ, δ_1	Actinomyces griseus	(55)
malonichromes	Fusarium roseum	(47), (56)
asperchromes A, B1, B2, B3, C, D1, D2, D3, E	Aspergillus ochraceous	(57)–(59)
fusarinine	Fusarium roseum	(57)–(59)
fusarinine A, B	Fusarium roseum	(60)
fusarinine C (fusigen)	Fusarium cubense	(61)
	Aspergillus nidulans	(61)
	Penicillium chrysogenum	(62)
N,N',N''-triacetylfusarinine	Penicillium sp.	(62)
	Aspergillus sp.	(63)

Table 1 (*continued*)

N-Hydroxy-amino acid	Natural product	Isolated from	Ref.
N^5-Hydroxy-L-ornithine	pyoverdine	*Pseudomonas aeruginose*	(64)
	rhodotorulic acid	*Rhodotorula pilimanae*	(65), (66)
	dimeric acid (verticillin S)	*Fusarium dimerum*	(67)
		Verticillium dahliae	(68)
	coprogen	*Epicoccum purpurascens*	(48)
		Pilobolus kleinii	(69)
		Penicillium sp.	(70)
		Neurospora crassa	(71)
	coprogen B	*Fusarium dimerum*	(67)
	neocoprogen I	*Curvularia sabulata*	(56)
	vanoxonin	*Saccharopolyspora hirsute*	(72)
N-Hydroxy-dehydrotyrosine	mycelianamide	*Penicillium griseofulvum*	(19), (20)
N^3-Hydroxy-2,3-diaminopropionic acid	alanosine	*Streptomyces alanosinicum*	(73), (74)
N^2-Hydroxy-2,2'-iminodipropionic acid	amavadine	*Amanita muscaria*	(75)
N^3-Hydroxyaminopropylphosphonic acid	FR-900098	*Streptomyces rubellomurinus*	(76)
	FR-31564	*Streptomyces lavendulae*	(77), (78)
N^3-Hydroxyamino-1-trans-propenyl-phosphonic acid	FR-32863	*Streptomyces lavendulae*	(77), (78)
2-Hydroxy-N^3-hydroxyaminopropyl-phosphonic acid	FR-33289	*Streptomyces rubellomurinus*	(78)

III. Physical and Chemical Properties

N-Hydroxyamino acids are colourless, solid, crystalline compounds only stable in the solid state. When heated they melt above 200 °C with decomposition (79). However the melting point is not a good criterion of purity (80). They are usually purified by recrystallization from hot water in which they are less soluble than amino acids. However, recrystallization in this case is accompanied by great losses due to decomposition (80). N-Hydroxyamino acids are slightly soluble in alcohol and sparingly soluble in ether, acetone and other organic solvents (1, 79, 82, 31). They are soluble in both acid and basic aqueous solution and form the corresponding salts (79, 81). Stable solid hydrochlorides are obtained in this way (1, 82).

Crystals of N^5-hydroxyarginine hydrobromide (83) and the synthetic 3,3-bis-(carboxyethyl)-3-hydroxyaminopropionic acid betaine (26) (84) (Scheme 3) were investigated by the X-ray technique. The N-O bond has a length of 1.40 Å and 1.42 Å, respectively, while for hydroxylamine a value of 1.46 Å is given (85).

A molecule of an α-N-hydroxyamino acid is a dipolar ion (27) (79, 81) as confirmed by the IR spectra which contain characteristic COO^- and $-NH_2^+$-OH group bands. The isoelectric point of these compounds lies within the 6–7 pH range. pK_1 constants have values between 2.1 and 2.3, while pK_2 lies between 5.7 and 5.9 (81).

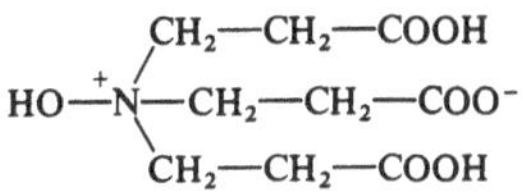

$$\underset{(26)}{HO-\overset{+}{N}\!\!\left\langle\begin{array}{l}CH_2-CH_2-COOH\\CH_2-CH_2-COO^-\\CH_2-CH_2-COOH\end{array}\right.} \qquad \underset{(27)}{HO-NH_2^+-CHR-COO^-}$$

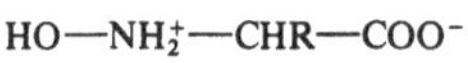

Scheme 3

In contrast to α-amino acids, N-hydroxyglycine (28) and N-hydroxyalanine (29) form copper complexes (1:1 and 1:2) at pH as low as 1.5–2.5 (86). In such complexes the cation is coordinated with the carboxylic group with participation of the nitrogen atom of the hydroxylamine group, with $\lambda_{max} = 620–630$ nm. The structure of N-hydroxyamino acids may be investigated by NMR using lanthanide shift reagents, as described in the case of N-hydroxyarginine (6).

It should be emphasized that the first naturally-occurring compound containing vanadium, with unspecified biological activity was the blue substance amavadine (16) isolated from a mushroom (*Amanita muscaria*) (75). ESR studies showed that it was a complex of N-hydroxy-α,α'-iminodipropionic acid with oxovanadium (IV) (*87, 88*). Also nickel and copper complexes of this and of a similar synthetic ligand (**30**) have been investigated (*89*).

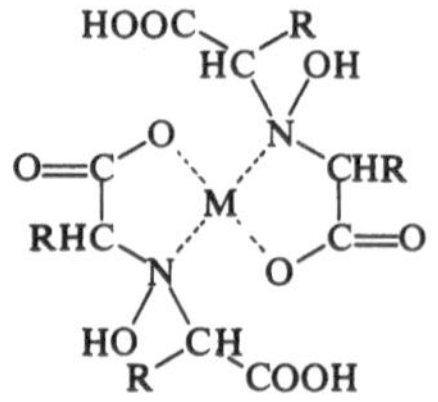

(**16**) R = Me   M = VO^{2+}
(**30**) R = Me   M = Ni^{2+} or Cu^{2+}
      H   M = VO^{2+}, Ni^{2+} or Cu^{2+}

Scheme 4

N-Hydroxyamino acids, just like N-hydroxyamines, are strong reducing agents. At room temperature they reduce solutions of salts of such metals as silver, copper, mercury and lead (*1*). They also reduce Fehling's reagent (*1*) and decolorize iodine solutions in neutral or alkaline but not acidic media (*36*). As the result of the above reactions compounds of type (**1**) are oxidized to the oximes of α-keto acids (**31**). Esters of N-hydroxyamino acids (*90*), like N-alkylhydroxylamines, undergo oxidation to dimers of *cis*-nitroso compounds with a characteristic absorbance at 264 nm, this being analytically important (*91*).

$$\text{HO—NH—CHR—COOH} \longrightarrow \text{HO—N=CR—COOH}$$

      (**1**)   (**31**)

R = Et, Pr, iBu, He

Scheme 5

N-Hydroxyamino acids readily undergo catalytic reduction, being transformed into the corresponding amino acids (**32**). Amino acids are also the main products formed when N-hydroxyamino acids are heated with hydrochloric acid (*82*) at elevated pressure (Scheme 6).

SPENSER and AHMAD (*81, 92*) found that disproportionation of N-hydroxyamino acids (**27**) occurs in aqueous solution. The keto acid oxime (**31**) decompose further to give the corresponding nitrile (**33**) (*93*) (Scheme 6). The above results are questionable as STEIGER (*94*) and MØLLER *et al.* (*95*) obtained amino acids and the corresponding aldoximes (**36**) from aromatic N-hydroxyamino acids (**34**) and (**35**) (Scheme 7).

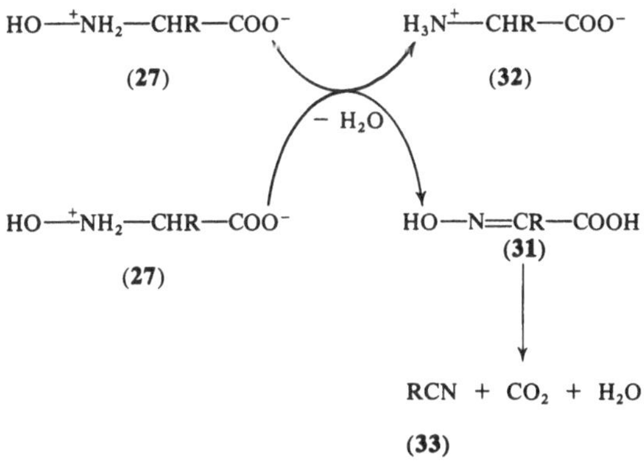

$$HO-^{+}_{-}NH_2-CHR-COO^- \qquad H_3N^{+}_{-}-CHR-COO^-$$
$$(27) \qquad\qquad (32)$$

$$-\ H_2O$$

$$HO-^{+}_{-}NH_2-CHR-COO^- \qquad HO-N{=}CR-COOH$$
$$(27) \qquad\qquad\qquad (31)$$

$$RCN\ +\ CO_2\ +\ H_2O$$
$$(33)$$

R = Me, Et, Pr, iPr

Scheme 6

$$HO-NH-CHR-COOH \xrightarrow{\ N_2\ } H_2N-CHR-COOH\ +\ HO-N{=}CHR$$

(34) R = Bzl   (32)   (36)
(35) R = *p*-HO—Bzl

Scheme 7

In view of these unfavourable chemical properties it is evident why no underivatized free N-hydroxyamino acid has been isolated from biological material so far.

N-Hydroxyamino acids have been, until now, rarely used as substrates for syntheses. However a general synthesis of 3-oxazolin-5-ones (**38**) from such substrates (**1**) and from carbonyl compounds (**37**) (*96*) has been described (Scheme 8).

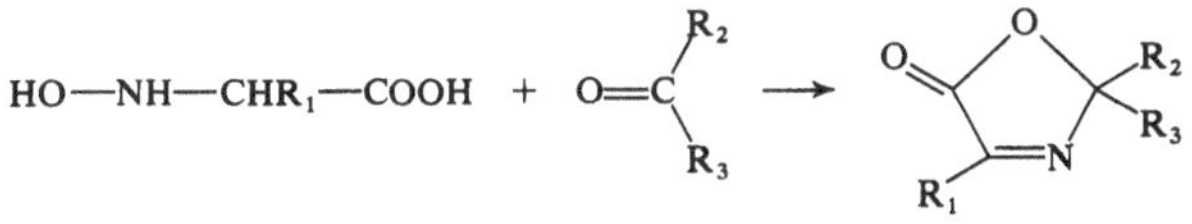

$$HO-NH-CHR_1-COOH\ +\ O{=}C\overset{R_2}{\underset{R_3}{\diagdown}} \longrightarrow$$

(1)   (37)   (38)

R_1 = Me or Ph R_2 = H or Me R_3 = Me, Et, iPr, Bu, iBu or Ph

Scheme 8

IV. Analysis

The reactions described in the previous section render difficult analyses of compounds containing N-hydroxyamino residues. The decompositions described above also proceed as the result of acid hydrolysis.

The N^5-hydroxyornithine residue frequently oxidises to the corresponding semialdehyde and then to glutamic acid (97). Thus misleading results were observed during the determination of the structure of the albomycines (98).

Only a few N-hydroxyamino acids have been analysed using a standard Durrum D-500 (99) amino acid analyser. The retention times for all compounds investigated so far (with norleucine as a standard), are given in Table 2 and are always lower than those of the corresponding amino acids.

Table 2. *Data of Amino Acid Analysis*
(Durrum – model D-500) (99)

Compound	Retention time	Colour yield %	Percent contamination by parent amino acid
norleucine (standard)	–	100	100
N-hydroxyglutamate (39)	21'35''	26	2,8
N-hydroxyaminobutyrate (40)	22'40''	18	1,8
N-hydroxyglycine (28)	21'13''	4	<1
glycine	33'30''	–	–
α-aminobutyrate	38'34''	–	–
glutamate	22'54''	–	–

Table 3. R_f *Value of some N-hydroxyamino Acids* (1) $HO-NR_1-CHR-COOH$

No	R	R_1	S_1^a	S_2^a	S_3^a	S_4^b	S_5^c
(28)	H	H	–	–	–	0.71	–
(40)	C_2H_5	H	0.54	0.20	0.50	0.89	–
(41)	n-C_3H_7	H	0.63	0.24	0.65	–	–
(42)	i-C_3H_7	H	0.58	0.19	0.65	–	–
(39)	$HOOC-(CH_2)_2$	H	–	–	–	0.76	–
(43)	H	CH_3	–	–	–	–	0.37
(44)	$-(CH_2)_3-$		–	–	–	–	0.42
(45)	$-(CH_2)_4-$		–	–	–	–	0.50

where:
S_1: nBuOH/AcOH/H_2O (4:1:1)   acc. (80)
S_2: iBuOH/3% aq NH_3 (3:1)   acc. (80)
S_3: $CH_3COC_2H_5$/Py/H_2O (60:15:25)   acc. (80)
S_4: $CH_3COC_2H_5$/tBuOH/HCOOH/
H_2O (40:30:15:15)   acc. (99)
S_5: MeOH/AcOH   (98:2)   acc. (101)

[a] TLC on silica gel-Kieselgel G, Merck;
[b] Chromatography on Whatman # 1 paper;
[c] TLC on silica gel F_{254}.

Chromatography of N-hydroxyamino acids was an object of controversy. Finally it was discovered that paper chromatography (99) and TLC (80) gave good results. Silver nitrate, ninhydrin and 2,3,5-triphenyl-2H-tetrazolium chloride were applied as detecting agents (80, 100).

R_f coefficients of several N-hydroxyamino acids are given in Table 3. Their values are higher than those of the corresponding amino acids.

V. Naturally-Occurring N-Hydroxypeptides

Up to the present only two naturally-occurring N-hydroxypeptides have been isolated: asparaginyl-L-α-N-hydroxyasparagine (46) (22) and aspartyl-L-α-N-hydroxyaspartyl-D-cycloserine (47) (23). The former was isolated from cultures of *Mycobacterium avium*, the latter (47) from *Corynebacterium kutscheri*. Both probably act as iron ionophores (siderophores) in these strains. It has also been reported (102) that in human brain tumor cell proteins one N-hydroxyamide bond occurs for every 300–500 amino acid residues. To date this has not been confirmed by other authors.

$$NH_2CO—CH_2—CH(NH_2)—CO—N(OH)—CH(CO_2H)—CH_2—CONH_2$$

(46)

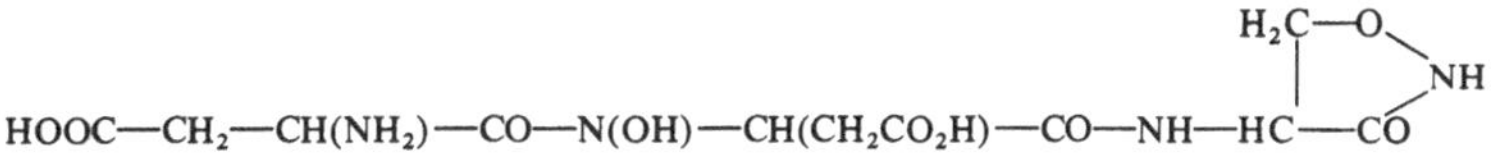

$$HOOC—CH_2—CH(NH_2)—CO—N(OH)—CH(CH_2CO_2H)—CO—NH—HC—CO$$

(47)

Scheme 9

VI. Biological Action of N-Hydroxyamino Acids and N-Hydroxypeptides

The specific activities and biological roles of naturally-occurring peptide derivatives of N-hydroxyamino acids of the siderophore class and their analogues are now well understood. The matter has been and still is being discussed in numerous reviews of which the latest have been cited in Section I. Therefore only the most important aspects of this topic will be summarized here.

In the course of evolution, about $2.8 \cdot 10^9$ years ago, during the build-up of free oxygen in the atmosphere, microorganisms began to adapt themselves to the acquisition of the less readily available iron (III). The method of iron transport, at present generally used by bacteria, fungi and algae to render their existence possible in environments of low iron levels, involves biosynthesis of low molecular (up to about 800 daltons) ferric-specific organic ligands named siderophores, their excretion outside the cell and the formation of complexes which can utilize iron contained in insoluble, difficulty accessible external compounds. Stability constants of iron complexes with siderophores, developed to perfection in the course of evolution, are typically of the order of 10^{30}. The complex undergoes active transport with participation of the outer membrane cell receptors; iron is released in the course of a complicated still, in part, controversial mechanism and is utilized by the cell, as one of the most essential ions, in the construction of many important metalloproteins. The problem, as a whole, is almost classical and is being introduced in some modern inorganic chemistry text-books (*103*).

In contrast to the above, the biological activity of other simple naturally-occurring and synthetic N-hydroxyamino acid derivatives has been investigated and described only rarely (*8*). Hadacidin, the sodium salt of N-formyl-N-hydroxyglycine (**13**), isolated from *Penicillium aurantio-violacium*, which acts as an antitumor and plant growth retardant factor, may be given as an example. Details concerning this compound are discussed by EMERY in a separate review (*117*). Hadacidin inhibits the *de novo* biosynthesis of adenylic acid in Ehrlich ascites tumor and in rat liver cells. The structural similarity of the hadacidin anion (**13**) to aspartate is responsible for the inhibition of adenylsuccinate synthetase (IMP: L-aspartate ligase) (*104, 105*). This effect may be competitively reversed by L-aspartate (*106*) (Scheme 10).

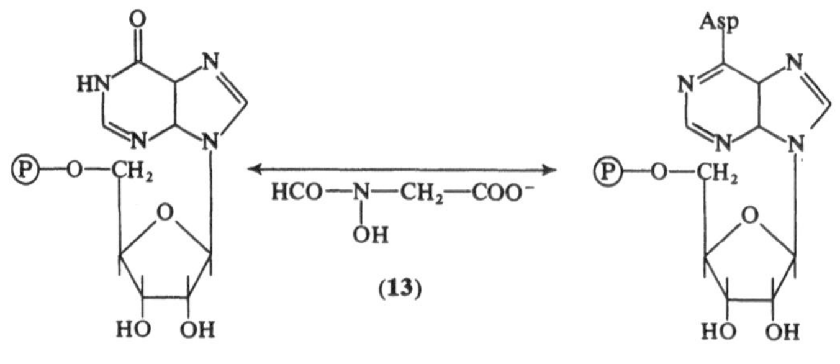

Scheme 10

The observation that N-formyl-N-hydroxyamino acetamide was markedly less active indicated that a negatively charged carboxyl group is essential for activity by complexation. Also formyl-N-hydroxyalanine (**48**) (Scheme 12), a homologue of hadacidin, acts as a strong inhibitor of the same enzyme (*107*). Only the L- form (**13**) is active because racemic compounds have only half of the activity. Both formyl and hydroxyl groups on the nitrogen atom are required for optimal activity, this being in accordance with previous considerations regarding the complexing action of N-hydroxyamino acids. Analogues of hadacidin containing a bulky group on the α-carbon atom were found to be inactive (*107*).

It was also found that N-unacylated N-hydroxyamino acids are enzyme inhibitors as well. N-Hydroxyglutamate (**39**), N-hydroxyaminobutyric acid (**40**) and N-hydroxyglycine (**28**) are strong irreversible inhibitors of pyridoxal 5′-phosphate dependent enzymes: glutamate-alanine transaminase, glutamate-aspartate transaminase and glutamate decarboxylase. These irreversible transaminase inhibitors interact with enzyme active sites and form a nitrone (**49**) (Scheme 11) as the ultimate

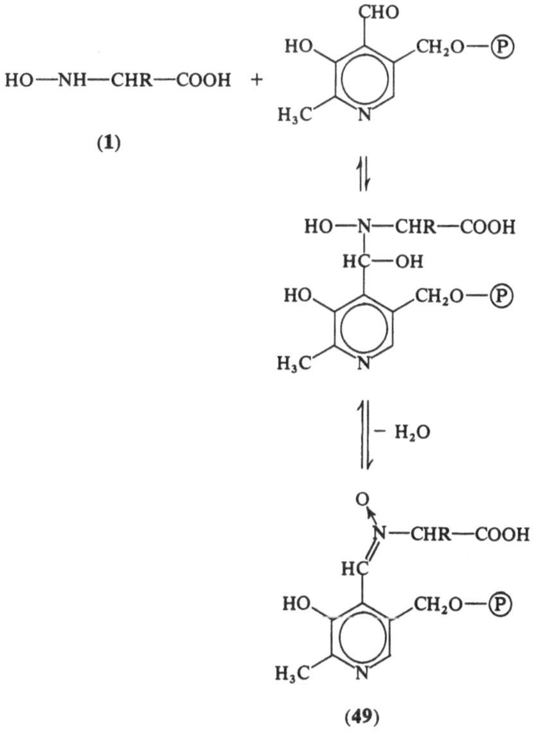

Scheme 11

product whose stable structure mimics the normal aldimine intermediate (*99*). N-Hydroxyamino acids also irreversibly inhibited two flavin-dependent enzymes: L- and D-amino acid oxidases, yielding the oxime of α-ketobutyrate as product in both cases (*99*). N-Hydroxyglutamate (**39**) is a strong competitive inhibitor of glutamate dehydrogenase but does not interact with glutamine or γ-glutamylcysteine synthetases.

The above facts strongly support the unusual role of N-hydroxyamino acids as amino acid antagonists. POWERS stated that the methyl ester of chloroacetyl-N-hydroxyleucine (**50**) (*108*) and the formyl-N-hydroxyleucyl-alanyl-glycine amide (**51**) (*109*) are the first specific non-reversible termolysine inhibitors. They act through coordination of the zinc atom of this metalloprotein, as has been determined by X-ray crystallography (*110*). Compound (**51**) is also an inhibitor of elastase (*111*).

$$HCO—N(OH)—CH(CH_3)—COO^-Na^+$$

(**48**)

$$ClCH_2—CO—N(OH)—CH(i\text{-}C_4H_9)—COOCH_3$$

(**50**)

$$HCO—N(OH)—CH(i\text{-}C_4H_9)—CONH—CH(CH_3)—CONH—CH_2—CONH_2$$

(**51**)

Scheme 12

N-Hydroxyamino acids, similar in action to N-hydroxyurea, some N-hydroxyamides and some aromatic hydroxyacids, are inhibitors of ribonucleotide reductases (*112*), a unique group of metalloenzymes, essential for cell proliferation. Inhibition of substrate reduction in vitro ($I_{50}=2.3\cdot 10^{-4}$) is accompanied by decay of the tyrosyl radical but not the iron atom from the *E. coli* subunit B2 of this enzyme. Inhibitors of the above type donate an electron to the enzyme's free radical, producing an inactive protein-enzyme with a still intact binuclear iron complex and a new more stable free radical of the nitroxide type (**52**)

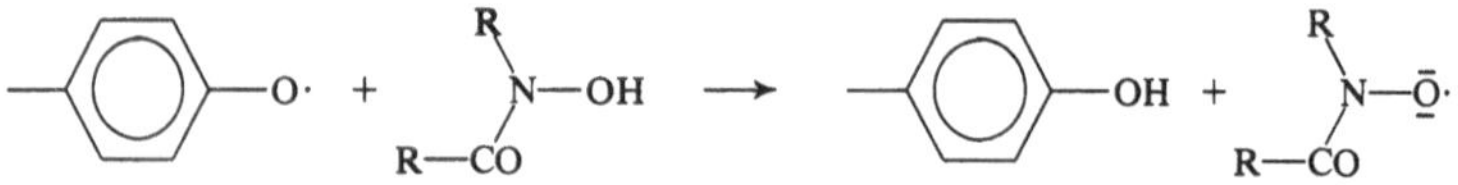

[Fe^{+++}] [Fe^{+++}] (**52**)

enzyme active centre — inactive enzyme

Scheme 13

(*112*). The reactions of stable nitroxide radicals have been reviewed and have been the subject of intensive investigations (*113*). It should be emphasized that ribonucleotide reductase inhibitors are of special importance because of the possibility of applying them as immunoregulatory drugs.

Comparison of the lability toward enzymatic splitting of the N-hydroxypeptide bond and the peptide bond is an interesting problem connected with biological action. It should be emphasized that the N-hydroxyamide bond in N-hydroxypeptides may be split by peptidases such as papaine, trypsine and leucine aminopeptidase (*102*). Lately AKIYAMA (*114*), while investigating synthetic N-hydroxypentapeptides, analogues of enkephalin, has stated that they are resistant to the action of aminopeptidase M. Carboxypeptidase Y degrades these compounds, splitting all carbon-nitrogen bonds. Of course, the biological activity of these synthetic compounds was also compared with that of natural products. A qualitative analgesia test shows that the N-hydroxypeptide enkephaline analogue possesses more lasting potency than Leu^5-enkephaline.

VII. Biogenesis of the N-Hydroxyamide Bond

The biogenesis of the N-hydroxyamide bond is an interesting problem (*115*). That it may involve oxidation of the α- or ω-amine group of an amino acid to a hydroxylamine derivative (Scheme 14, path a) was observed in the biosynthesis of ferrichrome (**5**) (*116*), hadacidin (**13**) (*25, 117*) and aerobactin (**8**) (*118*).

In EMERY's study of ferrichrome biosynthesis (*116*) incorporation of labelled N^5-hydroxyornithine was compared with incorporation of ornithine. In cultures of *Ustilago sphaerogena* incorporation of N-hydroxyamino acid was three times greater than that of ornithine. Initial work on hadacidin biosynthesis was performed by DULANEY (*119*), who found that neither glycine nor formylglycine stimulated production. STEVENS (*25*), using doubly labelled formylglycine, ruled out direct N-hydroxylation of the amide bond. Although glycine and formate are both excellent precursors of hadacidin, labelled N-hydroxyglycine (**28**) was an even better precursor than glycine itself. Molecular oxygen was found to be the source of oxygen in N-hydroxyglycine. When cells of *Penicillium* were grown in the presence of $^{18}O_2$, an excess of isotope was found in the hydroxylamino group. In the case of aerobactin (**8**) biosynthesis, conversion of L-lysine to its N^6-hydroxy derivative by cell-free extracts of *Aerobacter* was reported (*118*). The enzyme system is highly specific because D-lysine, N^2- or N^6-derivatives of lysine and α-amino acids were not hydroxylated.

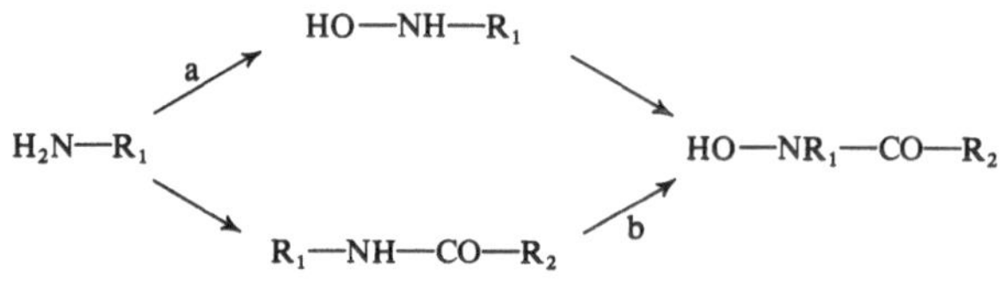
HO—NH—R₁
a
H₂N—R₁
HO—NR₁—CO—R₂
R₁—NH—CO—R₂
b

Scheme 14

That biogenesis of the N-hydroxyamide bond may also proceed by way of oxidation of the amide bond (Scheme 14, path b) was observed in the case of aspergillic acid (**21**) and neoaspergillic acid (*120*). The results with ^{14}C-labelled substrates support the hypothesis that aspergillic acid is synthesized by *Aspergillus flavus* from leucine and isoleucine. Oxidation of the amide bond during the biosynthesis of rhodotorulic acid (**6**) has been investigated more thoroughly by mass spectrometry (*121*). The presence of parent ions m/e 346 and 348 in acid isolated from *Rhodotorula pilimonae* grown in an atmosphere enriched with ^{18}O indicates that one or two oxygen atoms can be incorporated.

It is thought that N-hydroxyamino acids are intermediates in the biosynthesis of cyanoglucosides and glucosinolates (*4*, *122*, *123*) (Scheme 15). However, the assumption (*124*) that δ-(α-aminoadipoyl)-cysteinyl-N-hydroxyvaline (**53**) is an intermediate in the biosynthesis of the β-lactam ring of the penicillins has not yet been confirmed (*125*).

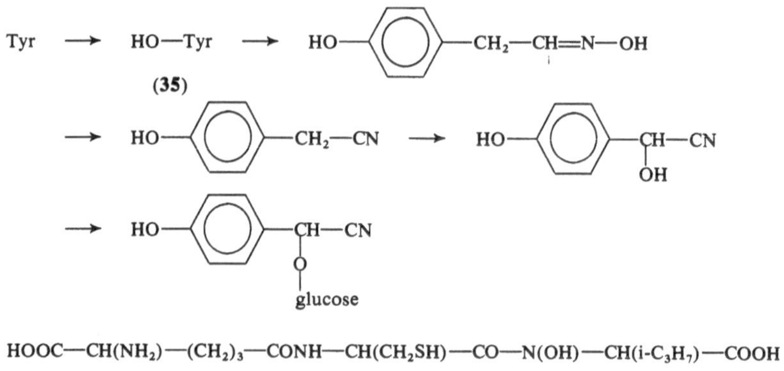
Tyr
HO—Tyr
(35)
HO
CH₂—CH=N—OH
HO
CH₂—CN
HO
CH—CN
OH
HO
CH—CN
O
glucose
HOOC—CH(NH₂)—(CH₂)₃—CONH—CH(CH₂SH)—CO—N(OH)—CH(i-C₃H₇)—COOH

(53)

Scheme 15

An important hypothesis concerning the function of N-hydroxy-amino acids has been formulated recently by UTTENHEIJM (126). There are indications that the metabolism of several types of non-protein amino acids such as N-hydroxyamino acids, dehydroamino acids and α-substituted amino acids is connected with the process of oxidation of amino acids or peptides. Hence it is suggested that N-hydroxyamino acids play an important role in the biosynthesis of dehydroamino acids (18), natural antiviral, antifungal or antibacterial products (as e.g. glio-toxin) (127) and some other fungal metabolites as sporidesmine (126), as sketched in Scheme 16.

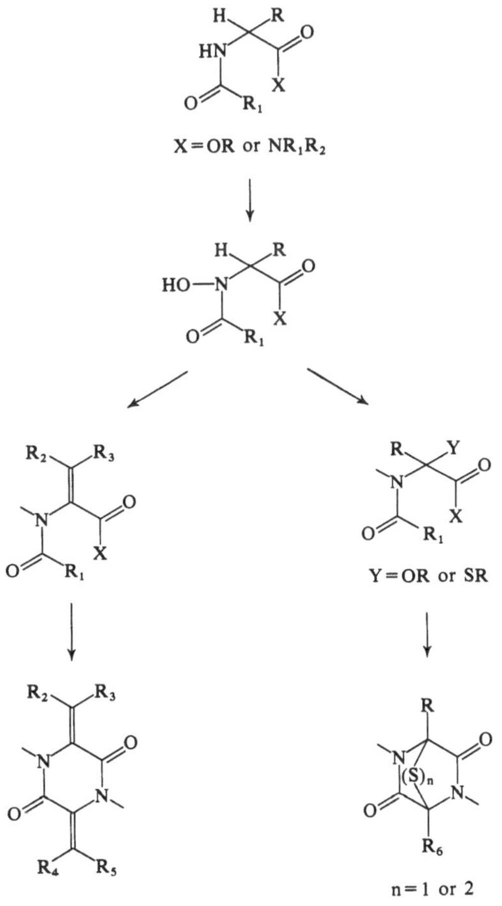

Scheme 16

VIII. Synthesis of N-Hydroxyamino Acids and Their Derivatives

The first methods for synthesis of N-hydroxyamino acid were developed as early as the end of the last century (*1*) but only after 1958 (*79*) did this subject again arouse interest. The difficulties caused by the instability of N-hydroxyamino acids were the reason for only occasional work of most authors in this field.

1. Reaction of Nitric Oxide with 1,3-Diketo Derivatives

The basis of this old method is the addition of nitric oxide to alcoholic solutions of sodium salts of C-alkyl derivatives of ethyl acetoacetate (**54**) (*128, 129*) (Scheme 17). Salts of alkyl derivatives of diethyl malonate (**55**) were used in NEELAKANTAN's modification (*79*). In both cases sodium salts of isonitramine (**56**), stable only in alkaline solution, were formed initially and subsequently transformed into N-hydroxyamino acids (**1**) by means of concentrated hydrochloric acid. This method using common substrates is quite general and may be used for the synthesis of N-hydroxyamino acids (**1**) of different types (*36, 79*). Usually it has been used for the synthesis of N-hydroxyglycine (**28**) (*129*) and N-hydroxyphenylalanine (**35**) (*79, 94, 129*).

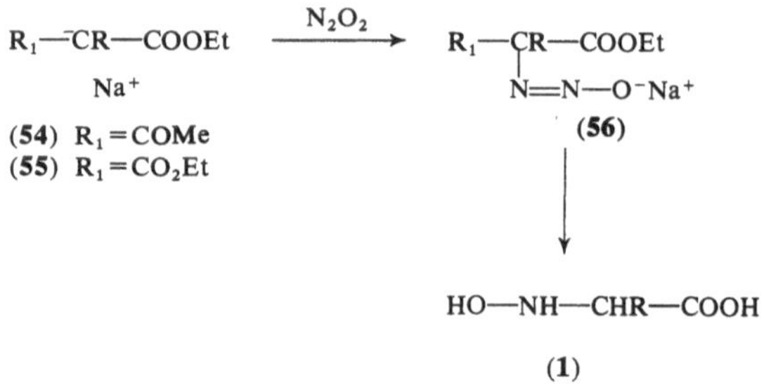

$$R_1-\!^-CR-COOEt \xrightarrow{\ N_2O_2\ } R_1-CR-COOEt$$

$$Na^+ \qquad\qquad N\!\!=\!\!N-O^-Na^+$$

(**54**) $R_1 = COMe$ (**56**)
(**55**) $R_1 = CO_2Et$

$$\downarrow$$

$$HO-NH-CHR-COOH$$

(**1**)

(**1**) R = H (**28**), Me (**29**), Et (**40**), nPr (**41**), nBu (**57**), Bzl (**35**)

Scheme 17

2. Strecker Synthesis

N-Hydroxyamino acids (**1**) were first obtained by adaptation of Strecker synthesis. Methods for obtaining the intermediate N-hydroxyamino nitriles (**58**) (Scheme 18) have been modified repeatedly

(130 133). Originally they were prepared by reaction of oxime with anhydrous hydrogen cyanide (82, 131, 132). Subsequently the oxime was added to an aqueous solution of hydroxylamine hydrochloride, sodium cyanide and sodium hydrogen phosphate (131). Lastly hydroxylamine hydrochloride, aldehyde and sodium cyanide may be mixed in a one-pot synthesis (79).

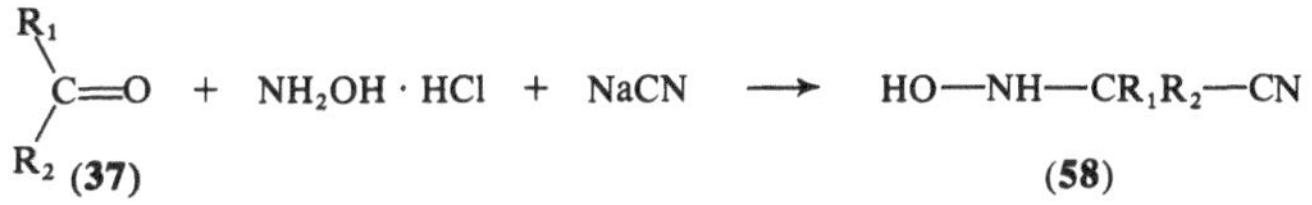

$$\begin{array}{c} R_1 \\ \diagdown \\ C{=}O \\ \diagup \\ R_2 \end{array} \; (37) \quad + \; NH_2OH \cdot HCl \; + \; NaCN \quad \longrightarrow \quad HO{-}NH{-}CR_1R_2{-}CN$$

(58)

$R_1 = H$ or Me $R_2 =$ Me, Et, nPr, iPr, nBu, iBu $R_1, R_2 = -(CH_2)_4-, \; -(CH_2)_5-, \; -(CH_2)_6-$

Scheme 18

The hydrolysis of nitriles (58) formed in this way is usually difficult (Scheme 19). In the presence of conc. sulfuric acid mainly amides of α-keto acid oximes (59) were obtained (1). Concentrated hydrochloric acid yields amides of N-hydroxyamino acids (60) (130) while in dilute acid the desired products (1) are formed (1, 79). Taking into account the instability of N-hydroxyamino nitrile (58) and the possibility of its polymerization, the first stage of hydrolysis is carried out at lower temperature (79). The Strecker method has also been applied to the synthesis of several cyclic N-hydroxyamino acids (70, 71) (134).

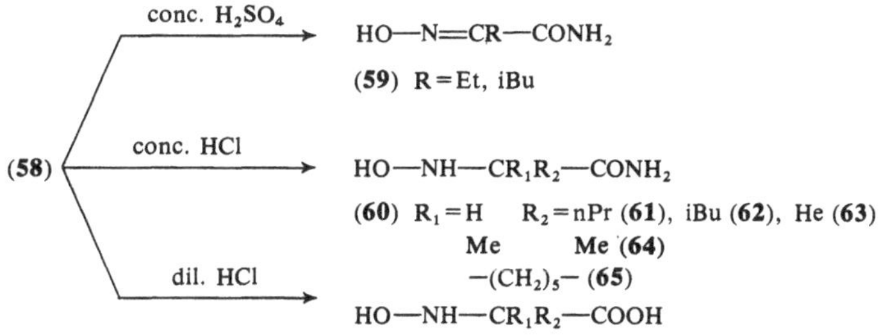

conc. H₂SO₄ ⟶ $HO{-}N{=}CR{-}CONH_2$

(59) R = Et, iBu

(58)

conc. HCl ⟶ $HO{-}NH{-}CR_1R_2{-}CONH_2$

(60) $R_1 = H$ $R_2 =$ nPr (61), iBu (62), He (63)
Me   Me (64)
$-(CH_2)_5-$ (65)

dil. HCl ⟶ $HO{-}NH{-}CR_1R_2{-}COOH$

(1)

(1) $R_1 = H$ $R_2 =$ Me (29), Et (40), nPr (41), iPr (42), nBu (57), iBu (66)
Me   Me (67), Et (68)
Et   Et (69)
$-(CH_2)_5-$ (70)
$-(CH_2)_6-$ (71)

Scheme 19

3. Reduction of Nitro Acids and Nitro Esters

This method was first applied by WEISBALT (135) to the synthesis of ethyl α-hydroxyamino-α-carboethoxy-β-(3-indole)-propionate. The reduction of nitrohydantoins (72) with zinc dust in the presence of ammonium chloride (136, 137) or isopropylamine hydrochloride (97) was the basis for the first syntheses of N^5-hydroxyornithine (73) and N^6-hydroxylysine (74), but gave only low yields (Scheme 20). Further attempts at synthesis of N^5-hydroxyornithine (73) (138) and N^6-hydroxylysine (74) (139) based on such reductions were unsuccessful. The conditions for reduction of nitroester (75) to the ester of N-hydroxyamino acid (76) were investigated by CHIMIAK (80), who obtained (78, 79) in aqueous methanol solution. SHIN (140) also reduced nitroesters (75) to (76, 77) using aluminium amalgam (Scheme 21).

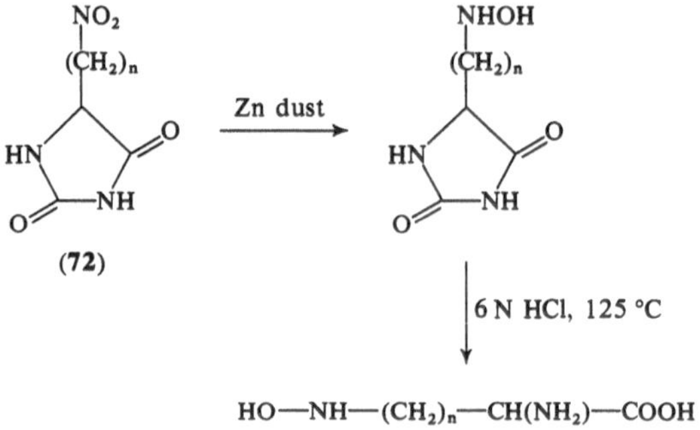

$$HO-NH-(CH_2)_n-CH(NH_2)-COOH$$

(2) n=3 (73)
    n=4 (74)

Scheme 20

(76) R=iPr (78) or C_3H_7—CH(OCH$_3$), i-C_3H_7—CH(OCH$_3$)
(77) R=nPr (79) or C_2H_5—CH(OCH$_3$), C_3H_7—CH(OCH$_3$), i-C_3H_7—CH(OCH$_3$)

Scheme 21

Finally KELLER-SCHIERLEIN et al. used reduction of appropriate nitro derivatives of cyclic peptides with subsequent acetylation for the synthesis of ferrichrome (141, 142) and rhodotorulic acid (143) (Scheme 22).

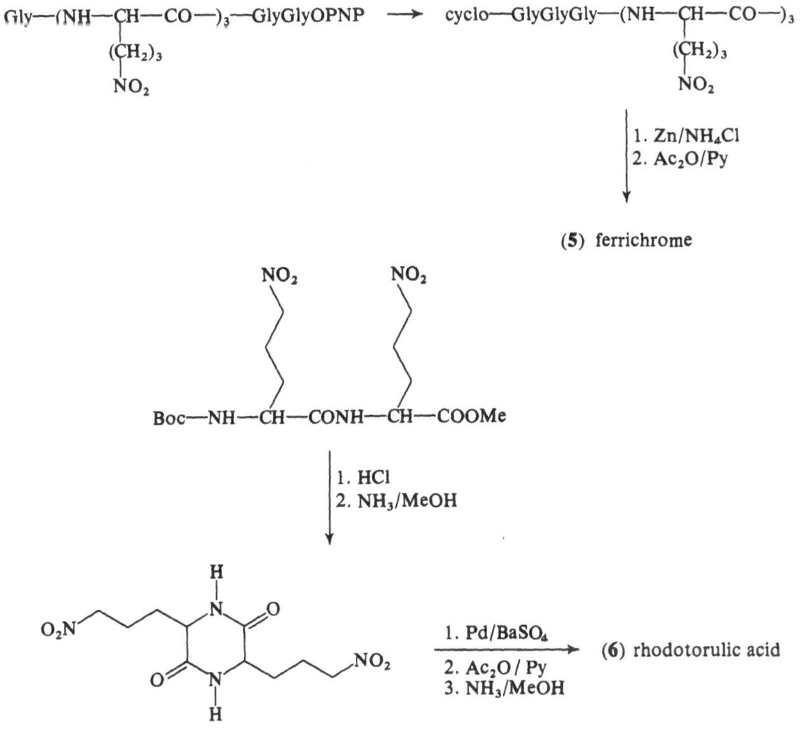

Gly—(NH—CH—CO—)$_3$—GlyGlyOPNP $\longrightarrow$ cyclo—GlyGlyGly—(NH—CH—CO—)$_3$
         | |
       (CH$_2$)$_3$ (CH$_2$)$_3$
         | |
        NO$_2$ NO$_2$

1. Zn/NH$_4$Cl
2. Ac$_2$O/Py

(5) ferrichrome

        NO$_2$ NO$_2$

Boc—NH—CH—CONH—CH—COOMe

1. HCl
2. NH$_3$/MeOH

O$_2$N ... N ... O
         |
         H

         NO$_2$

1. Pd/BaSO$_4$ (6) rhodotorulic acid
2. Ac$_2$O / Py
3. NH$_3$/MeOH

Scheme 22

4. Reduction of Oximes of α-Keto Acids and Substituted Aldoximes

This method (Scheme 23) was proposed in 1974 by AHMAD (*144*) who used excess lithium cyanoborohydride as the reducing agent. α-Oximino acids (**80**) or equimolar quantities of α-keto acid and hydroxylamine were reduced at pH 5 in aqueous solution. This procedure was also successfully applied to the synthesis of N-hydroxytyrosine (**35**) (*95*) (Scheme 23).

Since this method did not render good results in the reduction of esters and amides (**81, 82**), HERSCHEID and OTTENHEIJM (*145, 146*) used pyridine-borane as reducing agent and obtained esters (**77**) and esters of benzyloxyamino acids (**86**) in high yields. Esters (**86**) are good synthons because as derivatives of N-hydroxyamino acids they have a protected N-hydroxyamine function and therefore do not undergo redox reactions characteristic of hydroxylamines. In the case of amides (**82**) the reduction was not complete, but use of trimethylamine/borane complex as reducing agent was successful (*146*). OTTENHEIJM and co-

$$XO{-}N{=}CR_1{-}COR_2 \xrightarrow{\text{LiBH}_3\text{CN or Py}\cdot\text{BH}_3\text{ or TEA}\cdot\text{BH}_3} XO{-}NH{-}CHR_1{-}COR_2$$

(80) X = H R_2 = OH
(81) X = H R_2 = OEt
(82) X = H R_2 = NHMe
(83) X = Bzl R_2 = OEt
(84) X = Bzl R_2 = NHMe

(1) X = H R_2 = OH
(77) X = H R_2 = OEt
(85) X = H R_2 = NHMe
(86) X = Bzl R_2 = OEt
(87) X = Bzl R_2 = NHMe

(1) R_1 = H (28), Me (29), Et (40), iPr (42), nPr (41),
     HOOC—$(CH_2)_2$ (39), Ph (88), Bzl (34), p-HO—Bzl (35)

(77) R_1 = H (89), Me (90), Et (91), iPr (92), Bzl (93)

(85) R_1 = H (94), Me (95), Et (96), iPr (97), Bzl (98)

(86) R_1 = H (99), Me (100), Et (101), iPr (102), Ph (103), Bzl (104),

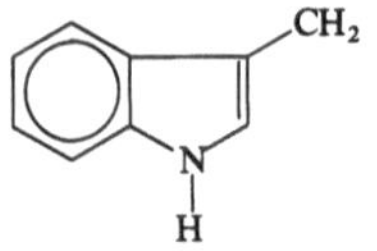

(105)

(87) R_1 = H (106), Me (107), Et (108), iPr (109), Ph (110), Bzl (111)

Scheme 23

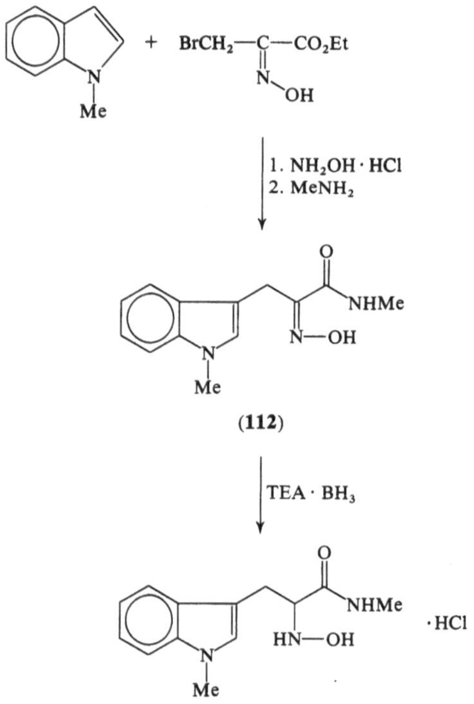

1. $NH_2OH \cdot HCl$
2. $MeNH_2$

(112)

TEA · BH_3

(113)

Scheme 24

workers (126) have used the hydride technique for preparation of the ester of benzyloxytryptophane (105) from the corresponding oxime and preparation of the N-methylamide of N-hydroxy-N'-methyl-tryptophane (113) from the appropriate oxime (112) (Scheme 24). LEE and MILLER have employed reduction of a O-benzyloxyimine for the synthesis of N-hydroxyornithine derivatives (114) from glutamic acid (147) (Scheme 25).

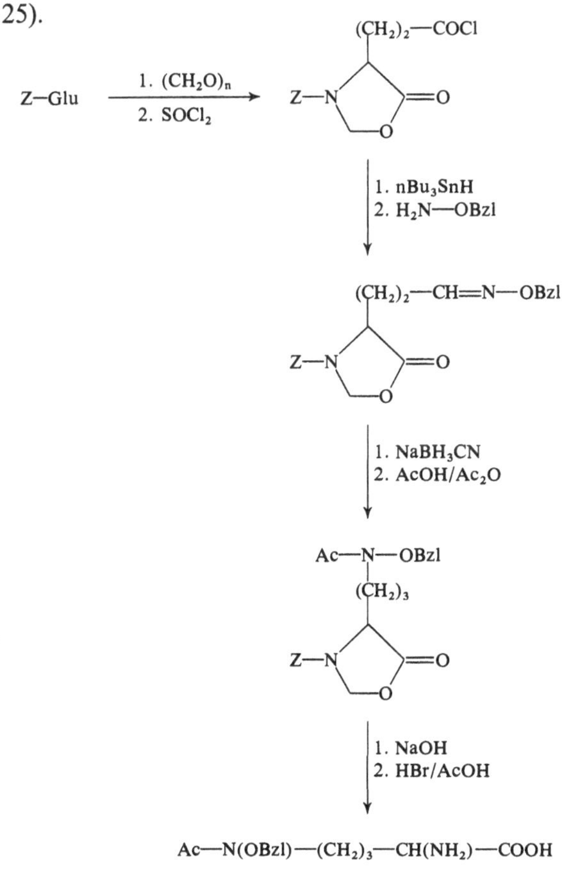

$$Z-Glu \xrightarrow[\text{2. SOCl}_2]{\text{1. (CH}_2\text{O)}_n}$$

$$\text{(CH}_2\text{)}_2-\text{COCl}$$

ring structure: Z—N, C=O, O

1. nBu$_3$SnH
2. H$_2$N—OBzl

$$\text{(CH}_2\text{)}_2-\text{CH}=\text{N}-\text{OBzl}$$

ring structure: Z—N, C=O, O

1. NaBH$_3$CN
2. AcOH/Ac$_2$O

$$\text{Ac}-\text{N}-\text{OBzl}$$
$$\text{(CH}_2\text{)}_3$$

ring structure: Z—N, C=O, O

1. NaOH
2. HBr/AcOH

$$\text{Ac}-\text{N(OBzl)}-\text{(CH}_2\text{)}_3-\text{CH(NH}_2\text{)}-\text{COOH}$$

(114)

Scheme 25

5. Addition of Hydroxylamine to α,β-Unsaturated Acids

Hydroxylamine reacts with the double bond of unsaturated acids (84, 151) such as acrylic acid (120), its esters (121) (152) or amide (122) (153) via the Michael addition, forming N-hydroxy-*sec*-amino

esters (124, 125), amide (126) or 3,3-bis-(2-carboxyethyl)-3-hydroxy-
aminopropionic acid betaine (26) (84) (Scheme 26).

The hydroxylamine addition reaction was used for the preparation
of β-N-hydroxyaminophenylpropionic acid (115) (Scheme 26) from
cinnamic acid (148). Prolonged heating of a mixture of (115) and hydroxylamine gives the appropriate β-amino acid (150).

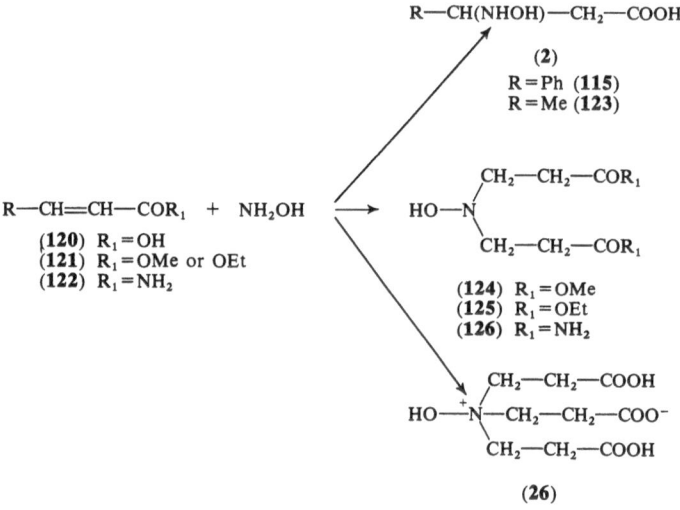

Scheme 26

FOUNTAIN (149) investigated the mechanism of the addition of N-
methylhydroxylamine to ester (116), obtaining oxazolone (117). This
compound was readily transformed into salt (118) or hydroxamic acid
of β-N-hydroxyamino acid (119) (Scheme 27).

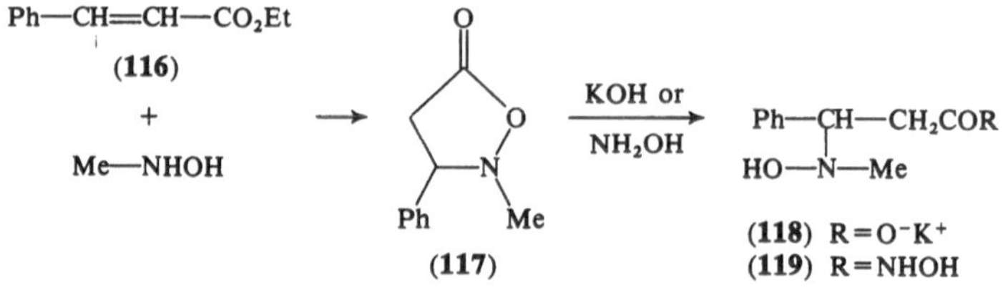

Ph—CH=CH—CO₂Et

(116)

+

Me—NHOH

→

O

(117)

$\xrightarrow[\text{NH}_2\text{OH}]{\text{KOH or}}$

Ph—CH—CH₂COR
HO—N—Me

(118) R=O⁻K⁺
(119) R=NHOH

Scheme 27

Addition of hydroxylamine to the double bond of fumaric acid
was accelerated by aspartase but unstable N-hydroxyaspartic acid was
not isolated (57, 154).

6. Reaction of α-Halo Acids with Hydroxylamine

Already in the last century HANTZSCH (*155*) observed that reaction of α-halo acids (**127**) with excess hydroxylamine leads to oximes of α-keto acids (**31**) as the result of an oxidation-reduction process (Scheme 28). This process, as a method of synthesis of (**31**), was later thoroughly studied by BARRY and HARTUNG (*156*). The oxidation-reduction reaction is responsible for the fact that later attempts to obtain the title compounds (**1**) from (**127**) by this route were never efficient (*81, 100*) although sometimes they were patented (*157*).

$$X—CHR—COOH \ + \ 2\,NH_2OH$$

(**127**) X = Cl or Br

$$\downarrow$$

$$HO—N = CR—COOH \ + \ HCl \ + \ NH_3 \ + \ H_2O$$

(**31**) R = H, Me or Et

Scheme 28

The fact that substitution of bromine in compounds of type (**127**) with hydroxylamine is accompanied by Walden inversion led NOCE *et al.* (*158*) to the first synthesis of optically active N-hydroxyamino acids L- and D-N-hydroxyleucine (**66**). The same method was later used to synthesize the aromatic D-amino acids: N-hydroxytyrosine (**35**) and N-hydroxyphenylalanine (**34**) (*159*). Reaction of hydroxylamine with α,α'-dibromopimelic acid gave 1-hydroxy-2,6-piperidine dicarboxylic acid (**134**) (*158*). SHIN (*160*) studied the reaction of esters (**128**) and (**129**) with hydroxylamine in the presence of triethylamine and obtained the ethyl (**77**) and *tert*-butyl (**130**) esters of N-hydroxyamino acids as oils.

$$Br—CHR—COOR_1 \ \xrightarrow{\ NH_2OH\ } \ HO—NH—CHR—COOR_1$$

(**127**) R_1 = H   (**1**) R_1 = H
(**128**) R_1 = Et   (**77**) R_1 = Et
(**129**) R_1 = tBu   (**130**) R_1 = tBu

(**1**) R = H (**28**), Me (**29**), Et (**49**), nPr (**41**), iPr (**42**), nBu (**57**), iBu (**66**), Ph (**88**), Bzl (**34**), $H_3C—S—(CH_2)_2$ (**131**), p-HO—Bzl (**35**)
(**77**) R = Et (**91**), nPr (**79**), Bzl (**93**)
(**130**) R = Me (**132**), Et (**133**), nPr (**134**)

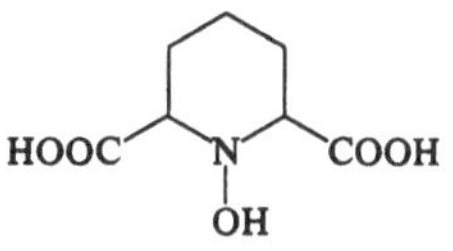

HOOC   COOH

N

OH

(**135**)
Scheme 29

Anhydrous hydroxylamine which is explosive was used for the synthesis of alanosine (18) (161), N-hydroxylysine (74), N-hydroxyornithine (73) and lower homologues (136, 137) (162). The ligand of amavadine was synthesized as a racemate using this method (87).

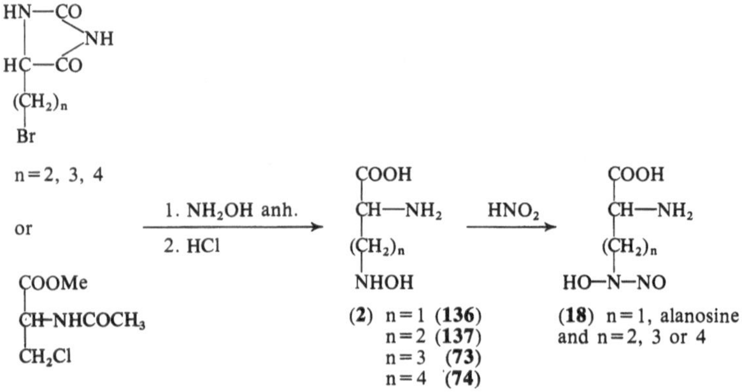

Scheme 30

7. Reaction of α-Halo Acids with Benzyloxyamine

As described in section 6 the reaction of hydroxylamine with bromo derivatives (127) is complex. Therefore it is better to use stable alkoxyamines for the substitution reaction. In a second step which used acids or boron trifluoroacetate the protecting O-alkyl group is removed to form the N-hydroxyamino acid (163) (Scheme 31).

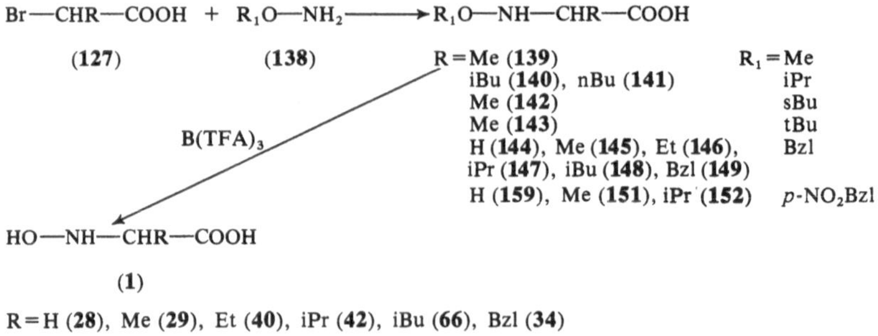

Scheme 31

The synthesis of N-benzyloxyaspartic acid (154) from bromoacid (153) and benzyloxyamine was complicated by formation of the β-N-benzyloxyamide of malic acid (156) as by-product (164) from the β-lactone intermediate (155) (Scheme 32). Reaction of bromosuccinic acid diesters (157) or O-tosyl malic diesters (158) with benzyloxyamine gave the desired esters of N-benzyloxyaspartic acid (159–161) (Scheme 33) but efforts to isolate and purify N-hydroxyaspartic acid were unsuccessful although its di-*tert*-butyl ester was obtained. Additionally *threo*-β-hydroxy-N-benzyloxyaspartic acid (163) was synthesized by reaction of *cis*-epoxysuccinic acid (162) with O-benzylhydroxylamine (*164*) (Scheme 34).

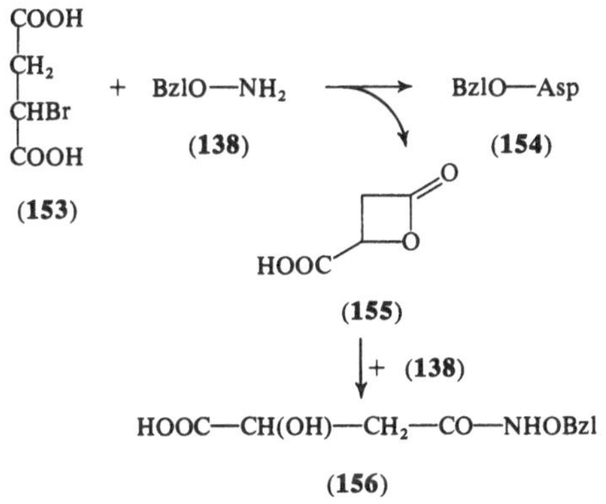

$$COOH$$
$$|$$
$$CH_2$$
$$|\qquad + \quad BzlO—NH_2 \quad \longrightarrow \quad BzlO—Asp$$
$$CHBr$$
$$|$$
$$COOH \qquad\qquad (138) \qquad\qquad\qquad (154)$$

(153)

HOOC / (155) lactone

$$\downarrow + \ (138)$$

$$HOOC—CH(OH)—CH_2—CO—NHOBzl$$

(156)

Scheme 32

$$ROOC—CH_2—CHX—COOR \ + \ (138) \ \longrightarrow \ ROOC—CH_2—CH(NHOBzl)—COOR$$

(157) X = Br   R = Me (159), Et (160), tBu (161)
(158) X = OTos

Scheme 33

$$Na^+{}^-OOC—CH—CH—COO^-Na^+ \ + \ (138)·HCl$$

(162)

$$\longrightarrow \ HOOC—CH(OH)—CH(NHOBzl)—COOH$$

(163)

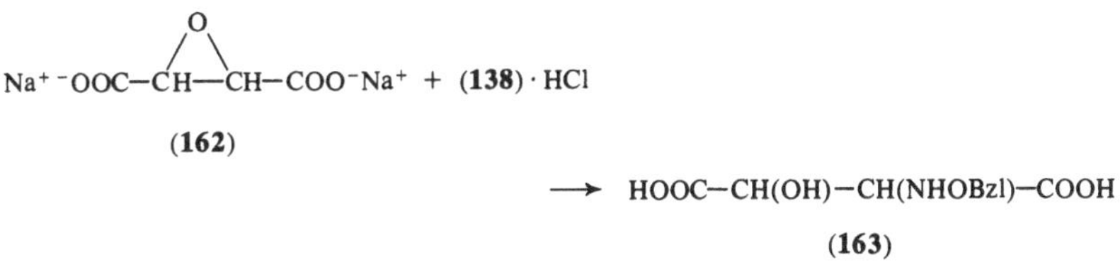

Scheme 34

Work at the Technical University of Gdańsk has shown that N-benzyloxyamino acids are the best substrates for the synthesis of N-hydroxypeptides with hydroxylamine-protected oxygen because they crystallize well and, what is most important, the benzyl group may be removed easily by hydrogenolysis (*165*). Several optically active esters of type (**159–161**) have been synthesized using this method (*163, 164*). However N^5-benzyloxyornithine (**164**) undergoes cyclization during acetylation to the N-acetyl-N-benzyloxy-lactam (**165**) (*166*) (Scheme 35).

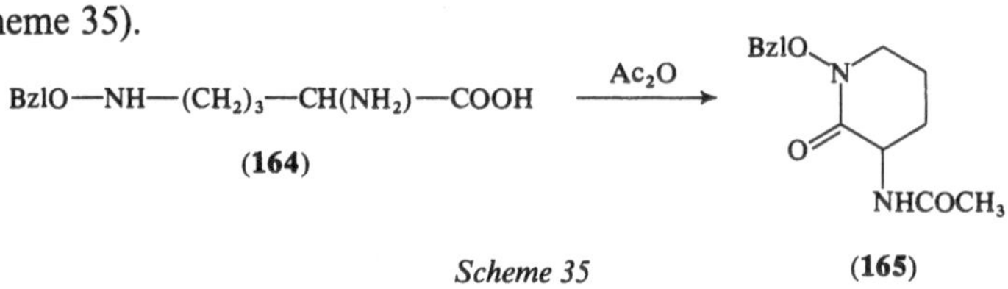

BzlO—NH—$(CH_2)_3$—CH(NH_2)—COOH $\xrightarrow{Ac_2O}$

(**164**)

Scheme 35　　　　　　(**165**)

8. From N-Tosyl-O-Benzylhydroxylamine

A procedure using two protecting groups on the hydroxyamine function was introduced by ISOWA for the synthesis of L- and D-N^5-hydroxyornithine (*167, 168*) and L-N^6-hydroxylysine (*169*) (Scheme 36). The key step was mono-substitution of a dihalide by tosylamide anion (**166**). After reaction of (**167**) with diethyl acetamidomalonate anion (**168**) and subsequent hydrolysis, the amino acids (**169**)

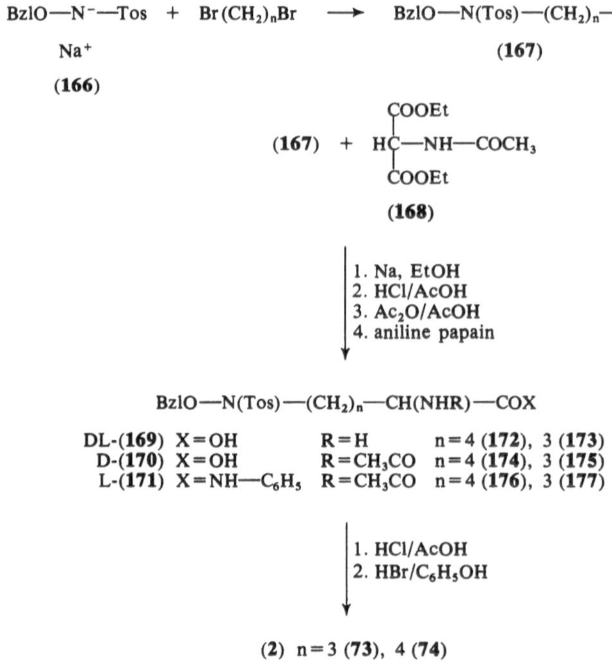

BzlO—N^-—Tos　+　Br$(CH_2)_n$Br　$\longrightarrow$　BzlO—N(Tos)—$(CH_2)_n$—Br

Na$^+$　　　　　　　　　　　　　　　　　　　　　　　　(**167**)

(**166**)

COOEt
|
(**167**)　+　HC—NH—$COCH_3$
|
COOEt

(**168**)

1. Na, EtOH
2. HCl/AcOH
3. Ac_2O/AcOH
4. aniline papain

BzlO—N(Tos)—$(CH_2)_n$—CH(NHR)—COX

DL-(**169**)	X=OH	R=H	n=4 (**172**), 3 (**173**)
D-(**170**)	X=OH	R=CH_3CO	n=4 (**174**), 3 (**175**)
L-(**171**)	X=NH—C_6H_5	R=CH_3CO	n=4 (**176**), 3 (**177**)

1. HCl/AcOH
2. HBr/C_6H_5OH

(**2**)　n=3 (**73**), 4 (**74**)

Scheme 36

were acetylated and the acetyl derivatives resolved with papaine. The resulting L-anilides (171) were hydrolysed to the N-hydroxyamino acids.

The synthesis of alanosine by ISOWA *et al.* (*170*) followed a similar course (Scheme 37). Resolution with papaine gave anilide (181), which was transformed into the benzoyl derivative (182). The O-benzyl group was hydrogenolytically removed from amide (182), giving L-β-N-hydroxyamino acid (136) which was finally nitrosated to L-alanosine (18). DL-N-Hydroxyamino acid (136) was similarly obtained from N-tosyl-2,4,6-trimethylbenzyloxyamine (178) (*171*). On the other hand FUJII (*172*), during the synthesis of rhodotorulic acid, resolved N-acetyl enantiomers by means of enzymatic selective deacetylation with diastase.

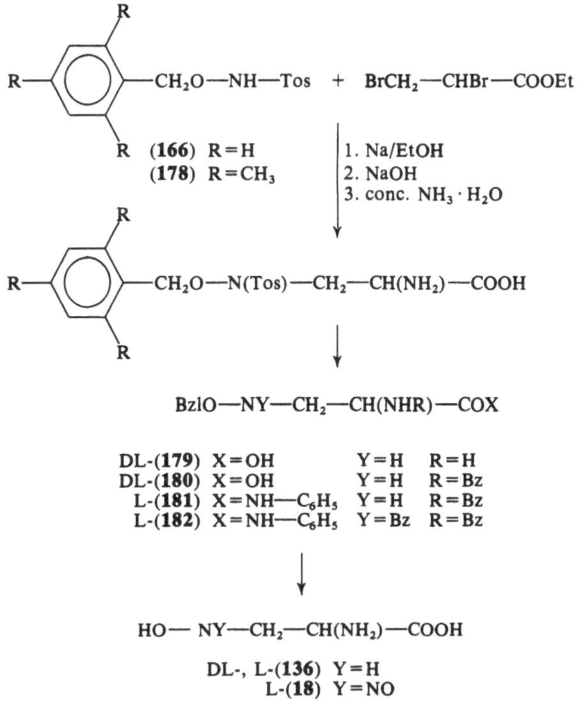

R—⟨O⟩—CH₂O—NH—Tos   +   BrCH₂—CHBr—COOEt

with R substituents labeled R on the ring.

(166) R = H
(178) R = CH₃

1. Na/EtOH
2. NaOH
3. conc. NH₃·H₂O

R—⟨O⟩—CH₂O—N(Tos)—CH₂—CH(NH₂)—COOH

BzlO—NY—CH₂—CH(NHR)—COX

DL-(179)	X = OH	Y = H	R = H
DL-(180)	X = OH	Y = H	R = Bz
L-(181)	X = NH—C₆H₅	Y = H	R = Bz
L-(182)	X = NH—C₆H₅	Y = Bz	R = Bz

HO— NY—CH₂—CH(NH₂)—COOH

DL-, L-(136) Y = H
L-(18) Y = NO

Scheme 37

This method was also used in the total synthesis of ferrichrome from N⁵-benzyloxyornithine (164) (*173*) and in syntheses of the phosphonic antibiotics (22–25) (*174, 175, 176*) from (167) using the Michaelis-Becker reaction (Scheme 38).

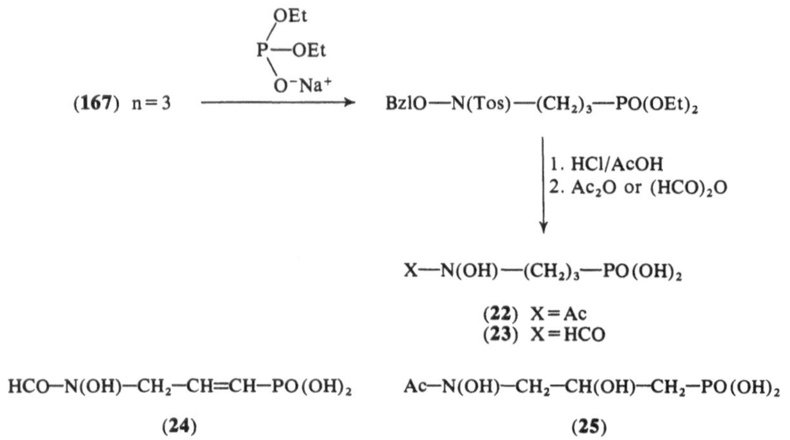

(167) n = 3 ⟶ BzlO—N(Tos)—(CH₂)₃—PO(OEt)₂

> 1. HCl/AcOH
> 2. Ac₂O or (HCO)₂O

X—N(OH)—(CH₂)₃—PO(OH)₂

(22) X = Ac
(23) X = HCO

HCO—N(OH)—CH₂—CH=CH—PO(OH)₂   Ac—N(OH)—CH₂—CH(OH)—CH₂—PO(OH)₂

(24)   **(25)**

Scheme 38

9. From Nitrones

9.1. Substitution

In 1896 Hantzsch (*155*) obtained N-hydroxyglycine (**28**) by reaction of benzaldoxime with chloroacetic acid and hydrolysis of the intermediate nitrone (**190**) with hydrochloric acid. Much later Buehler (*177*) observed that alkylation of Z-benzaldoxime (**187**) leads to nitrone formation and he generalized this method for synthesis of N-hydroxy-amino acids (**1**) (*178*) (Scheme 39). Sodium or potassium salts of Z-benzaldoxime (**187**) (*179*) as well as thallium salts (**180**), can be used

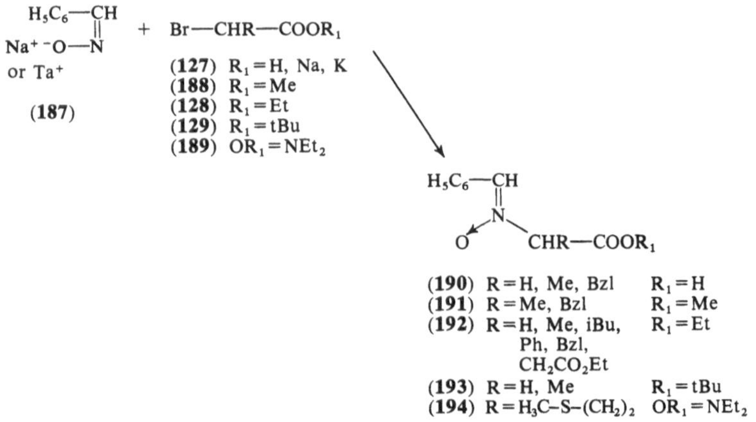

H₅C₆—CH
‖
Na⁺ ⁻O—N   +   Br—CHR—COOR₁
or Ta⁺
            (127) R₁ = H, Na, K
(187)   (188) R₁ = Me
            (128) R₁ = Et
            (129) R₁ = tBu
            (189) OR₁ = NEt₂

H₅C₆—CH
‖
N
O⁺   CHR—COOR₁

(190) R = H, Me, Bzl   R₁ = H
(191) R = Me, Bzl   R₁ = Me
(192) R = H, Me, iBu,   R₁ = Et
      Ph, Bzl,
      CH₂CO₂Et
(193) R = H, Me   R₁ = tBu
(194) R = H₃C–S–(CH₂)₂   OR₁ = NEt₂

Scheme 39

to give nitrones. Nitrones (190–194) were obtained in crystalline form and in good yields by alkylation of bromoacids (127) and their sodium (178) and potassium (181) salts. Amides such as (189) can also be used as substrates (182), but it is easier to alkylate esters (128, 129, 188) (178, 183).

As substitution of α-bromoesters with benzaldoxime (187) proceeds with inversion, D-N-hydroxyphenylalanine (34) had been obtained in this way (188). Benzyl and p-nitrobenzyl nitrone esters (212) may also be obtained (183) by alkylation of triethylammonium nitrone salts (190) (Scheme 40).

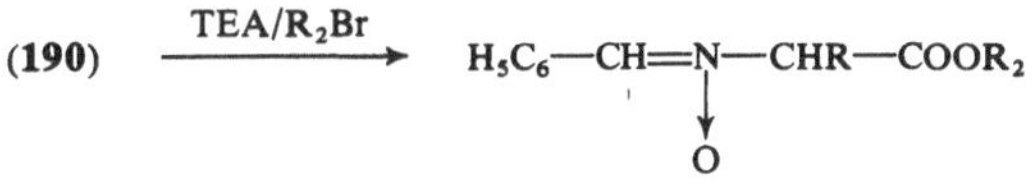

$$(190) \xrightarrow{\text{TEA}/R_2\text{Br}} H_5C_6-CH\!=\!\overset{\displaystyle \downarrow}{\underset{\displaystyle O}{N}}-CHR-COOR_2$$

(212) R = H or Me
       R_2 = Bzl or p-NO_2—Bzl

Scheme 40

Ross-Petersen and Hjeds (186) carried out this substitution reaction with 5-bromo-γ-butyrolactone obtaining cyclic nitrone (209) and from it by hydrazinolysis 5-hydroxyamino-butyrolactone (210) and N-hydroxyhomoserine hydrazide (211) (Scheme 41).

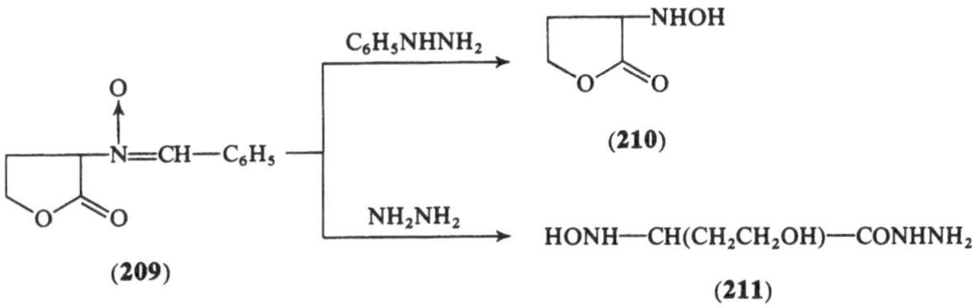

Scheme 41

Lau and Schöllkopf (184) prepared α-mono- or disubstituted nitrone esters (195) (Scheme 42) by metalation of (191) with potassium tert-butylate in THF and subsequent alkylation using alkyl halides. Hydrolysis of (195) then furnishes α,α-disubstituted N-hydroxyamino acids (67, 196) and their methyl esters (76).

Nitrones (192, 193) may also be alkylated with Meerwein's reagent and dimethyl sulfate to give esters of N-alkoxyamino acids (207, 208) after hydrolysis (185) (Scheme 43).

Finally it should be pointed out that β-N-hydroxyamino acids have also been prepared by the nitrone technique (187).

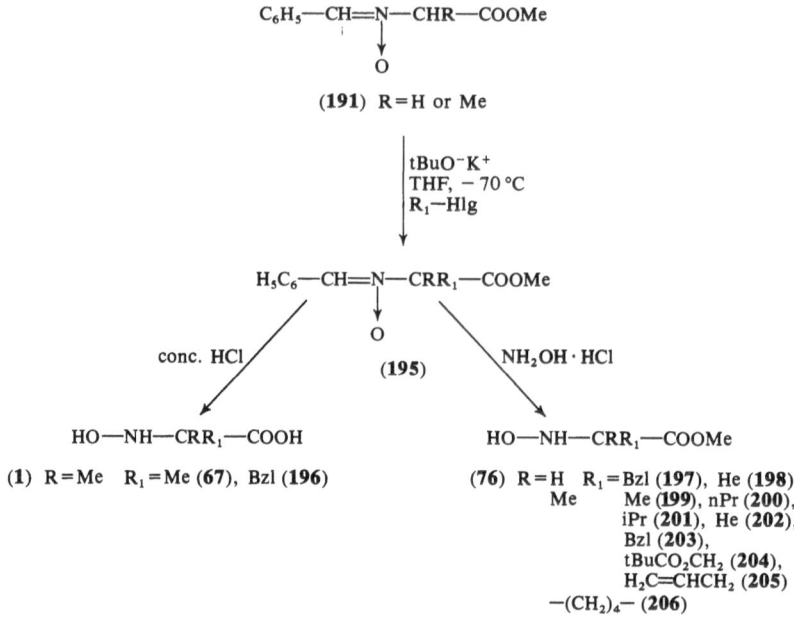

Scheme 42

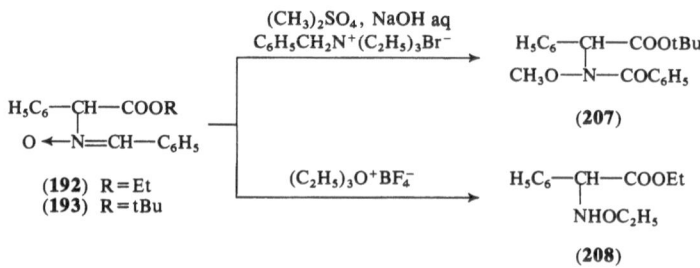

Scheme 43

The splitting of nitrones may be carried out by hydrolysis with hydrochloric acid (*155, 178, 182*), hydrazinolysis (*182, 186*) or most suitably hydroxylaminolysis (*183*) (Scheme 44). Esters (**191–193, 212**) are easily converted into N-hydroxyamino acid esters (**76, 77, 130, 213**) by action of hydroxylamine or its crystalline salts such as hydroxylamine hydrochloride (*184*), oxalate or toluenosulfonate (*183*) without disturbing the ester function. The salts of the last type are easily crystallized.

Nitrone intermediates have been used in syntheses of hadacidin (**13**) (*189*), alanosine (**18**) (*190*), N^5-hydroxyornithine (**73**) and N^5-hydroxyarginine (**224**) (*180*), as well as 2-methyl-N^5-hydroxyornithine (**225**)

(190), (192) $\xrightarrow{\text{conc. HCl or NH}_2\text{NH}_2}$ HO—NH—(CH$_2$)$_n$—CHR—COOH

(1), (2)

(1) R=H (28), Me (29), Ph (88), Bzl (34), (CH$_2$)$_2$COOH (39)
(2) R=H   n=1 (214), 2 (215)
(2) R=Ph n=1 (216)

(191), (192), (193), (212) $\xrightarrow{\text{NH}_2\text{OH or A}^-\ ^+\text{NH}_3\text{OH}}$ HO—NH—CHR—COOR$_1$

(76), (77), (130), (213)

(76) R=Me (217), Bzl (197)   R$_1$=Me
(77) R=Me (90), iBu (218)   R$_1$=Et
(130) R=H (219), Me (132)   R$_1$=tBu
(213) R=H (220), Me (221)   R$_1$=Bzl

Scheme 44

and 2-methyl-N^5-hydroxyarginine (226) (*83*). α,α′-Dibromocarboxylic acids give symmetrical dinitrones (227), and thence N,N′-dihydroxy-aminodicarboxylic acids (228) (*191*) (Scheme 45). Thus, use of α,α′-dibromopimelic acid finally gave di-N-hydroxyaminopimelic acid (*182*).

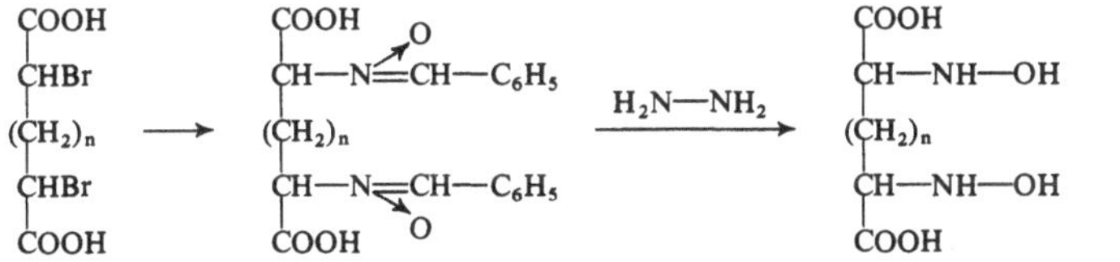

(227)   (228) n=2, 3, 4, 5, 6

Scheme 45

9.2. Addition

TESTA *et al.* (*187*) added benzaldehyde oxime (187) to α,β-unsaturated esters (121) and diethylamides (229) obtaining nitrones (191, 192, 194) and from them esters of β-N-hydroxyamino acids (230) and their derivatives (231) (Schema 46).

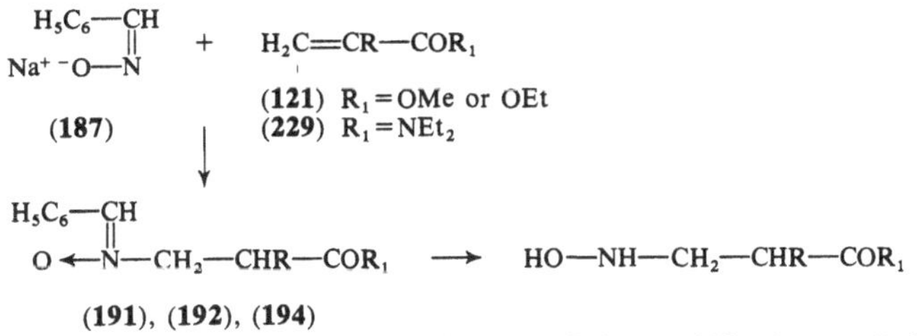

(187)

(121) R$_1$=OMe or OEt
(229) R$_1$=NEt$_2$

(191), (192), (194)

(230) R=H (232), Cl (233), CH$_3$CONH (234) R$_1$=OMe
          Ph (235)   OEt
(231) R=H (236), Ph (237)   R$_1$=NEt$_2$

Scheme 46

10. From Oxaziridines

This method was introduced by Połoński and Chimiak (*192, 193*) in 1974 (Scheme 47). It is based on the oxidation of Schiff bases (**239**) to appropriate oxaziridines (**240**) in ether using monoperphtalic acid (MPP). Bases (**239**) are obtained from esters of amino acids and anisyl aldehyde (**238**) and are oxidised without isolation. Oxaziridines (**240**) are next hydrolyzed with hydrochloric acid to N-hydroxyamino acids (**1**) or give *p*-toluenesulfonates of (**76, 213**), which crystallize readily, by splitting with hydroxylamine *p*-toluenesulfonates in alcohol. Use of benzaldehyde is unfavourable and leads to nitrones. Use of mono-perphthalic acid permits one to follow the progress of the reaction due to precipitation of phthalic acid. This method is general. Because bases (**239**) racemize only very slowly it is possible to obtain (*193*) optically active compounds (**1, 76, 213**).

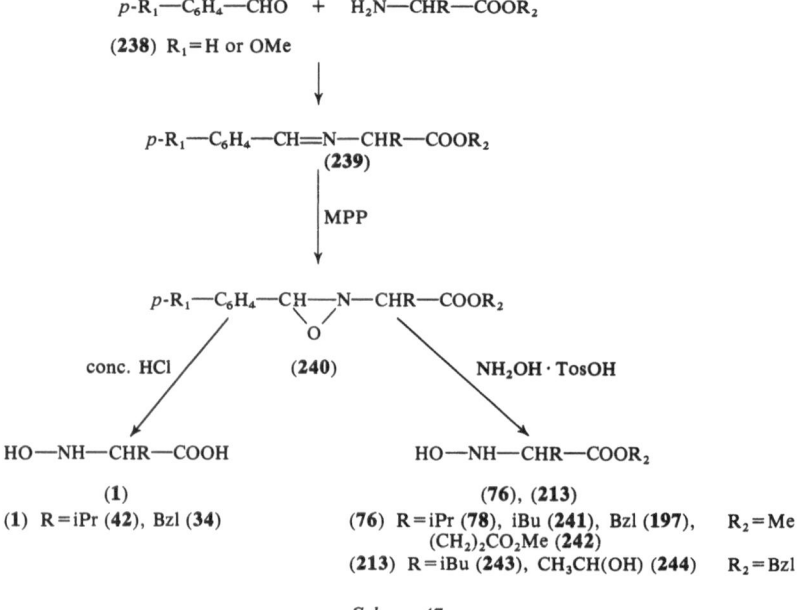

p-R₁—C₆H₄—CHO + H₂N—CHR—COOR₂

(**238**) R₁ = H or OMe

p-R₁—C₆H₄—CH═N—CHR—COOR₂
(**239**)

MPP

p-R₁—C₆H₄—CH—N—CHR—COOR₂
\ /
O

conc. HCl   (**240**)   NH₂OH · TosOH

HO—NH—CHR—COOH   HO—NH—CHR—COOR₂

(**1**)   (**76**), (**213**)
(1) R = iPr (**42**), Bzl (**34**)   (**76**) R = iPr (**78**), iBu (**241**), Bzl (**197**), R₂ = Me
                                      (CH₂)₂CO₂Me (**242**)
                                      (**213**) R = iBu (**243**), CH₃CH(OH) (**244**) R₂ = Bzl

Scheme 47

In the same year Widmer and Keller-Schierlein (*194*) used an oxaziridine for the synthesis of N⁵-hydroxyarginine (**224**) (Scheme 48). However, in their synthesis *m*-chloroperbenzoic acid was used for oxidation of Schiff base and Dowex 50 X2 for cleavage of the oxaziridine. The resulting derivative of N⁵-hydroxyornithine (**245**) was transformed

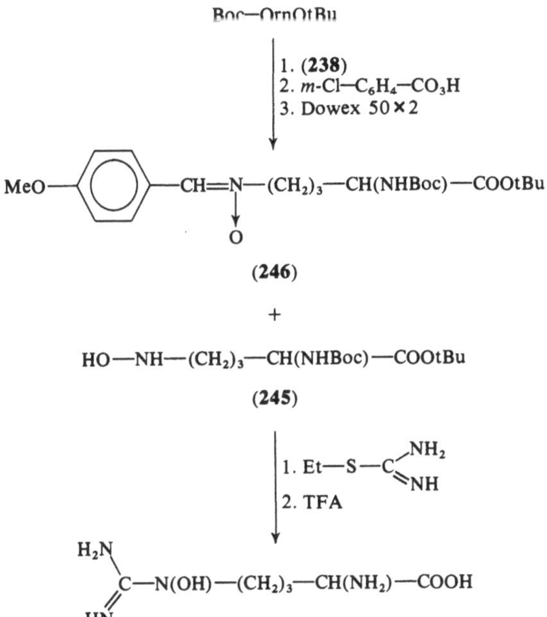

Boc—OrnOtBu

1. **(238)**
2. m-Cl—C_6H_4—CO_3H
3. Dowex 50 × 2

MeO—⟨benzene ring⟩—CH=N—(CH$_2$)$_3$—CH(NHBoc)—COOtBu
                                      |
                                      O

(246)

+

HO—NH—(CH$_2$)$_3$—CH(NHBoc)—COOtBu

(245)

1. Et—S—C(NH$_2$)=NH
2. TFA

H$_2$N\
    C—N(OH)—(CH$_2$)$_3$—CH(NH$_2$)—COOH
HN⁄

(224)

Scheme 48

into a derivative of N^5-hydroxyarginine by action of S-ethylisothio-urea; subsequent removal of protective groups was carried out with trifluoroacetic acid. Oxidation of the Schiff base also produces nitrone **(246)**, which has been utilized to prepare N^5-hydroxyornithine (*194*). The oxaziridine method of obtaining N^5-hydroxyornithine residues was also applied to preparation of cyclo-tri(D-N-hydroxyornithyl)-trigly-cine during the synthesis of enantioferrichrome **(247)** (*195*) (Scheme 49).

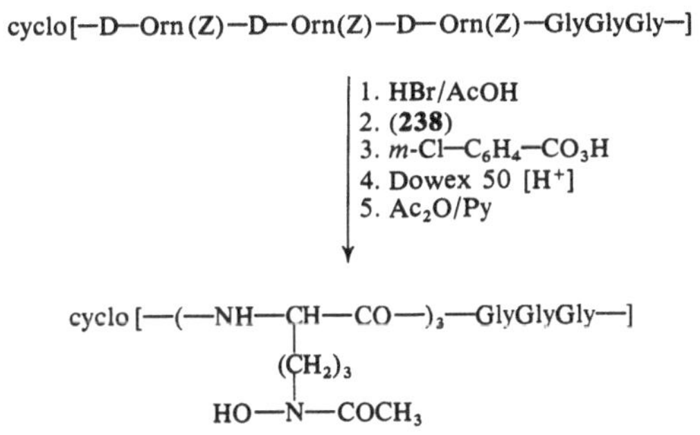

cyclo[−D−Orn(Z)−D−Orn(Z)−D−Orn(Z)−GlyGlyGly−]

1. HBr/AcOH
2. **(238)**
3. m-Cl—C_6H_4—CO_3H
4. Dowex 50 [H⁺]
5. Ac$_2$O/Py

cyclo [—(—NH—CH—CO—)$_3$—GlyGlyGly—]
                   |
                (CH$_2$)$_3$
                   |
         HO—N—COCH$_3$

(247) enantioferrichrome

Scheme 49

11. Alkylation of Hydroxamic Acids

Alkylation of hydroxamic acids as a method of ω-N-hydroxyamino acids (**2**) synthesis was introduced by MAURER and MILLER (*196*). When N-*tert*-butoxycarbonyl-6-hydroxynorleucine benzylhydroxamate (**248**) or a homologue was treated with triphenylphosphine and diethylazodicarboxylate (DEAD) under Mitsunobu conditions (*197*), intramolecular alkylation took place leading to N-hydroxylactams (**249**) or (**250**) as well as lesser amounts of hydroximates *Z*-(**251**) and *E*-(**252**) (Scheme 50). The products were separated and distinguished by NMR spectrometry (*196, 198, 199*). Derivatives of the seven-membered N-hydroxylactam (**253**) were applied for the total synthesis of mycobactin S2 (**254**) (*199*) (Scheme 51).

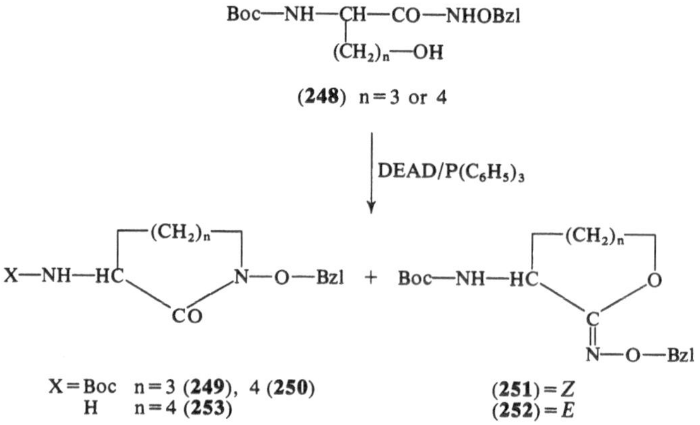
Boc—NH—CH—CO—NHOBzl
(CH2)n—OH
(248) n=3 or 4
DEAD/P(C6H5)3
(CH2)n
X—NH—HC
N—O—Bzl
CO
+
Boc—NH—HC
(CH2)n
O
C
N—O—Bzl
X=Boc n=3 (249), 4 (250)
H n=4 (253)
(251)=Z
(252)=E

Scheme 50

In the alkylation reaction appropriate bromo derivatives (**256**), obtained directly from N^6-hydroxynorleucine (**255**) or in several steps from glutamic acid, may be used (Scheme 52). From acetylbenzyloxyamine in the presence of potassium carbonate a mixture of N-(**257–259**) and O-alkylderivatives (**260, 261**) (*Z* and *E* isomers) was formed. After separation the products were used for synthesis of N^6-hydroxylysine (*200*), N^5-hydroxyornithine (*198, 201*) and their derivatives (**257, 258, 259**).

In the latest of a series of papers MILLER and coworkers (*202*) described the alkylation of hydroxamic acids (**263**) with alcohol (**262**) under Mitsunobu conditions (*197*). In this manner, starting from glutamic acid they obtained several derivatives of N^5-hydroxyornithine (**264–269**) which were used subsequently in the synthesis of rhodotorulic acid (*202*) (Scheme 53).

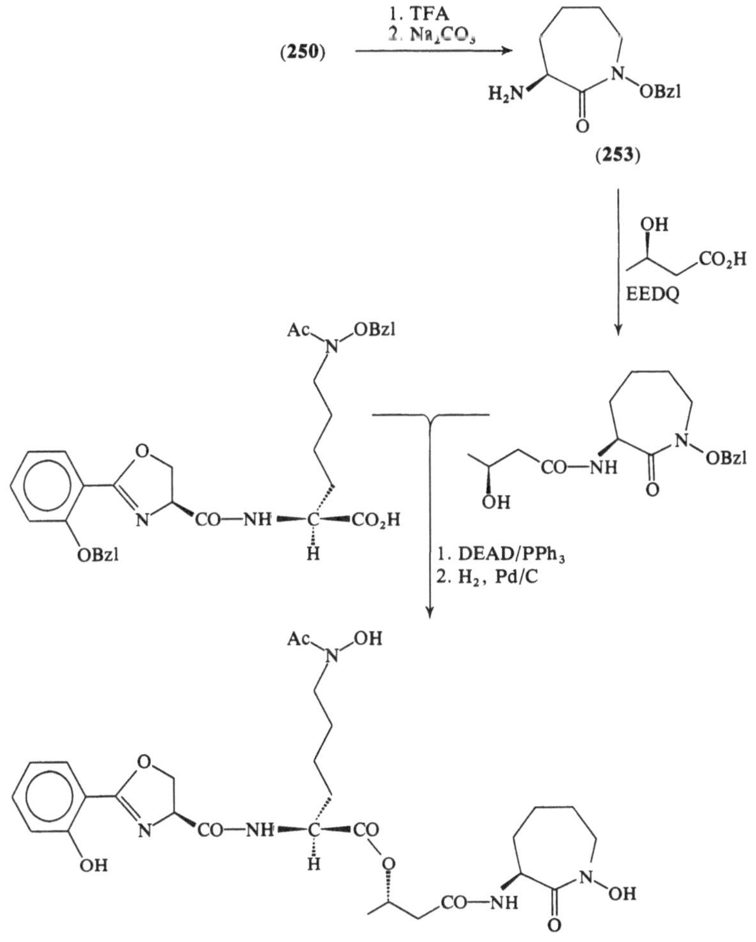

(**254**) mycobactin S2

Scheme 51

$HO—(CH_2)_n—CH(NHX)—COOR$

(**255**)

↓ $P(C_6H_5)$, CBr_4

$Br—(CH_2)_n—CH(NHX)—COOR$

(**256**)

Scheme 52

(256)

$$\Big\downarrow \text{K}_2\text{CO}_3, \ \text{BzlO—NH—Ac}$$

BzlO—N(Ac)—(CH$_2$)$_n$—CH(NHX)—COOR

(257) X = Boc　R = Me　n = 4
(258) X = Z　　R = Bzl　n = 4
(259) X = Boc　R = tBu　n = 3

+

BzlO~N≡C(Me)—O—(CH$_2$)$_n$—CH(NHX)—COOR

(260) = Z
(261) = E

Scheme 52 (continued)

Boc—L—GluOtBu　$\xrightarrow[\text{2. NaBH}_4]{\text{1. EtOCOCl/TEA}}$　HO—(CH$_2$)$_3$—CH(NHBoc)—COOtBu

(262)

$$\Big\downarrow \begin{array}{l} \text{DEAD, PPh}_3, \\ \text{BzlO—NH—Y} \\ \textbf{(263)} \ \text{Y = Z, Ac or trOC} \end{array}$$

BzlO—NY—(CH$_2$)$_3$—CH(NHX)—COOR

Y = Z　　X = Boc　R = tBu (264)
Y = trOC　X = Boc　R = tBu (265), H (266)
Y = Ac　　X = Boc　R = H (267)
Y = trOC　X = H　　R = Me (268)
Y = Ac　　X = H　　R = Me (269)

(267) + (269)

$$\Big\downarrow \text{EEDQ}$$

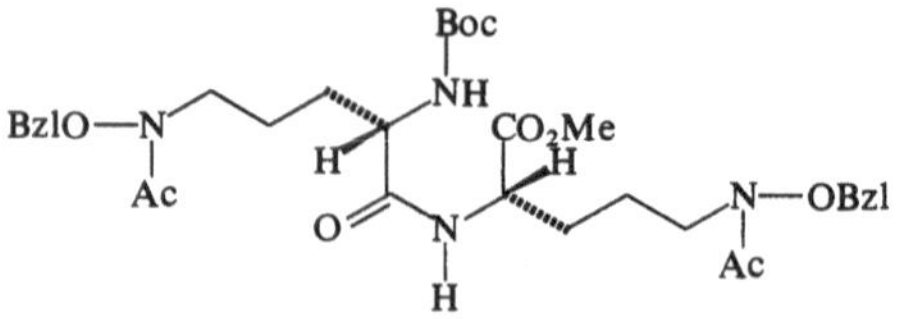

1. TFA
2. MeOH
3. H$_2$, Pd/C

(6) rhodotorulic acid

Scheme 53

12. Synthesis of N-Hydroxy-*sec*-amino Acids

As shown in Scheme 54, proline was transformed into the N-cyano-ethyl derivative (270) and oxidised with *m*-chloroperbenzoic acid to the *trans*-N-oxide (271). Cope elimination in boiling acetone gave N-hydroxyproline (44). N-Hydroxypipecolic acid (45) and N-hydroxysarcosine (43) were obtained analogously (*101*).

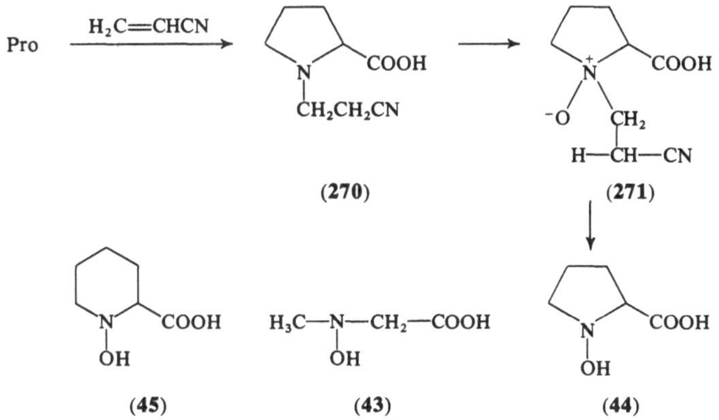

Pro $\xrightarrow{\text{H}_2\text{C}=\text{CHCN}}$

(270)

(271)

(45)   (43)   (44)

Scheme 54

As already mentioned N-hydroxy-*sec*-amino acids (118) may also be obtained by addition of the N-substituted hydroxylamine to a double bond (*149*).

13. Esterification of N-Hydroxyamino Acids and N-Alkoxyamino Acids

Apart from the synthetic methods for esters described in sections: VIII.3–VIII.7 and VIII.9–VIII.11, N-hydroxyamino acid esters (76, 77, 130, 213) may also be obtained by acid-catalysed esterification of N-hydroxyamino acids (*80, 100, 160*) although the best reagent for quantitative esterification is diazomethane (*80*) (Scheme 55). When this reagent is used N-methyl derivatives are not formed although this had been suspected previously (*100*).

N-Benzyloxyamino acids have also been esterified (*203*) to give O-protected substrates (275–277) for the synthesis of N-hydroxypeptides (Scheme 56). To obtain methyl (275), *tert*-butyl (276) and benzyl (277) esters diazomethane, *tert*-butyl acetate with perchloric acid and benzyl chloride with acetoacetate were used, respectively. N-Isopropoxyamino acid esters (291) have also been prepared (*203*).

N-Benzyloxyamino acid *p*-nitrophenyl esters (297) have been obtained from α-keto acids *via* benzyloxyimino acids (295), their esterifica-

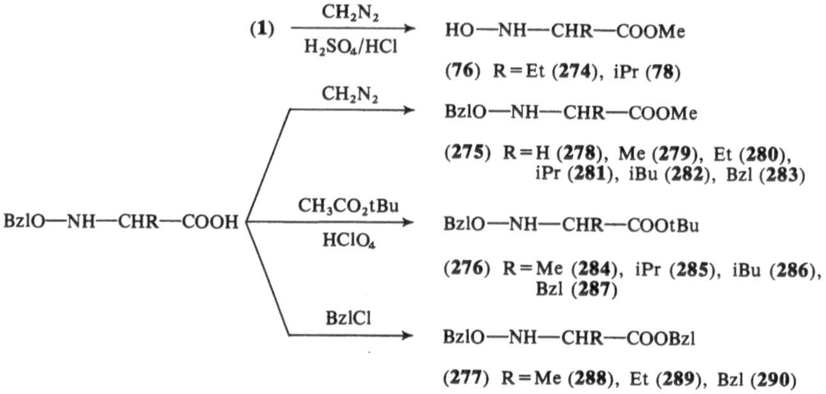

$$(1) \xrightarrow[\text{H}_2\text{SO}_4/\text{HCl}]{\text{CH}_2\text{N}_2} \quad \text{HO—NH—CHR—COOMe}$$

$$(76) \quad R = \text{Et } (274), \text{ ipr } (78)$$

$$\xrightarrow{\text{CH}_2\text{N}_2} \quad \text{BzlO—NH—CHR—COOMe}$$

$$(275) \quad R = H \ (278), \ Me \ (279), \ Et \ (280),$$
$$\text{ipr } (281), \text{ iBu } (282), \text{ Bzl } (283)$$

$$\text{BzlO—NH—CHR—COOH} \xleftarrow[\text{HClO}_4]{\text{CH}_3\text{CO}_2\text{tBu}} \quad \text{BzlO—NH—CHR—COOtBu}$$

$$(276) \quad R = \text{Me } (284), \text{ ipr } (285), \text{ iBu } (286),$$
$$\text{Bzl } (287)$$

$$\xrightarrow{\text{BzlCl}} \quad \text{BzlO—NH—CHR—COOBzl}$$

$$(277) \quad R = \text{Me } (288), \text{ Et } (289), \text{ Bzl } (290)$$

Scheme 55

$$\text{iPrO—NH—CHR—COOH} \xrightarrow[\text{CH}_3\text{CO}_2\text{tBu, HClO}_4]{\text{CH}_2\text{N}_2 \text{ or}} \quad \text{iPrO—NH—CHR—COOR}_1$$

$$(291) \quad R = \text{iBu} \quad R_1 = \text{Me } (292),$$
$$\text{tBu } (293)$$
$$\text{ipr} \qquad \text{Me } (294)$$

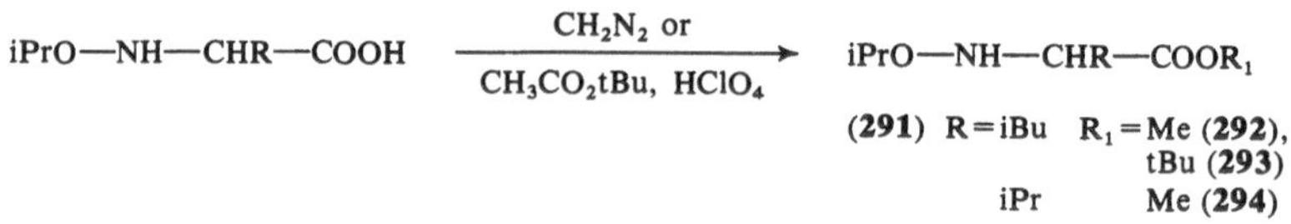

Scheme 56

tion in the presence of dicyclohexylcarbodiimide (DCCI), and finally by reduction with borohydride (*204*) (Scheme 57).

After describing the various methods of preparing N-hydroxyamino acids and their derivatives we have listed in Table 4 all the compounds of this class synthesized so far together with literature references to the original papers.

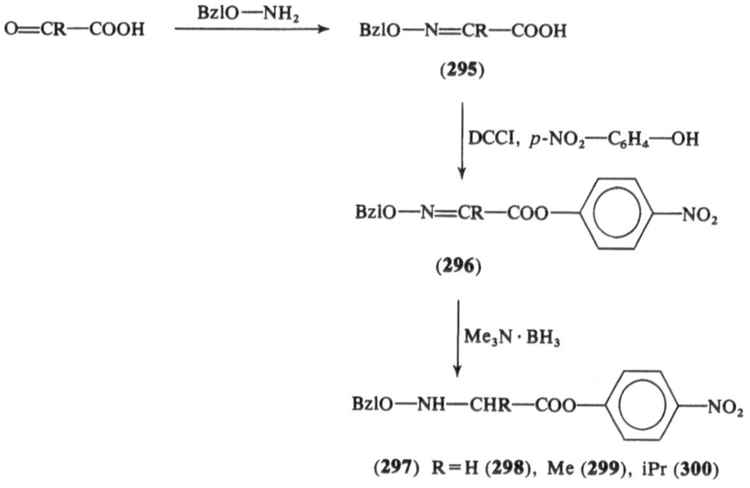

$$\text{O}{=}\text{CR—COOH} \xrightarrow{\text{BzlO—NH}_2} \quad \text{BzlO—N}{=}\text{CR—COOH}$$

$$(295)$$

$$\Big\downarrow \text{DCCI, } p\text{-NO}_2\text{—C}_6\text{H}_4\text{—OH}$$

$$\text{BzlO—N}{=}\text{CR—COO—}⟨○⟩\text{—NO}_2$$

$$(296)$$

$$\Big\downarrow \text{Me}_3\text{N} \cdot \text{BH}_3$$

$$\text{BzlO—NH—CHR—COO—}⟨○⟩\text{—NO}_2$$

$$(297) \quad R = H \ (298), \ Me \ (299), \ \text{ipr } (300)$$

Scheme 57

Table 4. *Synthetic N-Hydroxyamino Acids and Their Derivatives*

1. N²-Hydroxyamino Acids R—CH—COR₂ with NHOR₃

N-Hydroxyamino acid	Config.	No.	R or n	R_1	R_2	R_3	R_4	Synthetic method VIII.
N-Hydroxyglycine	—	(28)	H	—	HO	H	—	4 (144), 6 (158), 9.1 (178)
	—	(89)		—	EtO	H	—	4 (145, 146)
	—	(219)		—	tBuO	H	—	6 (160), 9.1 (183)
	—	(220)		—	BzlO	H	—	9.1 (183)
	—	(222)		—	p-NO$_2$—BzlO	H	—	9.1 (183)
	—	(94)		—	MeNH	H	—	4 (146)
	—	(144)		—	HO	Bzl	—	7 (163)
	—	(278)		—	MeO	Bzl	—	7 (203)
	—	(99)		—	EtO	Bzl	—	4 (145, 146)
	—			—	tBuO	Bzl	—	7 (165)
	—	(298)		—	p-NO$_2$—C$_6$H$_4$O	Bzl	—	4 (204)
	—	(106)		—	MeNH	Bzl	—	4 (146)
	—	(150)		—	HO	p-NO$_2$—Bzl	—	7 (163)
	—			—	MeO	p-NO$_2$—Bzl	—	7 (203)
N-Hydroxyalanine	DL	(29)	CH$_3$	—	HO	H	—	2 (79), 4 (144), 6 (158), 9.1 (178)
	DL	(217)		—	MeO	H	—	9.1 (183)
	DL	(90)		—	EtO	H	—	3 (80), 4 (145, 146), 6 (160), 9.1 (183)
	DL	(132)		—	tBuO	H	—	6 (160), 9.1 (183)
	DL	(221)		—	BzlO	H	—	9.1 (183), 10 (192)
	DL	(223)		—	p-NO$_2$—BzlO	H	—	9.1 (183)

A. CHIMIAK and MARIA J. MILEWSKA:

Table 4 *(continued)*

N-Hydroxyamino acid	Config.	No.	R or n	R₁	R₂	R₃	R₄	Synthetic method VIII.
N-Hydroxyalanine	DL	(95)		—	MeNH	H	—	4 (146)
	DL	(145)		—	HO	Bzl	—	7 (163)
	D, L	(279)		—	MeO	Bzl	—	7 (165, 203)
	DL	(100)		—	EtO	Bzl	—	4 (145, 146)
	L	(284)		—	tBuO	Bzl	—	7 (165, 203)
	D	(288)		—	BzlO	Bzl	—	7 (165, 203)
	DL	(299)		—	$p\text{-}NO_2\text{-}C_6H_4O$	Bzl	—	4 (204)
	DL	(107)		—	MeNH	Bzl	—	4 (146)
	DL	(151)		—	HO	$p\text{-}NO_2 - Bzl$	—	7 (163)
	DL	(142)		—	HO	sBu	—	7 (163)
	DL	(143)		—	HO	tBu	—	7 (163)
N-Hydroxy-2-aminobutanoic acid	DL	(40)	CH₃CH₂	—	HO	H	—	2 (79, 80), 4 (144), 6 (158)
	DL	(274)		—	MeO	H	—	2 (80)
	DL	(91)		—	EtO	H	—	3 (80), 4 (146), 6 (160)
	DL	(133)		—	tBuO	H	—	6 (160)
	DL	(96)		—	MeNH	H	—	4 (146)
	DL	(146)		—	HO	Bzl	—	7 (163)
	DL	(280)		—	MeO	Bzl	—	7 (203)
	DL	(101)		—	EtO	Bzl	—	4 (146)
	DL	(289)		—	BzlO	Bzl	—	7 (203)
	DL	(108)		—	MeNH	Bzl	—	4 (146)
N-Hydroxynorvaline	DL	(41)	CH₃CH₂CH₂	—	HO	H	—	1 (79), 2 (1, 79, 80, 130, 134)

	Config	(No.)	R		O-subst.	N-subst.		References
N-Hydroxynorvaline	DL	(79)		—	EtO	H	—	3 (80), 6 (160)
	DL	(134)		—	tBuO	H	—	6 (160)
	DL	(61)		—	NH₂	H	—	2 (130)
N-Hydroxyvaline	DL	(42)	(CH₃)₂CH	—	HO	H	—	2 (79, 80), 4 (144), 6 (158), 10 (192, 193)
	L	(78)		—	MeO	H	—	2 (80), 3 (80), 10 (192, 193)
	DL	(92)		—	EtO	H	—	4 (146)
	DL	(97)		—	MeNH	H	—	4 (146)
	D, L	(147)		—	HO	Bzl	—	7 (163)
	DL	(281)		—	MeO	Bzl	—	7 (203)
	DL	(102)		—	EtO	Bzl	—	4 (146)
	DL	(285)		—	tBuO	Bzl	—	7 (193)
	DL	(300)		—	p-NO₂—C₆H₄O	Bzl	—	4 (204)
	DL	(109)		—	MeO	Bzl	—	4 (146)
	DL	(294)		—	MeO	iPr	—	7 (193)
	DL	(152)		—	HO	p-NO₂ — Bzl	—	7 (163)
N-Hydroxyisoleucine	DL	(57)	CH₃(CH₂)₃	—	HO	H	—	1 (79), 2 (79, 82)
	DL	(141)		—	HO	iPr	—	7 (163)
N-Hydroxyleucine	DL	(66)	(CH₃)₂CHCH₂	—	HO	H	—	2 (79)
	L, D	(241)		—	MeO	H	—	6 (158)
	L	(218)		—	EtO	H	—	10 (192, 193)
	DL	(243)		—	BzlO	H	—	9.1 (183)
	L	(62)		—	NH₂	H	—	10 (193)
	DL	(148)		—	HO	Bzl	—	2 (130)
	L, D	(282)		—	MeO	Bzl	—	7 (163), 7 (203)
	L, D	(286)		—	tBuO	Bzl	—	7 (163), 7 (165)

Table 4 (*continued*)

N-Hydroxyamino acid	Config.	No.	R or n	R_1	R_2	R_3	R_4	Synthetic method VIII.
N-Hydroxyleucine	DL	(140)		—	HO	iPr	—	7 (163)
	D					iPr	—	7 (163)
	DL	(292)			MeO	iPr	—	7 (203)
	D					iPr	—	7 (203)
	D	(293)			tBuO	iPr	—	7 (203)
N-Hydroxy-2-aminooctanoic acid	DL	(198)	$CH_3(CH_2)_5$	—	MeO	H	—	9.1 (184)
	DL	(63)			NH_2	H	—	2 (130)
N-Hydroxyphenylglycine	DL	(88)	C_6H_5	—	HO	H	—	9.1 (185, 178)
	DL	—			MeO	H	—	6 (158)
	DL	—			EtO	H	—	6 (160), 9.1 (185)
	DL	(335)			tBuO	H	—	9.1 (185)
	DL	(336)			ACA	H	—	9.1 (185)
	DL	(104)			APA	H	—	9.1 (185)
	DL	(110)			EtO	Bzl	—	4 (146)
	DL	—			MeNH	Bzl	—	4 (146)
	DL	—			HO	Me	—	9.1 (185)
	DL				HO	Et	—	9.1 (185)
N-Hydroxyphenylalanine	DL	(208)	$C_6H_5CH_2$	—	EtO	Et	—	9.1 (185)
	DL	(34)			HO	H	—	1 (79, 206), 3 (206), 4 (144), 6 (158), 9.1 (178)
	D							6 (159)
	L							7 (192, 193)
	—	(197)			MeO	H	—	6 (158), 9.1 (183, 184), 10 (192, 193)
	—	(93)			EtO	H	—	6 (158, 160), 4 (145, 146)

Compound	Config	(No.)	R	X	N-subst.		Ref.
N-Hydroxyphenylalanine	–	(98)		MeNH	H	—	4 (146)
	DL	(149)		HO	Bzl	—	7 (163)
	D				Bzl	—	7 (163)
	D	(283)		MeO	Bzl	—	7 (203)
	DL	(103)		EtO	Bzl	—	4 (145, 146)
	D	(287)		tBuO	Bzl	—	7 (203)
	D	(290)		BzlO	Bzl	—	7 (203)
	DL	(111)		MeNH	Bzl	—	4 (146)
N-Hydroxyaspartic acid	DL	—	$CH_2COOtBu$	tBuO	H	—	7 (164)
	DL	(161)		tBuO	Bzl	—	7 (164)
	D, L	(159)	CH_2COOMe	MeO	Bzl	—	7 (164)
	D, L	(160)	CH_2COOEt	EtO	Bzl	—	7 (164)
treo-β-Hydroxy-N-hydroxy-aspartic acid	DL	(163)	$CH(OH)COOH$	HO	Bzl	—	7 (164)
N-Hydroxyglutamic acid	DL	(39)	CH_2CH_2COOH	HO	H	—	4 (144), 9.1 (178)
	L	(242)	CH_2CH_2COOMe	MeO	H	—	10 (192, 193)
N-Hydroxyhomoserine	DL	(211)	CH_2CH_2OH	$NHNH_2$	H	—	9.1 (186)
N-Hydroxythreonine	DL	(244)	$CH_3CH(OH)$	$p\text{-}NO_2\text{–}BzlO$	H	—	10 (192)
N-Hydroxytyrosine	DL	(35)	$p\text{-}HO\text{–}C_6H_4CH_2$	HO	H	—	4 (95)
	D					—	6 (159)
N-Hydroxymethionine	DL	(131)	$CH_3SCH_2CH_2$	HO	H	—	6 (158)
	–			MeO	H	—	6 (158)
N-Hydroxytryptophan	DL	(105)	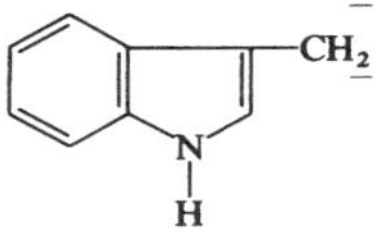 CH_2–(indol-3-yl, N–H)	EtO	Bzl	—	4 (126)

Table 4 *(continued)*

N-Hydroxyamino acid	Config.	No.	R or n	R_1	R_2	R_3	R_4	Synthetic method VIII.
N-Hydroxyamino-3-(N-methyl-indol-3-yl)propionic acid	DL	(113)	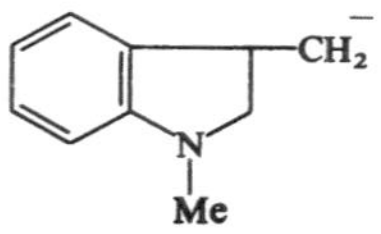 [indolyl-CH₂ structure, N-Me]		MeNH	Bzl	—	4 *(126)*

2. 2-Alkyl-N^2-hydroxyamino acids $R—CR_1—COR_2$ / NHOH

N-Hydroxyamino acid	Config.	No.	R or n	R_1	R_2	R_3	R_4	Synthetic method VIII.
2-Methyl-N^2-hydroxyamino-propionic acid	DL	(67)	CH_3	CH_3	HO	—	—	2 *(79, 134)*, 9.1 *(184)*
	DL	(199)			MeO	—	—	9.1 *(184)*
	DL	(64)			NH_2	—	—	2 *(130)*
2-Benzyl-N^2-hydroxyamino-propionic acid	DL	(196)	CH_3	$C_6H_5CH_2$	HO	—	—	9.1 *(184)*
	DL	(203)			MeO	—	—	9.1 *(184)*
2-Pivaloiloxymethyl-N^2-hydroxyaminopropionic acid	DL	(204)	CH_3	$tBuCO_2CH_2$	MeO	—	—	9.1 *(184)*
2-Methyl-N^2-hydroxyamino-butanoic acid	DL	(68)	CH_3CH_2	CH_3	HO	—	—	2 *(79)*
2-Ethyl-N^2-hydroxyamino-butanoic acid	DL	(69)	CH_3CH_2	CH_3CH_2	HO	—	—	1 *(129)*, 2 *(79)*
2,3-Dimethyl-N^2-hydroxy-aminobutanoic acid	DL	(201)	$(CH_3)_2CH$	CH_3	MeO	—	—	9.1 *(184)*
2-Methyl-N^2-hydroxyamino-pentenoic acid	DL	(205)	$CH_2=CHCH_2$	CH_3	MeO	—	—	9.1 *(184)*

Compound	Config.	No.	R	R_1				Ref.
2-Methyl-N²-hydroxyamino-pentanoic acid	DL	(200)	$CH_3CH_2CH_2$	CH_3	MeO		—	9.1 (184)
2-Methyl-N²-hydroxyamino-octanoic acid	DL	(202)	$CH_3(CH_2)_5$	CH_3	MeO		—	9.1 (184)
1-Hydroxyaminocyclo-pentanoic acid	DL	(206)	$-(CH_2)_4-$		MeO		—	9.1 (184)
1-Hydroxyaminocyclo-hexanoic acid	DL	(70)	$-(CH_2)_5-$		HO		—	2 (134)
	DL	(65)			NH_2		—	2 (134)
1-Hydroxyaminocyclo-heptanoic acid	DL	(71)	$-(CH_2)_6-$		HO		—	2 (134)

3. N-Hydroxy-sec-amino Acids R—CH—COOH

$$R_1N-OH$$

Compound	Config.	No.	R	R_1				Ref.
N-Hydroxysarcosine	DL	(43)	H	CH_3			—	12 (101)
N-Hydroxyproline	DL	(44)	$-(CH_2)_3-$				—	12 (101)
	D, L						—	12 (101)
N-Hydroxypipecolic acid	DL	(45)	$-(CH_2)_4-$				—	12 (101)
N-Hydroxyimino-diacetic acid	DL	(30)	H	CH_2COOH			—	6 (89)
N-Hydroxyimino-di-propionic acid	DL	(16)	$CH(CH_3)COOH$	$CH(CH_3)COOH$			—	6 (89)

4. N³-Hydroxyamino Acids R—CH—(CH₂)ₙ—CH—COR₂

$$R_3-N-OH \qquad R_1$$

Compound	Config.	No.	R	R_1				Ref.
N³-Hydroxy-3-aminopropionic acid (N-hydroxy-β-alanine)	DL	(214)	H, $n = 0$	H	HO	H		9.1 (178), 9.2 (187)
	DL	(232)	MeO		H			9.2 (187)
	DL	(236)			NEt_2	H	—	9.2 (187)
	DL	—			NH_2NH	H	—	9.2 (187)

Table 4 (continued)

N-Hydroxyamino acid	Config.	No.	R or n	R_1	R_2	R_3	R_4	Synthetic method VIII.
N^4-Hydroxy-4-aminobutanoic acid (N-hydroxy-γ-amino-butyric acid)	DL	(215)	H $n=1$	H	HO	H	—	9.1 (178)
N^3-Hydroxy-3-amino-2-phenylpropionic acid	DL	(216)	H $n=0$	C_6H_5	HO	H	—	9.2 (187)
	DL	(235)			EtO	H	—	9.2 (187)
	DL	(237)			Et_2N	H	—	9.2 (187)
N^3-Hydroxy-3-amino-butanoic acid	DL	(123)	CH_3 $n=0$	H	HO	H	—	5 (148)
N^3-Hydroxy-3-amino-3-phenylpropionic acid	DL	(115)	C_6H_5 $n=0$	H	HO	H	—	5 (148)
N^3-Hydroxyimino-di-propionic acid	DL	(124)	H $n=0$	H	MeO	$(CH_2)_2COOH$	—	5 (152)
	DL	(125)			EtO	$(CH_2)_2COOH$	—	5 (152)
	DL	(126)			NH_2	$(CH_2)_2COOH$	—	5 (153)
N^3-Hydroxy-N^3-methyl-3-amino-3-phenylpropionic acid	DL	(118)	C_6H_5 $n=0$	H	HO	Me	—	5 (149)
acid	DL	(119)			NHOH	Me	—	5 (149)
5. N^3-Hydroxydiamino Acids R_3—N—$(CH_2)_n$—$\underset{NHR_4}{CR}$—COR_2, OR_1								
N^3-Hydroxy-2,3-diamino-propionic acid	D, L	(137)	H $n=1$	H	HO	H	H	6 (162)

Compound	Config	(No.)	Side		Ester			Yields (Refs.)
N³-Hydroxy-2,3-diamino-propionic acid	DL	(234)		H	MeO	H	Ac	9.1 (190)
	DL	(179)		Bzl	HO	H	H	9.2 (187)
	DL	(180)		Bzl	HO	H	Bz	8 (165)
	DL	—		Bzl	HO	Tos	H	8 (165)
	DL	(181)		Bzl	C₆H₅NH	H	Bz	8 (170, 198)
	L	(182)		Bzl	C₆H₅NH	Bz	Bz	8 (165)
N⁴-Hydroxy-2,4-diamino-butanoic acid	DL	(136)	H, n=2	H	HO	H	H	6 (162)
	L			H	HO	H	H	6 (162)
N⁵-Hydroxyornithine	DL	(73)	H, n=3	H	HO	H	H	3 (97, 137, 138), 6 (162)
	L	(245)		H	tBuO	Ac	Boc	8 (167), 9.1 (180, 194), 10 (194)
	D	—		H	HO	H	H	10 (194)
	D	(164)		Bzl	HO	Tos	H	8 (166)
	DL	(173)		Bzl	HO	Ac	H	8 (167)
	L	(114)		Bzl	HO	Ac	H	8 (166, 198)
	L	(269)		Bzl	MeO	Ac	H	4 (147), 8 (167), 11 (202)
	L	(268)		Bzl	MeO	trOC	H	8 (172), 11 (202)
	L	—		Bzl	tBuO	Ac	Z	11 (202)
	D	(175)		Bzl	HO	Ac	Ac	4 (205)
	L	(267)		Bzl	HO	Tos	Boc	4 (147)
	L	(266)		Bzl	HO	Ac	Boc	8 (166)
	L	(259)		Bzl	tBuO	trOC	Boc	11 (198)
	L	(265)		Bzl	tBuO	Ac	Boc	11 (202)
	L	(264)		Bzl	tBuO	trOC	Boc	11 (202)
	L	(177)		Bzl	C₆H₅NH	Z	Boc	11 (202)
						Tos	Ac	11 (202)
N⁵-Hydroxy-2-methylornithine	DL	(225)	CH₃, n=3	H	HO	H	H	9.1 (83)

Table 4 (continued)

N-Hydroxyamino acid	Config.	No.	R or n	R_1	R_2	R_3	R_4	Synthetic method VIII.
N^5-Hydroxyarginine	L	(224)	H, n=3	H	HO	$\overset{NH}{CONH_2}$	H	9.1 (180), 10 (194)
N^5-Hydroxy-2-methylarginine	DL	(226)	CH_3, n=3	H	HO	$\overset{NH}{CONH_2}$	H	9.1 (83)
N^6-Hydroxylysine	DL	(74)	H, n=4	H	HO	H	H	3 (137), 6 (162), 8 (169)
	L	–		H	HO	Ac	H	11 (200)
	L	–		Bzl	HO	H	H	8 (169)
	DL	(172)		Bzl	HO	Tos	H	8 (169)
	D	(174)		Bzl	HO	Tos	Ac	8 (166)
	L	–		Bzl	MeO	Ac	H	11 (199, 200)
	L	(257)		Bzl	MeO	Ac	Boc	11 (200)
	L	(258)		Bzl	BzlO	Ac	Z	11 (200)
	L	(176)		Bzl	C_6H_5NH	Tos	Ac	8 (166)

6. N-Hydroxylactams

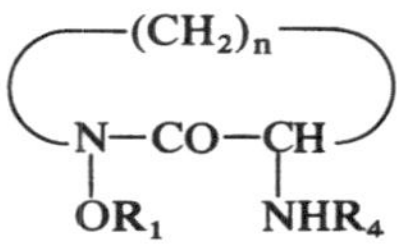

	Config.	No.	R or n	R_1	R_2	R_3	R_4	Synthetic method VIII.
N-Hydroxy-3-amino-2-piperidone	L	–	n=3	H	–	–	H	– (66, 116)
	D	–			–	–	H	8 (166)
	D	(249)		Bzl	–	–	H	8 (166)
	L	(165)		Bzl	–	–	Boc	11 (198)
	DL			Bzl	–	–	Ac	8 (166)
N-Hydroxy-3-amino-2-azepidone	L	(253)	n=4	Bzl	–	–	H	11 (196, 199)
	L	(250)		Bzl	–	–	Boc	11 (196, 199)

IX. Synthesis of N-Hydroxypeptides

1. N-Hydroxydiketopiperazines

Derivatives of N-hydroxydiketopiperazines are found among natural products (*vide infra*) and therefore present important problems to the synthetic chemist. The first synthesis of these heterocycles was carried out by COOK and SLATER in 1956 (*100*) (Scheme 58). Cyclization of methyl esters of chloroacyl-N-hydroxyamino acids (**301**) with ammonia or hydroxylamine furnished 1-hydroxy- or 1,4-dihydroxy-2,5-diketopiperazines (**302, 303**) in very low yields.

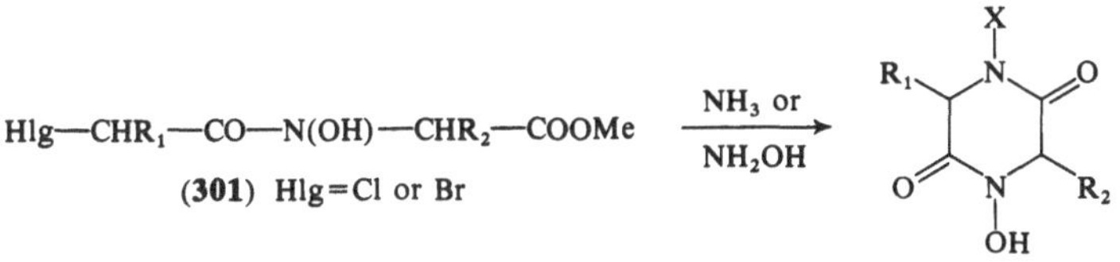

Hlg—CHR₁—CO—N(OH)—CHR₂—COOMe $\xrightarrow{\begin{array}{c} NH_3 \text{ or} \\ NH_2OH \end{array}}$

(**301**) Hlg = Cl or Br

(**302**), (**303**)

(**302**) X=H   R₁=nBu   R₂=nBu,
(**303**) X=OH   R₁=H   R₂=nBu, iBu
                    iPr   iBu

Scheme 58

Cyclization of esters of type (**304**) is sometimes accompanied by elimination and leads to cyclic derivatives of dehydro-N-hydroxyamino acids, as happens in the case (**304**) itself which results in formation of 1,4-dihydroxy-3,6-diisobutylidene-2,5-dioxopiperazine (**305**) (*140*) (Scheme 59). SHIN and coworkers (*207*) obtained unsaturated N-hydroxydiketopiperazines (**307**) from α,β-dehydro-halogenoacylamino acids (**306**) by cyclization with hydroxylamine in ethanol in the presence of sodium alcoholate (Scheme 60).

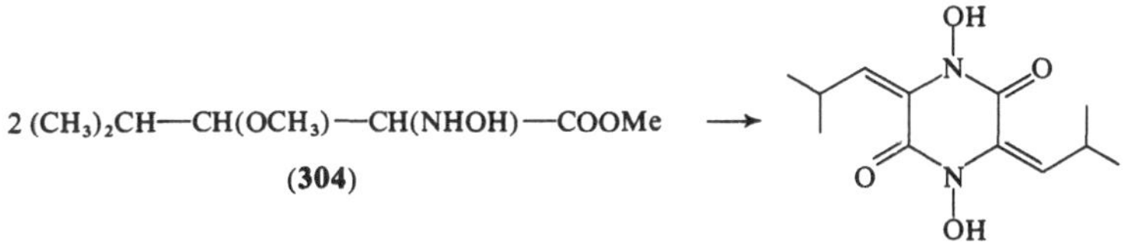

2 (CH₃)₂CH—CH(OCH₃)—CH(NHOH)—COOMe →

(**304**)

(**305**)

Scheme 59

During such a cyclization of ester (**308**) followed by elimination, they also observed (*208, 209*) the formation of imidazolidone (**310**) (Scheme 61). The cyclization of bromoacetyl derivatives (**309**) finally gives the desired 1,4-dihydroxy-2,5-diketopiperazines (312) in good yields (209).

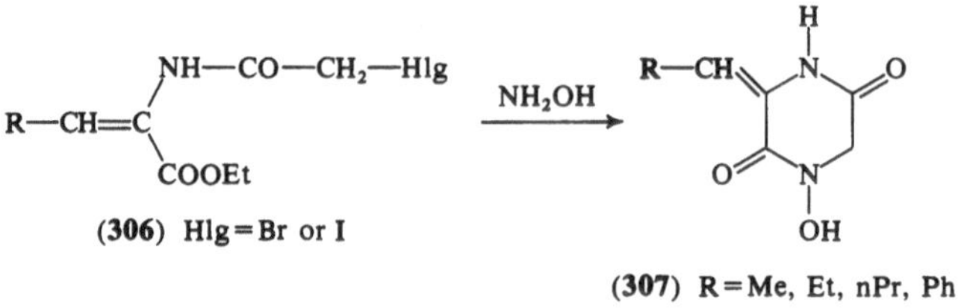

(306) Hlg = Br or I

(307) R = Me, Et, nPr, Ph

Scheme 60

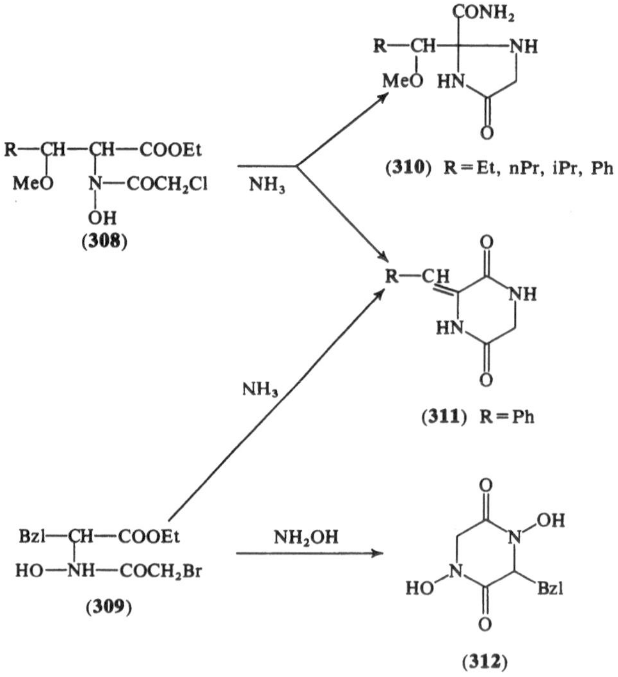

(308)

(310) R = Et, nPr, iPr, Ph

(311) R = Ph

(309)

(312)

Scheme 61

The synthesis of mycelianamide (Scheme 62) (14) (210), a naturally occurring benzylidene-2,5-dioxopiperazine residue, with protection of the olefinic bond as well as the N-hydroxyamide function, is an interesting solution.

OTTENHEIJM and coworkers (204) carried out successful cyclizations of p-nitrophenyl esters of N,N'-bis-(benzyloxy)-dipeptides (313) in pyridine to appropriate 1,4-dibenzyloxy-2,5-dioxopiperazines (314) (Scheme 63). Compounds (315) were also obtained in good yields (211)

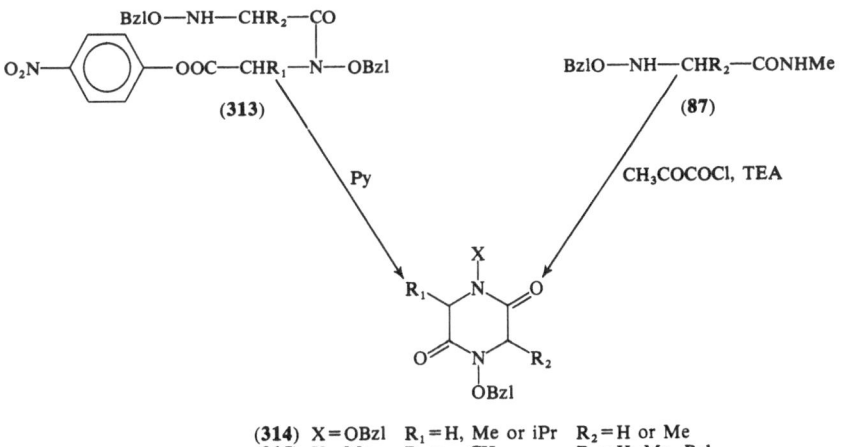

Scheme 62

Scheme 63

by reaction of pyruvoyl chloride with N-benzyloxyamino acid amides. Catalytic hydrogenolysis of (314) gave N-hydroxydioxopiperazines (303). They declared their intention to use N-hydroxydiketopiperazine derivatives of N-hydroxytryptophan (126) in a new synthesis of the fungal metabolites sporidesmine B and neoechinulin B.

2. Linear N-Hydroxypeptides

The first synthesis of a linear N-hydroxypeptide – glycyl-N-hydroxy-glycine (319) – was carried out by NEUNHOEFFER et al. (181) by acylation of the sodium salt of N-hydroxyglycine with phtalylglycine chloride (316) and hydroxylaminolysis of the phtalyl protective group in 14% overal yield (Scheme 64). The results were improved by acylation of ethyl (140, 160) and tert-butyl esters of N-hydroxyamino acids (77, 130) (160, 212) in the same manner. Acylation (212) by 1-ethoxycarbonyl-2-ethoxy-1,2-dihydroquinoline (EEDQ) and DCCI gave only low yields. Use of p-nitrophenyl and N-hydroxyphthalimidyl esters (318) resulted mainly in the formation of O-acyl derivatives (321). This result is not surprising because hydroxylamine derivatives are ambident reagents and undergo acylation reactions giving both N- and O-acyl mixtures (213).

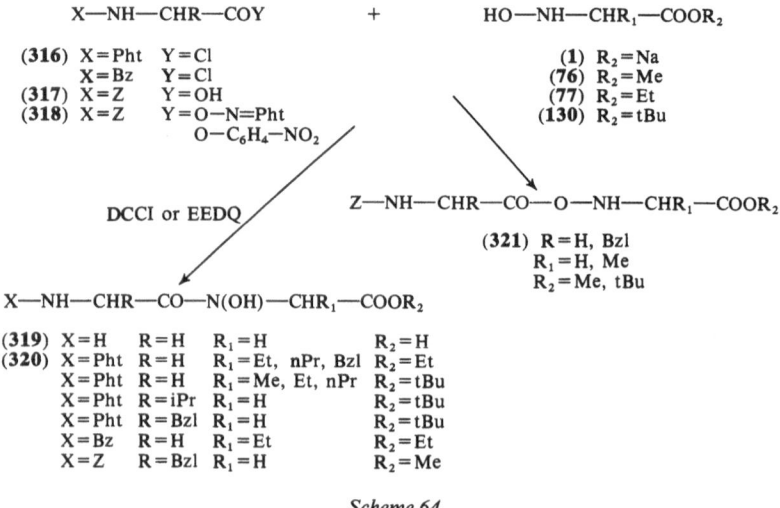

Scheme 64

The method was therefore improved by acylating N-hydroxyglycine and several derivatives (Scheme 65) with N-carboxy anhydrides (322) of amino acids, the reactions being carried out in acidic medium to prevent polymerization. The o-nitrophenylsulfenyl-N-carboxy anhydrides (323), the acylation reagents of choice because of their stability, were used by CHIMIAK and POŁOŃSKI (212) for the synthesis of several N-hydroxypeptides (325).

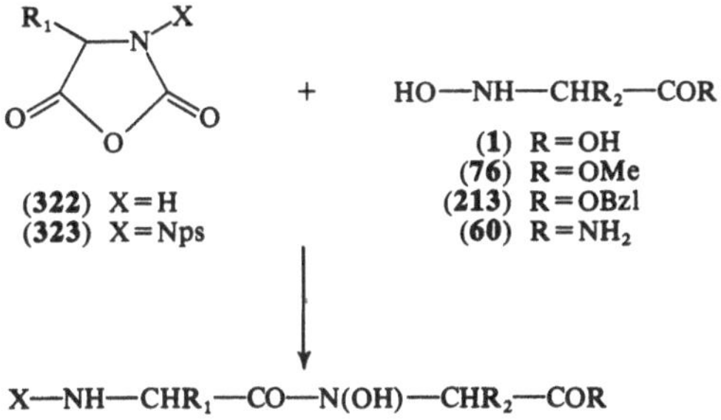

		+	HO—NH—CHR$_2$—COR
			(1) R=OH
			(76) R=OMe
(322)	X=H		(213) R=OBzl
(323)	X=Nps		(60) R=NH$_2$

X—NH—CHR$_1$—CO—N(OH)—CHR$_2$—COR

(324)	X=H	R=OH	R$_1$=H, Me, Bzl, (CH$_2$)$_2$CO$_2$Bzl	R$_2$=H
(325)	X=Nps	R=OH	R$_1$=Me	R$_2$=Bzl
	X=Nps	R=NH$_2$	R$_1$=iBu	R$_2$=Bzl
	X=Nps	R=OMe	R$_1$=Me, iBu	R$_2$=Bzl
	X=Nps	R=OMe	R$_1$=Bzl	R$_2$=H

Scheme 65

N-Benzyloxy-N-carboxy anhydrides (326) have also been used for the synthesis of poly-N-hydroxyamino acids (327) (215) (Scheme 66).

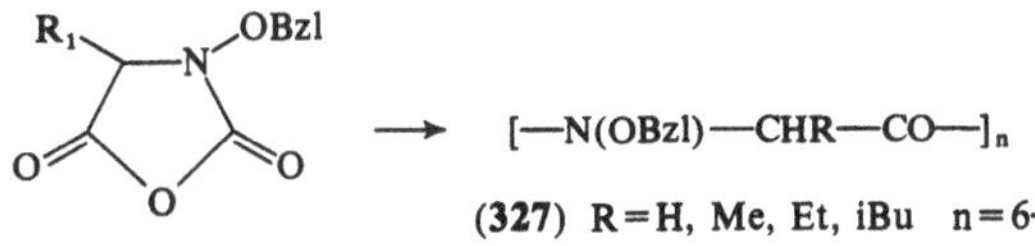

$\longrightarrow$ [—N(OBzl)—CHR—CO—]$_n$

(327) R=H, Me, Et, iBu n=6—30

(326)

Scheme 66

Esters of N-hydroxyamino acids, as unstable derivatives of hydroxylamine, are inconvenient substrates because they have strong reducing properties. In addition, their acylation, as already mentioned, always results in mixtures of N- and O-acyl derivatives. Taking this into account (163, 165), an unambiguous synthetic path to N-hydroxypeptides was developed at the Technical University of Gdańsk which employs acylation of N-benzyloxyamino acid esters (Scheme 67). Because of the relatively low nitrogen basicity in (275–277) acylation proceeds with difficulty and only low yields of benzyloxypeptide esters (328) were isolated using EEDQ and DCCI. However, use of strong activators such as acyl chlorides or mixed anhydrides improved the yield. Also use of iminochlorides (333) as activating agents afforded the N-benzyloxypeptides in good yields (164). The last method was used for preparing peptides of N-hydroxyaspartic acid (Scheme 68). Protective groups may be removed selectively from (**328**) to yield (**329, 330**) or (**331**). Hydrogenolysis or boron trifluoroacetate removes the benzyl group from the N-hydroxyamide bond, thus forming N-hydroxyamides (**322**) (*165, 216*). It is also possible to carry out, without elimination, hydrazinolysis of the phthalyl group and basic hydrolysis of N-hydroxypeptide esters (**331**) (*212*).

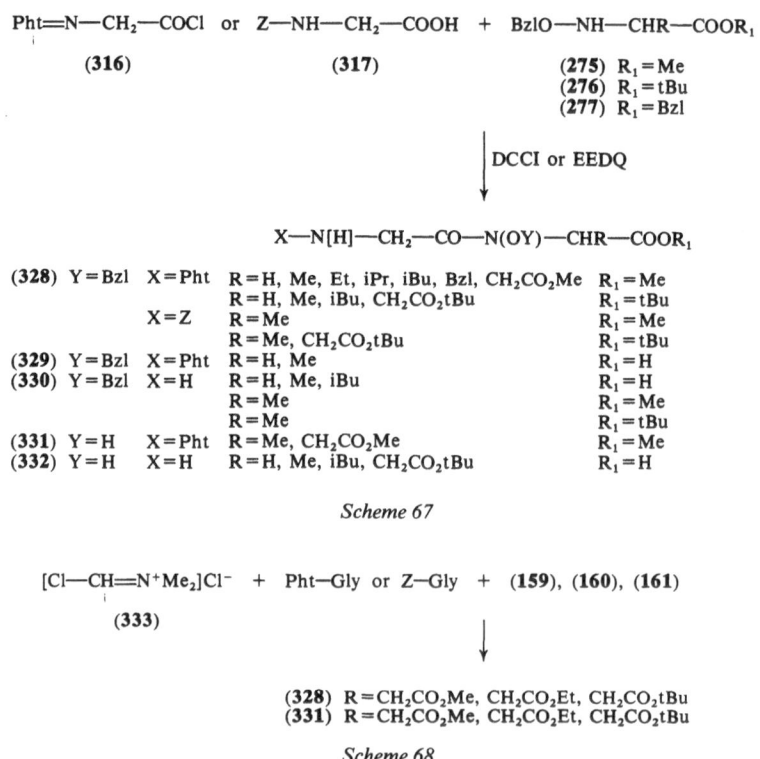

Pht≡N—CH₂—COCl   or   Z—NH—CH₂—COOH   +   BzlO—NH—CHR—COOR₁

(**316**)	(**317**)	(**275**) R₁ = Me
		(**276**) R₁ = tBu
		(**277**) R₁ = Bzl

DCCI or EEDQ

X—N[H]—CH₂—CO—N(OY)—CHR—COOR₁

(**328**) Y = Bzl	X = Pht	R = H, Me, Et, iPr, iBu, Bzl, CH₂CO₂Me	R₁ = Me
		R = H, Me, iBu, CH₂CO₂tBu	R₁ = tBu
	X = Z	R = Me	R₁ = Me
		R = Me, CH₂CO₂tBu	R₁ = tBu
(**329**) Y = Bzl	X = Pht	R = H, Me	R₁ = H
(**330**) Y = Bzl	X = H	R = H, Me, iBu	R₁ = H
		R = Me	R₁ = Me
		R = Me	R₁ = tBu
(**331**) Y = H	X = Pht	R = Me, CH₂CO₂Me	R₁ = Me
(**332**) Y = H	X = H	R = H, Me, iBu, CH₂CO₂tBu	R₁ = H

Scheme 67

[Cl—CH≡N⁺Me₂]Cl⁻   +   Pht—Gly or Z—Gly   +   (**159**), (**160**), (**161**)

(**333**)

(**328**) R = CH₂CO₂Me, CH₂CO₂Et, CH₂CO₂tBu
(**331**) R = CH₂CO₂Me, CH₂CO₂Et, CH₂CO₂tBu

Scheme 68

N-Hydroxypeptides with a N-terminal N-hydroxyamino acid residue were also obtained from appropriate nitrones (**334**) (Scheme 69), among them some interesting N-hydroxypeptides (**335, 336**) of 7-aminocephalosporanic acid and 6-aminopenicillanic acid (*185*). The mixed anhydride method was applied in the last case to form the amide bond in the presence of the nitrone group. Contrary to the above results PALACZ (*217*) observed that nitrone (**190**) easily rearranges into isomeric N-benzoyl derivatives (**337**) during activation and aminolysis.

N-Hydroxypeptides with a N-terminal hydroxyamine group were synthesized by AKIYAMA and coworkers (*218*) who used N-benzyloxy-

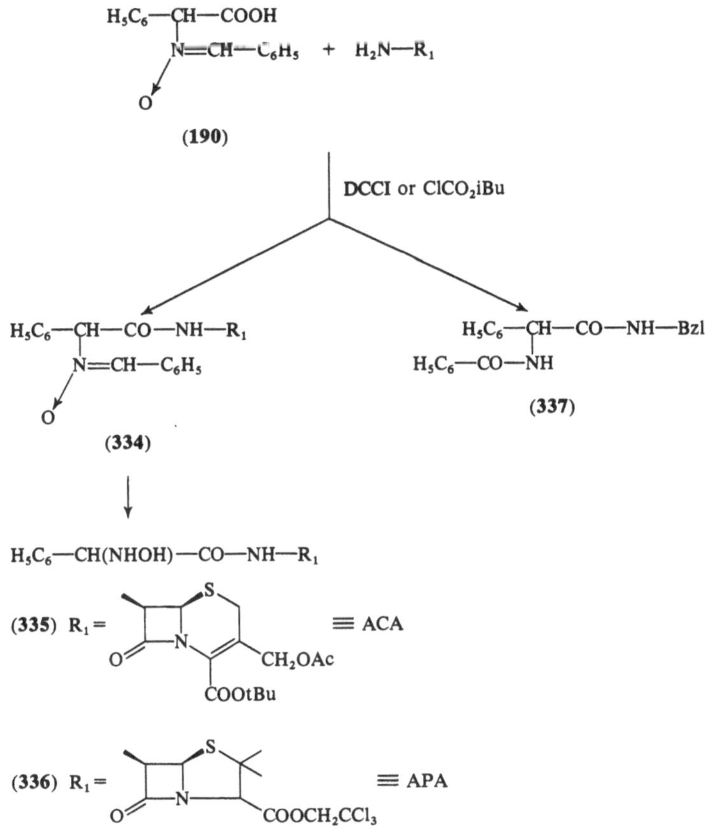

H₅C₆—CH—COOH
          |
          N=CH—C₆H₅   +   H₂N—R₁
          |
          O

(190)

DCCI or ClCO₂iBu

H₅C₆—CH—CO—NH—R₁   H₅C₆—CH—CO—NH—Bzl
          | |
          N=CH—C₆H₅   H₅C₆—CO—NH
          |
          O   **(337)**

(334)

H₅C₆—CH(NHOH)—CO—NH—R₁

(335) R₁ =

$\equiv$ ACA

CH₂OAc
COOtBu

(336) R₁ =

$\equiv$ APA

COOCH₂CCl₃

Scheme 69

(326) or BzlO—NH—CHR—COOSu   +   GlyOMe

(338)

BzlO—NH—CHR—CONH—CH₂—COOCH₃

(339) R = H, Me, Et, iPr, iBu, Bzl

Ac—[N(OH)—CH(CH₃)—CONH—CH₂—CO]₃—NH—C₆H₅

(340)

Scheme 70

N-carboxy anhydrides of amino acids (326), obtained from N-benzyloxy-amino acids as crystalline derivatives by reaction with phosgene. Acylation of glycine methyl ester by (326) gave the corresponding peptides

(339) in quantitative yield (Scheme 70). The same workers also obtained N-hydroxytripeptide (342) by the method of mixed anhydrides and from it, employing again (326), the tetrapeptide (343) with two protected N-hydroxyl functions (Scheme 71). The observation that the benzyloxyamino group did not react with N-hydroxysuccinimide ester prompted AKIYAMA et al. (216) to employ N-hydroxysuccinimide esters of benzyloxyamino acids without N-protection (338). With such substrates they obtained a N-hydroxyhexapeptide (340) with three N-hydroxyamide bonds (Scheme 70).

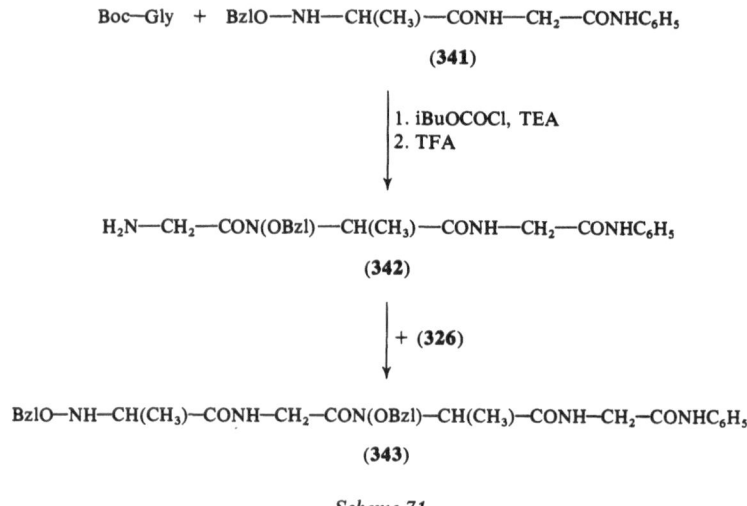

Boc—Gly + BzlO—NH—CH(CH₃)—CONH—CH₂—CONHC₆H₅

(341)

1. iBuOCOCl, TEA
2. TFA

H₂N—CH₂—CON(OBzl)—CH(CH₃)—CONH—CH₂—CONHC₆H₅

(342)

+ (326)

BzlO—NH—CH(CH₃)—CONH—CH₂—CON(OBzl)—CH(CH₃)—CONH—CH₂—CONHC₆H₅

(343)

Scheme 71

X. Reactions of N-Hydroxypeptides

N-Hydroxydiketopiperazines as well as linear N-hydroxypeptides undergo elimination under the influence of bases to α,β-unsaturated compounds (160). For best preparative results, appropriate tosyl derivatives (345, 347) are formed in the first stage by the action of toluenesulfonyl chloride and triethylamine; these then undergo elimination under the influence of potassium *tert*-butoxide to (348) (Scheme 73) or (346) (Scheme 72) (126, 127, 211, 219). In this manner N-hydroxypeptides can be used for the convenient synthesis of dehydropeptides (219). The elimination reaction of N-hydroxydiketopiperazines was employed in a biomimetic synthesis of the fungal metabolite neoechinulin B (126) (Scheme 73).

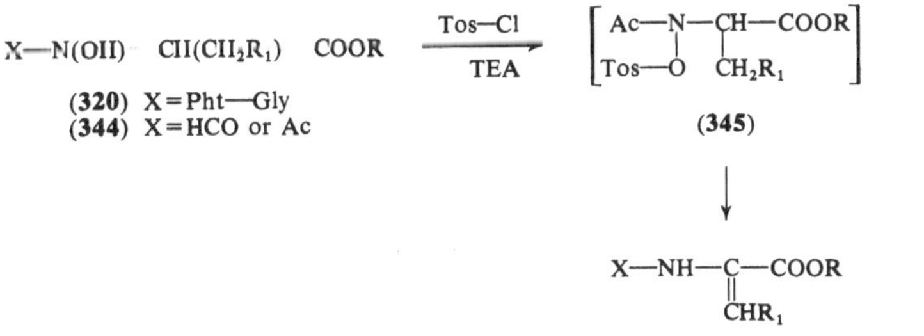

$$X-N(OH) \quad CH(CH_2R_1) \quad COOR \xrightarrow[\text{TEA}]{\text{Tos—Cl}} \begin{bmatrix} Ac-N-CH-COOR \\ Tos-O \quad CH_2R_1 \end{bmatrix}$$

(320) X = Pht—Gly
(344) X = HCO or Ac

(345)

$$X-NH-\underset{\underset{CHR_1}{\|}}{C}-COOR$$

(346) X = Pht—Gly, Ac or HCO
R = Me, Et, tBu
R_1 = H, Me, CO_2Me, CO_2Et

Scheme 72

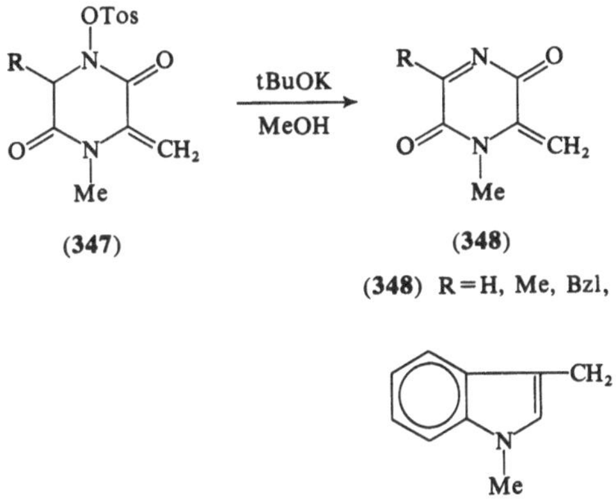

(347)　　(348)

(348) R = H, Me, Bzl,

Scheme 73

XI. Final Remarks

The chemistry of N-hydroxyamino acids and their derivatives deals with an important area lying between the chemistry of hydroxylamine, the chemistry of amino acids and the chemistry of dehydroamino acids, thus linking these fields. As it has not been satisfactorily and fully investigated it constitutes a good area for further synthetic and bio-chemical research. It is hoped that this review will stimulate further development in this field.

Addendum

After completion of this review another independent review appeared covering only a part of the subject discussed here, *i.e.* α-N-hydroxyamino acids and their chemistry (*220*). In addition a few papers on related subjects have been published recently and deserve to be mentioned.

VAN DER HELM's group elucidated the structure of des(diserylglycyl)-ferrirhodin – a novel siderophore isolated from *Aspergillus ochraceous* (*221*). This is the first natural linear N^5-hydroxyornithine tripeptide to be isolated. Evidence for the structure was derived from ^{1}H- and ^{13}C-NMR spectra of its deferri form and the gallium complex and is also based on the synthesis of N-acetyl and methyl esters.

OTTENHEIJM's group has recently improved procedure VIII.7 (*164*) (Scheme 33) for the preparation of optically active N-hydroxyamino acid derivatives by a substitution reaction. N-Benzyloxyamino acid esters were obtained from triflates of α-hydroxy esters and O-benzylhydroxylamine (*222*). The optical purity of the products was very high.

MILEWSKA and CHIMIAK utilized oxidation of the ω-amine group of amino acid derivatives with benzoyl peroxide in a new synthesis of useful derivatives of N^5-hydroxyornithine and N^6-hydroxylysine (*223*). The desired O-benzoyl hydroxylamine derivatives were the main products formed on oxidation. The O-benzoyl group and other protective groups may be removed easily after acetylation. It should be added that oxidation of α-amino acid esters under the same conditions leads to N-benzoylamino acid esters only (*224*).

In continuation of their research on N-hydroxypeptides AKIYAMA's group obtained the anilide hexapeptide with a 6-aminohexanoyl-3-(hydroxyamino)propanoyl sequence (*225*). This linear tri-N-hydroxyamide bears a structural resemblance to the natural siderophores ferrioxamines. Cyclic voltametry and UV spectrometry were used for the study of its iron complex. The key substrate was obtained by the addition of benzyloxyamine to *p*-nitrophenyl acrylate.

Acknowledgements

This work was supported in part by Grants CPBR 03.13 and CPBP 01.13. We thank M Sc. A. Cygan for his considerable help in translating this review.

References

1. VON MILLER, W., and J. PLÖCHL: Ueber Amidoxylsäuren. Ber. dtsch. chem. Ges. **26**, 1545 (1893).
2. IUPAC-IUB Commission on Biochemical Nomenclature. Symbols for Amino Acid Derivatives and Peptides. Recommendations 1971. J. Biol. Chem. **247**, 977 (1972).
3. CHIMIAK, A.: N-Hydroxyamino Acids and Their Derivatives. Wiad. Chem. **19**, 803 (1965); Chem. Abstr. **64**, 4881 (1966).
4. MØLLER, B.L.: The Involvement of N-Hydroxyamino Acids as Intermediates in Metabolic Transformations. In: Cyanide Biol. Academic: London U.K., **1981**, p. 197; Chem. Abstr. **96**, 176229 c (1982).

5. AKIYAMA, M.: Syntheses and Reactions of Compounds Containing the N-Hydroxy-amide Functionality. I. Synth. Org. Chem. Japan **40**, 1189 (1982).
6. FISCHER, B.,   W. KELLER-SCHIERLEIN,   H. KNEIFEL,   W.A. KÖNIG,   W. LOEFFLER, A. MÜLLER, R. MUNTWYLER, and H. ZÄHNER: Stoffwechselprodukte von Mikroorganismen. 118. δ-N-Hydroxy-L-arginin, ein Aminosäure-Antagonist aus *Nannizzia grypsea*. Arch. Mikrobiol. **91**, 203 (1973).
7. MAEHR, H., J.F. BLOUNT, D.L. PRUESS, L. YARMCHUK, and M. KELLETT: Antimetabolites Produced by Microorganisms. VIII. N^5-Hydroxy-L-arginine, a New Naturally Occuring Amino Acid. J. Antibiotics **26**, 284 (1973).
8. MAEHR, H.: Antibiotics and Other Naturally Occurring Hydroxamic Acid and Hydroxamates. Pure Appl. Chem. **28**, 603 (1971).
9. NEILANDS, J.B.: Microbial Iron Transport Compounds. In: Inorganic Biochemistry, **1**, p. 167. Amsterdam: Elsevier Scientific Pub. Comp. 1978.
10. – Hydroxamic Acids in Nature. Sophisticated Ligands Play a Role in Iron Metabolism and Possibly in Other Processes in Microorganisms. Science **156**, 1443 (1967).
11. – Microbial Iron Compounds. Annu. Rev. Biochem. **50**, 715 (1981).
12. – Iron Absorption and Transport in Microorganisms. Ann. Rev. Nutr. **1981**, 27.
13. – Methodology of Siderophores. Structure and Bonding (Berlin) **58**, 1 (1984).
14. BERGERON, R.J.: Synthesis and Solution Structure of Microbial Siderophores. Chem. Rev. **84**, 587 (1984).
15. CHIMIAK, A.: Siderophores – Carriers of Ferric Ion. Postępy Biochemii **30**, 435 (1984); Chem. Abstr. **104**, 144041 c (1986).
16. HIDER, R.C.: Siderophore Mediated Absorption of Iron. Structure and Bonding (Berlin) **58**, 25 (1984).
17. SNOW, C.A.: Mycobactins: Iron-chelating Growth Factors from *Mycobacteria*. Bacteriol. Rev. **34**, 99 (1970).
18. SCHMIDT, U., J. HÄUSLER, E. ÖHLER, and H. POISEL: Dehydroamino Acids, α-Hydroxy-α-amino Acids and α-Mercapto-α-amino Acids. Fortschr. Chem. organ. Naturstoffe **37**, 251 (1979).
19. BIRCH, A.J., R.A. MASSAY-WESTROPP, and R.W. RICKARDS: Studies in Relation to Biosynthesis. Part VIII. The Structure of Mycelianamide. J. Chem. Soc. (London) **1956**, 3717.
20. OXFORD, A.E., and H. RAISTRICK: Studies in the Biochemistry of Microorganisms. 76. Mycelianamide, $C_{22}H_{28}O_5N_2$ a Metabolic Product of *Penicillium griseofulvum* Dierck X. Part 1. Preparation, Properties and Breakdown Products. Biochem. J. **42**, 323 (1948).
21. PERLMAN, D., A.J. VLIETINCK, H.W. MATTHEWS, and F.F. LO: Microbial Production of Vitamin B_{12} Antimetabolites. I. N^5-Hydroxy-L-arginine from *Bacillus cereus* 439. J. Antibiotics **27**, 826 (1974).
22. MCCULLOUGH, W.G., and R.S. MERKAL: Iron – chelating Compound from *Mycobacterium avium*. J. Bacteriol. **128**, 15 (1976).
23. – – Tripeptide Hydroxamate from *Corynebacterium kutscheri*. J. Bacteriol. **137**, 243 (1979).
24. FUGMANN, B., and W. STEGLICH: Unusual Components of the Toadstool *Lyophyllum connatum* (Agaricales). Angew. Chem., Int. Ed. Engl. **23**, 72 (1984).
25. STEVENS, R.L., and T.F. EMERY: The Biosynthesis of Hadacidin. Biochemistry **5**, 74 (1966).
26. DULANEY, E.L., and R.A. GRAY: Penicillia that Make N-Formyl Hydroxyaminoacetic Acid, a New Fungal Product. Mycologia **54**, 476 (1962).
27. GITTERMAN, C.O., E.L. DULANEY, E.A. KACZKA, D. HENDLIN, and H.B. WOODRUFF: The Human Tumor-egg Host System. II. Discovery and Properties of a New Antitumor Agent, Hadacidin. Proc. Soc. Exptl. Biol. Med. **109**, 852 (1962).

28. GRAY, R.A., G.W. GAUGER, E.L. DULANEY, E.A. KACZKA, and H.B. WOODRUFF: Hadacidin, a New Plant Growth Regulator Produced by Fermentation. Plant Physiol. 39, 204 (1964).

29. KACZKA, E.A., C.O. GITTERMAN, E.L. DULANEY, and K. FOLKERS: Hadacidin, a New Growth – Inhibitory Substance in Human Tumor Systems. Biochemistry 1, 340 (1962).

30. DUTCHER, J.: Aspergillic Acid: an Antibiotic Substance Produced by *Aspergillus flavus*. I. General Properties; Formation of Desoxyaspergillic Acid; Structural Conclusions. J. Biol. Chem. 171, 321 (1947).
– Aspergillic Acid: an Antibiotic Substance Produced by *Aspergillus flavus*. II. Bromination Reactions and Reduction with Sodium and Alcohol. J. Biol. Chem. 171, 341 (1947).
– Aspergillic Acid: an Antibiotic Substance Produced by *Aspergillus flavus*. III. The Structure of Hydroxyaspergillic Acid. J. Biol. Chem. 232, 785 (1958).

31. COOK, A.H., and C.A. SLATER: The Structure of Pulcherrimin. J. Chem. Soc. (London) 1956, 4133.

32. – – Metabolism of "Wild" Yeasts. I. The Chemical Nature of Pulcherrimin. J. Inst. Brew. 60, 213 (1954).

33. KLUYVER, A.J., J.P. VAN DER WALT, and A.J. VAN TRIET: Pulcherrimin, the Pigment of *Candida pulcherrima*. Proc. Nat. Acad. Sci. (USA) 39, 583 (1953).

34. GIBSON, F., and D.J. MAGRATH: Isolation and Characterization of a Hydroxamic Acid (Aerobactin) Formed by *Aerobacter aerogenes 62-1*. Biochim. Biophys. Acta 192, 175 (1969).

35. MACHAM, L.P., C. RATLEDGE, and J.C. NOCTON: Extracellular Iron Acquisition by *Mycobacteria*: Role of the Exochelins and Evidence Against the Participation of Mycobactin. Infect. Immun. 12, 1242 (1975).
MACHAM, L.P., and C. RATLEDGE: A New Group of Water-soluble Iron-binding Compounds from *Mycobacteria*: the Exochelins. J. Gen. Microbiol. 89, 379 (1975).

36. SNOW, G.A.: Mycobactin. A Growth Factor for *Mycobacterium johnei*. II. Degradation, and Identification of Fragments. J. Chem. Soc. (London) 1954, 2588.

37. RATLEDGE, C., and G.A. SNOW: Isolation and Structure of Nocobactin NA, a Lipid-soluble Iron-binding Compound from *Nocardia asteroides*. Biochem. J. 139, 407 (1974).

38. ENG-WILMOT, D.L., A. RAHMAN, J.V. MENDENHALL, S.L. GRAYSON, and D. VAN DER HELM: Molecular Structure of Ferric Neurosporin, a Minor Siderophore-like Compound Containing N^δ-Hydroxy-D-ornithine. J. Amer. Chem. Soc. 106, 1285 (1984).

39. YANG, CH.CH., and J. LEONG: Structure of Pseudobactin 7SR1, a Siderophore from a Plant-Deleterius *Pseudomonas*. Biochemistry 23, 3534 (1984).

40. TEINTZE, M., M.B. HOSSAIN, C.L. BARENS, J. LEONG, and D. VAN DER HELM: Structure of Ferric Pseudobactin: a Siderophore from a Plant Growth Promoting *Pseudomonas*. Biochemistry 20, 6446 (1981).

41. MAURER, B., A. MULLER, W. KELLER-SCHIERLEIN, and H. ZÄHNER: Metabolic Products of Microorganisms. LXI. Ferribactin, a Siderochrome from *Pseudomonas fluorescens*. Arch. Mikrobiol. 60, 326 (1968).

42. PHILSON, S.B., and M. LLINAS: Siderochromes from *Pseudomonas fluorescens*. I. Isolation and Characterization. J. Biol. Chem. 257, 8081 (1982).

43. NEILANDS, J.B.: A Crystalline Organo-iron Pigment from a Rust Fungus (*Ustalago sphaerogena*). J. Amer. Chem. Soc. 74, 4846 (1952).

44. GARIBALDI, J.A., and J.B. NEILANDS: Isolation and Properties of Ferrichrome A. J. Amer. Chem. Soc. 77, 2429 (1955).

45. ATKIN, C.L., J.B. NEILANDS, and H. PAFF: Rhodotorulic Acid from Species of *Leucosporidium*, *Rhodosporidium*, *Rhodotorula*, *Sporidiobolus*, and *Sporobolomyces*, and

a New Alanine-containing Ferrichrome from *Cryptococcus melibiosum*. J. Bacteriol. 103, 722 (1970).

46. KELLER-SCHIERLEIN, W., and A. DEER: 215. Stoffwechselprodukte von Mikroorganismen. 44 Mitt. Zur Konstitution von Ferrichrysin und Ferricrocin. Helv. Chim. Acta 46, 1907 (1963).

47. JALAL, M.A.F., R. MOCHARLA, C.L. BARNES, M.B. HOSSAIN, D.R. POWELL, D.L. ENG-WILMOT, S.L. GRAYSON, B.A. BENSON, and D. VAN DER HELM: Extracellular Siderophores from *Aspergillus ochraceus*. J. Bacteriol. 158, 683 (1984).

48. FREDERICK, C.B., P.J. SZANISZLO, P.E. VICKREY, M.D. BENTLEY, and W. SHIVE: Production and Isolation of Siderophores from the Soil Fungus *Epicoccum purpurascens*. Biochemistry 20, 2432 (1981).

   FREDERICK, C.B., M.D. BENTLEY, and W. SHIVE: Structure of Triornicin, a New Siderophore. Biochemistry 20, 2436 (1981).

   – – – The Structure of the Fungal Siderophore Isotriornicin. Biochem. Biophys. Res. Comm. 105, 133 (1982).

49. CHARLANG, G., N.G. BRADFORD, N.H. HOROWITZ, and R.M. HOROWITZ: Cellular and Extracellular Siderophores of *Aspergillus nidulans* and *Penicillium chrysogenum*. Mol. Cell. Biol. 1, 94 (1981).

50. KELLER-SCHIERLEIN, W.: 216. Stoffwechselprodukte von Mikroorganismen. 45 Mitt. Über die Konstitution von Ferrirubin, Ferrirhodin und Ferrichrom A. Helv. Chim. Acta 46, 1920 (1963).

51. DEML, G., K. VOGES, G. JUNG, and G. WINKELMANN: Tetraglycylferrichrome – the First Heptapeptide Ferrichrome. FEBS Letters 173, 53 (1984).

52. TADENUMA, M., and S. SATO: Studies on the Colorants in Sake. The Presence of Ferrichrysin as Iron Containing Colorant in Sake. Agr. Biol. Chem. 31, 1482 (1967).

53. GAUZE, G.F., and M.G. BRAZNIKOVA: The Action of Albomycin on Bacteria. Novosti Med. Acad. Med. Sci. USSR 23, 3 (1951).

54. BENZ, G., T. SCHRÖDER, J. KÜRZ, C. WÜNSCHE, W. KARL, G. STEFFENS, J. PFITZNER, and D. SCHMIDT: Constitution of the Deferriform of Albomycins δ_1, δ_2, and ϵ. Angew. Chem., Int. Ed. Engl. 21, 527 (1982).

55. EMERY, T.: Malonichrome, a New Iron Chelate from *Fusarium roseum*. Biochim. Biophys. Acta 629, 382 (1980).

56. JALAL, M.A.F., R. MOCHARLA, and D. VAN DER HELM: Separation of Ferrichromes and Other Hydroxamate Siderophores of Fungal Origin by Reversed-phase Chromatography. J. Chromatogr. 301, 247 (1984).

57. EMERY, T.: Aspartase-catalyzed Synthesis of N-Hydroxyaspartic acid. Biochemistry 2, 1041 (1963).

58. EMERY, T.F.: Isolation, Characterization, and Properties of Fusarinine, a δ-Hydroxamic Acid Derivative of Ornithine. Biochemistry 4, 1410 (1965).

59. SAYER, J.M., and T.F. EMERY: Structures of the Naturally Occurring Hydroxamic Acids, Fusarinines A and B. Biochemistry 7, 184 (1968).

60. DIEKMANN, H.: Stoffwechselprodukte von Mikroorganismen. 56 Mitt. Fusigen – ein neues Sideramin aus Pilzen. Arch. Mikrobiol. 58, 1 (1967).

61. CHARLANG, G., R.M. HOROWITZ, P.H. LOWY, N.G. BRADFORD, S.M. POLING, and N.H. HOROWITZ: Extracellular Siderophores of Rapidly Growing *Aspergillus nidulans* and *Penicillium chrysogenum*. J. Bacteriol. 150, 785 (1982).

62. MOORE, R.E., and T. EMERY: N^α-Acetylfusarinines: Isolation, Characterization, and Properties. Biochemistry 15, 2719 (1976).

63. ANKE, H.: Metabolic Products of Microorganisms. 163. Desferritriacetylfusigen, an Antibiotics from *Aspergillus deflectus*. J. Antibiotics 30, 125 (1977).

64. WENDENBAUM, S., P. DEMANGE, A. DELL, J.M. MEYER, and M.A. ABDALLAH: The Structure of Pyoverdine Pa, the Siderophore *Pseudomonas aeruginosa*. Tetrahedron Letters 24, 4877 (1983).

65. ATKIN, C.L., and J.B. NEILANDS: Rhodotorulic Acid, a Diketopiperazine Dihydrox-amic Acid with Growth Factor Activity. I. Isolation and Characterization. Biochem-istry 7, 3734 (1968).
66. AKERS, H.A., and J.B. NEILANDS: A Hydroxamic Acid Present in *Rhodotorula pili-manae*. Cultures Growth at Low pH and Its Metabolic Relation to Rhodotorulic Acid. Biochemistry 12, 1006 (1973).
67. DIEKMANN, H.: Metabolic Products of Microorganisms. 81. Occurrence and Struc-ture of Coprogen and Dimerum Acid. Arch. Mikrobiol. 73, 65 (1970).
68. HARRINGTON, G.J., and J.B. NEILANDS: Isolation and Characterization of a Dihyd-roxamic Acid Siderophore from *Verticillium dahliae*. Int. Symp. Iron Nutrition and Interaction in Plants; Brigham Young University, 1981.
69. HESSELTINE, C.W., C. PIDACKS, A.R. WHITEHILL, N. BOHONOS, B.L. HUTCHINGS, and J.H. WILLIAMS: Coprogen, a New Growth Factor for Coprophilic Fungi. J. Amer. Chem. Soc. 74, 1362 (1952).
70. PIDACKS, C., A.R. WHITEHILL, L.M. PRUESS, C.W. HESSELTINE, B.L. HUTCHINGS, N. BOHONOS, and J.H. WILLIAMS: Coprogen, the Isolation of a New Growth Factor Required *Pilobolus* Species. J. Amer. Chem. Soc. 75, 6064 (1953).
71. HOROWITZ, N.H., G. CHARLANG, G. HORN, N.P. WILLIAMS: Isolation and Identifica-tion of the Conidial Germination Factor of *Neurospora crassa*. J. Bacteriol. 127, 135 (1976).
72. KANAI, F., T. SAWA, M. HAMADA, H. NAGANAWA, T. TAKEUCHI, and H. UMEZAWA: Vanoxonin, a New Inhibitor of Thimidylate Synthetase. J. Antibiotics 36, 656 (1983).
73. CORONELLI, C., C.R. PASQUALUCCI, G. TAMONI, and G.G. GALLO: Isolation and Structure of Alanosine, a New Antibiotic. Il Farmaco, Ed. Sci. 21, 269 (1966).
74. MURTHY, Y.K.S., J.E. THIEMANN, C. CORONELLI, and P. SENSI: Alanosine, a New Antiviral and Antitumour Agent Isolated From a *Streptomyces*. Nature 211, 1198 (1966).
75. BAYER, E., and H. KNEIFEL: Isolation of Amavadin, a Vanadium Compound Occur-ing in *Amanita muscaria*. Z. Naturforsch. 27b, 207 (1972).
76. OKUHARA, M., Y. KURODA, T. GOTO, M. OKAMOTO, H. TERANO, M. KOHSAKA, H. AOKI, and H. IMANAKA: Studies on New Phosphonic Acid Antibiotics. 1. FR-900098, Isolation and Characterization. J. Antibiotics 33, 13 (1980).
77. KURODA, Y., M. OKUHARA, T. GOTO, M. OKAMOTO, H. TERANO, M. KOHSAKA, H. AOKI, and H. IMANAKA: Studies on New Phosphonic Acid Antibiotics. IV. Struc-ture Determination of FR-33289, FR-31564 and FR-32863. J. Antibiotics 33, 29 (1980).
78. OKUHARA, M., Y. KURODA, T. GOTO, M. OKAMOTO, H. TERANO, M. KOHSAKA, H. AOKI, and H. IMANAKA: Studies on New Phosphonic Acid Antibiotics. III. Isola-tion and Characterization of FR-31564, FR-32863 and FR-33289. J. Antibiotics 33, 24 (1980).
79. NEELAKANTAN, L., and W.H. HARTUNG: α-Hydroxylamino Nitriles and α-Hydroxyl-amino Acids. J. Organ. Chem. (USA) 23, 964 (1958).
80. CHIMIAK, A.: Esters of N-Hydroxyamino Acids. Polish J. Chem. 42, 225 (1968).
81. SPENSER, I.D., and A. AHMAD: α-Amino Acids and α-Keto Acid Oximes from α-Hydroxylamino Acids: a New Disproportionation Reaction. Proc. Chem. Soc. (Lon-don) 1961, 375.
82. VON MILLER W., and J. PLÖCHL: Ueber Schiff'sche Basen. Ber. dtsch. chem. Ges. 25, 2020 (1892).
83. MAEHR, H., and M. LEACH: Antimetabolites Produced by Microorganisms. XVI. Synthesis of N^5-Hydroxy-2-methylarginine and N^5-Hydroxy-2-methylornithine. J. Antibiotics 31, 165 (1978).

84. BRACHTEL, G., and M. JANSEN: On the Reaction Between Hydroxylamine and Acrylic Acid. Z. Naturforsch. **40b**, 574 (1985).

85. GIGUERE, P.A., and J.D. LIU: Infrared Spectrum, Molecular Structure, and Thermodynamic Functions of Hydroxylamine. Canad. J. Chem. **30**, 948 (1952).

86. KARCZYŃSKI, F., B. LIBEREK, and Z. PALACZ: Complex of Compounds of N-Hydroxyglycine and N-Hydroxy-β-alanine with Cu(II). Zeszyty Naukowe Wydz. Mat. Fiz. Chem. UG **1**, 77 (1971).

87. KNEIFEL, H., and E. BAYER: Determination of the Structure of the Vanadium Compound, Amavadine, from Fly Agaric. Angew. Chem., Int. Ed. Engl. **12**, 508 (1973).

88. KRAUSS, P., E. BAYER, and H. KNEIFEL: Electron Spin Resonance Investigations of Amavadine, a Natural Product Containing Vanadium. Z. Naturforsch. **39b**, 829 (1984).

89. FELCMAN, J., M. CÂNDIDA, T.A. VAZ, and J.J.R. FRAÙSTO DA SILVA: Metal Complexes of N-Hydroxy-imino-di-α-propionic Acid and Related Ligands. Inorg. Chim. Acta **93**, 101 (1984).

90. DUYNSTEE, E.F.J., and M.E.A.H. MEVIS: Decomposition of Dimeric α-Nitrosocarboxylic Acids in Sulfuric Acid. Rec. trav. chim. Pay-Bas **86**, 715 (1967).

91. EMERY, T.F., and J.B. NEILANDS: Further Observations Concerning the Periodic Acid Oxidation of Hydroxylamine Derivatives. J. Organ. Chem. (USA) **27**, 1075 (1962).

92. AHMAD, A.: The Chemistry of α-N-Hydroxyamino Acids. Bull. Chem. Soc. Japan **47**, 2583 (1974).

93. AHMAD, A., and I.D. SPENSER: The Conversion of α-Keto Acids and of α-Keto Acid Oximes to Nitriles in Aqueous Solution. Canad. J. Chem. **39**, 1340 (1961).

94. STEIGER, R.E.: N-Hydroxy-α-amino Acid as Possible Intermediates in the Oxidative Degradation of α-Amino Acids. J. Biol. Chem. **153**, 691 (1944).

95. MØLLER, B.L., I.J. McFARLANE, and E.E. CONN: Chemical Synthesis and Disproportionation of N-Hydroxytyrosine. Acta Chem. Scand. **B 31**, 343 (1977).

96. PINZA, M., G. PIFFERI, and F. NASI: A New Synthesis of 3-Oxazolin-5-ones (5-oxo-2,5-dihydro-1,3-oxazoles). Synthesis **1980**, 55.

97. TOMLINSON, G., and T. VISWANATHA: Synthesis and Properties of δ-N-Hydroxyornithine. Canad. J. Biochem. **51**, 754 (1973).

98. TURKOWA, J., O. MIKES, and F. ŠORM: Chemical Composition of the Antibiotic Albomycin. VI. Determination of the Structure of the Peptide Moiety of the Antibiotic Albomycin. Collect. Czech. Chem. Comm. **29**, 280 (1964).

99. COOPER, A.J.L., and O.W. GRIFFITH: N-Hydroxyamino Acids. Irreversible Inhibitors of Pyridoxal 5' Phosphate Enzymes and Substrates of D- and L-Amino Acids Oxidases. J. Biol. Chem. **254**, 2748 (1979).

100. COOK, A.H., and C.A. SLATER: Pulcherrimin: A Synthesis of 1,4-Dihydroxy-2,5-dioxopiperazines. J. Chem. Soc. (London) **1956**, 4130.

101. NAGASAWA, M.T., J.G. KOHLHOFF, P.S. FRASER, and A.A. MIKHAIL: Synthesis of 1-Hydroxy-L-proline and Related Cyclic N-Hydroxyamino Acids. Metabolic Disposition of ^{14}C-Labeled 1-Hydroxy-L-proline in Rodents. J. Med. Chem. **15**, 483 (1972).

102. NEUNHOEFFER, O.: Proof of N-Hydroxypeptide Groups in the Protein of Cancer. Z. Naturforsch. **25b**, 299 (1970).

103. HUHEEY, J.E.: The Biochemistry of Iron. In: Inorganic Chemistry: Principles of Structure and Reactivity, p. 895. New York: Harper & Row, Pub., Inc., 1984.

104. Report of the Commission on Enzymes of the International Union of Biochemistry, p. 137. New York: Pergamon Press, 1962.

105. MICHAL, G.: Table. Biochemical Pathways. Boehringer Mannheim, 1982.

106. Shigeura, H.T., and C.N. Gordon: Hadacidin, a New Inhibitor of Purine Biosynthesis. J. Biol. Chem. 237, 1932 (1962).
– – The Mechanism of Action of Hadacidin. J. Biol. Chem. 237, 1937 (1962).
107. Shigeura, H.T.: Structural Modifications of Hadacidin and Their Effects on the Activity of Adenylosuccinate Synthetase. J. Biol. Chem. 238, 3999 (1963).
108. Rasnick, D., and J.C. Powers: Active Site Directed Irreversible Inhibition of Thermolysin. Design of Potent Reversible Inhibitors for Thermolysin. Biochemistry 17, 4363 (1978).
109. Nishino, N., and J.C. Powers: Peptides Containing Zinc Coordinating Ligands and Their Use in Affinity Chromatography. Biochemistry 18, 4340 (1979).
110. Holmes, M.A., D.E. Tronrud, and B.W. Matthews: Structural Analysis of the Inhibition of Thermolysin by an Active-Site-Directed Irreversible Inhibitor. Biochemistry 22, 236 (1983).
111. Nishino, N., and J.C. Powers: *Pseudomonas aeruginosa* Elastase. Development of a New Substrate, Inhibitors, and an Affinity Ligand. J. Biol. Chem. 255, 3482 (1980).
112. Lammers, M., and H. Follmann: The Ribonucleotide Reductases a Unique Group of Metalloenzymes Essential for Cell Proliferation. Structure and Bonding (Berlin) 54, 27 (1983).
113. Rozantsev, E.G., and V.D. Sholle: Synthesis and Reactions of Stable Nitroxyl Radicals. I. Synthesis. Synthesis 1971, 191.
114. Akiyama, M., K. Iesaki, A. Katoh, and K. Shimizu: N-Hydroxy Amides. Part 5. Synthesis and Properties of N-Hydroxypeptides Having Leucine Enkephalin Sequences. J. Chem. Soc. Perkin Trans. I (London) 1986, 851.
115. Emery, T.: Biosynthesis and Mechanism of Action of Hydroxamate-type Siderochromes. In: Microbial Iron Metabolism. (Neilands, J.B., ed.). New York: Academic Press 1974, p. 107.
116. – Initial Steps in the Biosynthesis of Ferrichrome Incorporation of δ-N-Hydroxyornithine and δ-N-Acetyl-δ-N-hydroxyornithine. Biochemistry 5, 3694 (1966).
117. Emery, T.F.: Hadacidin. In: Antibiotics. (Gottlieb, D., and P.D. Shaw, eds.,) Berlin: Springer 1967, vol. 2, p. 17.
118. Parniak, M.A., G.E.D. Jackson, G.J. Murray, and T. Viswanatha: Studies on the Formation of N^6-Hydroxylysine in Cell-Free Extracts of *Aerobacter aerogenes* 62-1. Biochim. Biophys. Acta 569, 99 (1979).
119. Dulaney, E.L.: Formation of N-Formylhydroxyaminoacetic Acid by *Penicillium*. Mycologia 55, 211 (1963).
120. Mac Donald, J.C.: Biosynthesis of Aspergillic Acid. J. Biol. Chem. 236, 512 (1961). Micetich, R.G., and J.C. Mac Donald: Biosynthesis of Neoaspergillic and Neohydroxyaspergillic Acids. J. Biol. Chem. 240, 1692 (1965).
121. Akers, H.A., and J.B. Neilands: Biosynthesis of Rhodotorulic Acid and Other Hydroxamate Type Siderophores. In: Biological Oxidation of Nitrogen. (Gorrad, J.W., ed.). Amsterdam: Biomedical Press 1978, p. 429.
122. Møller, B.L., and E.E. Conn: N-Hydroxyamino Acids as Intermediates in the Biosynthesis of Cyanogenic Glucosides in Plants. In: Biological Oxidation of Nitrogen. (Gorrad, J.W., ed.). Amsterdam: Biomedical Press 1978, p. 437.
123. – – The Biosynthesis of Cyanogenic Glucosides in Higher Plants. N-Hydroxytyrosine as an Intermediate in the Biosynthesis of Dhurrin by *Sorghum bicolor* (Linn) Moench. J. Biol. Chem. 254, 8575 (1979).
124. Scott, A.I., S.E. Yoo, S.K. Chung, and J.A. Lacadie: Reactivity of Peptide Hydroxamates: a Model for the Biosynthesis of β-Lactam Antibiotics. Tetrahedron Letters 1976, 1137.
125. Baxter, R.L., G. A. Thomson, and A.I. Scott: Synthesis and Biological Activity of δ-(L-α-Aminoadipoyl)-L-cysteinyl-N-hydroxy-D-valine: a Proposed in the Biosynthesis of the Penicillins. J. Chem. Soc. Chem. Commun. (London) 1984, 32.

126. OTTENHEIJM, H.C.J., R. PLATE, J.H. NOORDIK, and J.D.M. HERSCHEID: Synthesis of Natural Products Containing Oxidized Dioxopiperazines. An Approach to the Neoechinulin and Sporidesmin Series. J. Organ. Chem. (USA) 47, 2147 (1982).

127. HERSCHEID, J.D.M., R.J.F. NIVARD, M.W. TIJHUIS, and H.C.J. OTTENHEIJM: Biosynthesis of Gliotoxin. Synthesis of Sulfur – Bridged Dioxopiperazines from N-Hydroxyamino Acids. J. Organ. Chem. (USA) 45, 1885 (1980).

128. TRAUBE, W.: Ueber Isonitramine. Ber. dtsch. chem. Ges. 27, 1507 (1894).
     – Ueber Isonitramin- und Oxazo-Fettsäuren. Ber. dtsch. chem. Ges. 28, 1785 (1895).

129. – Ueber die Constitution der Isonitramine. Ber. dtsch. chem. Ges. 28, 2297 (1895).

130. HURD, CH.D., and J.M. LONGFELLOW: Preparation and Reactions of α-Hydroxylamino Nitriles. J. Organ. Chem. (USA) 16, 761 (1951).

131. PORTER, C.C., and L. HELKERMAN: α-Hydroxylaminoisobutylonitrile an Intermediate in the Synthesis of Porphyrexide and Porphyrindine. J. Amer. Chem. Soc. 61, 754 (1939).

132. – – Porphyrexide and Porphyrindine Analogs Derived from 1-Hydroxyaminocyclohexyl Cyanide. J. Amer. Chem. Soc. 66, 1652 (1944).

133. LILLEVIK, H.A., R.L. HOSSFELD, H.V. LINDSTROM, R.T. ARNOLD, and R.A. GORTNER: Technics in the Synthesis of Porphyrindin. J. Organ. Chem. (USA) 7, 164 (1942).

134. DUYNSTEE, E.F.J., J.L.J.P. HENNEKENS, and M.E.A.H. MEVIS: Decarbonylation of α-Hydroxylaminocarboxylic Acids. Rec. trav. chim. Pays-Bas 84, 1442 (1965).

135. WEISBALT, D.I., and D.A. LYTTLE: Synthesis of Ethyl α-Nitro-β-(3-indole)-propionate from Gramine and Ethyl Nitromalonate. J. Amer. Chem. Soc. 71, 3079 (1949).

136. EMERY, T., and J.B. NEILANDS: Contribution to the Structure of the Ferrichrome Compounds: Characterization of the Acyl Moieties of the Hydroxamate Functions. J. Amer. Chem. Soc. 82, 3658 (1960).
     – – Structure of the Ferrichrome Compounds. J. Amer. Chem. Soc. 83, 1626 (1961).

137. ROGERS, S., and J.B. NEILANDS: The α-Amino-ω-Hydroxyamino Acids. Biochemistry 2, 6 (1963).

138. MAKAJEJENINA, L.G., and N.A. PODDUBNAJA: Studies on the Structure of the Antibiotic Albomycin. XIII. Synthesis of δ-N-Hydroxyornithine. Zhurn. Obshchei. Khimii 36, 1755 (1966).

139. BLACK, D.S.C., R.F.C. BROWN, and A.M. WADE: Synthetic Studies Related to Mycobactins. II. Formation of Mycobactin Hydroxamic Acid Units by Malonic Esters Synthesis. Austral. J. Chem. 25, 2155 (1972).

140. SHIN, C., M. MASAKI, and M. OHTA: The Synthesis and Reaction of α,β-Unsaturated α-Nitrocarboxylic Esters. Bull. Chem. Soc. Japan 43, 3219 (1970).

141. MAURER, B., and W. KELLER-SCHIERLEIN: Stoffwechselprodukte von Mikroorganismen 74. Mitt. Synthese des Ferrichroms; 1 Teil. (S)-α-Amino-δ-nitrovaleriansäure (δ-Nitro-L-norvalin). Helv. Chim. Acta 52, 388 (1969).

142. KELLER-SCHIERLEIN, W., and B. MAURER: Stoffwechselprodukte von Mikroorganismen 74. Mitt. Synthese des Ferrichroms; 2. Teil. Helv. Chim. Acta 52, 603 (1969).

143. WIDMER, J., and W. KELLER-SCHIERLEIN: Synthesen in der Sideramin-Reine: Rhodotorulasäure und Dimerumsäure. Helv. Chim. Acta 57, 1904 (1974).

144. AHMAD, A.: Syntheses of α-Hydroxyamino Acids from α-Keto Acids. Bull. Chem. Soc. Japan 47, 1819 (1974).

145. HERSCHEID, J.D.M., and H.C.J. OTTENHEIJM: A Practical Synthesis of N-Hydroxy-α-amino Acid Esters. Tetrahedron Letters 1978, 5143.

146. TIJHUIS, M.W., J.D.M. HERSCHEID, and H.C.J. OTTENHEIJM: A Practical Synthesis of N-Hydroxy-α-amino Acid Derivatives. Synthesis 1980, 890.

147. LEE, B.H., and M.J. MILLER: Constituents of Microbial Iron Chelators. The Synthesis of Optically Active Derivatives of δ-N-Hydroxy-L-ornithine. Tetrahedron Letters 25, 927 (1984).

274 A. CHIMIAK and MARIA J. MILEWSKA:

148. POSNER, T.: Beiträge zur Kenntnis der ungesättigten Verbindungen. IX. Ueber die Addition von Hydroxylamin an ungesättigte Säuren und Ester der Zimmtsäurereine sowie an analoge Verbindungen. Liebigs Ann. Chem. **389**, 1 (1912).
149. FOUNTAIN, K.R., R. ERWIN, T. EARLY, and H. KEHL: The Mechanism of Hydroxylamine Addition to α,β-Unsaturated Esters. Tetrahedron Letters **1975**, 3027.
150. STEIGER, R.E.: dl-β-Amino-β-phenylpropionic Acid. Organic Syntheses; Wiley: New York, 3, p. 91 (1955).
151. POSNER, T.: Beiträge zur Kenntnis der ungesättigten Verbindungen. II. Ueber die Anlagerung von freiem Hydroxylamin an Zimmtsäure. Constitution und Derivate der α-Hydroxylamino-β-phenylopropionsäure. Ber. dtsch. chem. Ges. **39**, 3519 (1906).
- Beiträge zur Kenntnis der ungesättigten Verbindungen. I. Ueber die Einwirkung von freiem Hydroxylamin auf ungesättigte Säuren. Ber. dtsch. chem. Ges. **36**, 4314 (1903).
152. HARRIES, C., and W. HAARMANN: Ueber die Einwirkung von Hydroxylamin auf ungesättigte Säureester. Ber. dtsch. chem. Ges. **37**, 253 (1904).
BECKE, F., and G. MUTZ: Über die Reaktion von Hydroxylamin mit Acrylsäurederivaten. Chem. Ber. **98**, 1322 (1965).
153. SAYIGH, A.A.R., H. ULRICH, and M. GREEN: Michael Addition of Hydroxylamines to Activated Double Bonds. A Convenient Synthesis of N,N-Dialkyl Hydroxylamines. J. Organ. Chem. (USA) **29**, 2042 (1964).
154. GROSSOWICZ, N., and Y. LICHTENSTEIN: Enzymic Binding of Hydroxylamine by Fumaric Acid. Nature **191**, 412 (1961).
155. HANTZSCH, A., and W. WILD: Ueber Oxime aus α-halogenisierten Aldehyden, Ketonen und Säuren sowie über Oximessigsäuren. Liebigs Ann. Chem. **289**, 285 (1896).
156. BARRY, R.H., and W.H. HARTUNG: α-Oximino Acid Intermediates for the Synthesis of α-Amino Acids. J. Organ. Chem. (USA) **12**, 460 (1947).
157. ORGANON, N.V.: Hydroxyamine Solutions and Their Use for Hydroxyamino Acids Preparation. Belgium Patent 660,703, **1965**; Chem. Abstr. **64**, 2164a (1966).
158. LA NOCE, T., E. BELLASIO, and E. TESTA: Synthesis of α-N-Hydroxyamino Acids. Part 1. Ann. chimie **58**, 393 (1968).
159. KAMIŃSKI, K., and T. SOKOŁOWSKA: Synthesis of Optically Active, Aromatic N-Hydroxyamino Acids, Analogues of Phenylalanine and Tyrosine. Polish. J. Chem. **47**, 653 (1973).
160. SHIN, C., K. NANJO, E. ANDO, and J. YOSHIMURA: α,β-Unsaturated Carboxylic Acid Derivatives. VI. New Synthesis of N-Acyl-α-dehydroamino Acid Esters. Bull. Chem. Soc. Japan **47**, 3109 (1974).
161. LANCINI, G.C., A. DIENA, and E. LAZZARI: The Synthesis of Alanosine [L-2-Amino-3-(N-nitrosohydroxylamino)propionic Acid]. Tetrahedron Letters **1966**, 1769.
162. LANCINI, G.C., E. LAZZARI, and A. DIENA: Synthesis of Homologues of the Antibiotic Alanosine. Il Farmaco **24**, 169 (1969).
163. KOLASA, T., and A. CHIMIAK: O-Protected Derivatives of N-Hydroxyamino Acids. Tetrahedron **30**, 3591 (1974).
CHIMIAK, A., and T. KOLASA: N-Alkoxyamino Acids – Key Substrates for N-Hydroxypeptides. In: Peptides 1972. p. 118. (HANSON, H., and H.D. JAKUBKE, eds.). 12th European Peptide Symposium. Reinhardsbrunn Castle 1972. Amsterdam: Elsevier 1973.
164. KOLASA, T.: Synthetic Study of N-Hydroxyaspartic Acid. Canad. J. Chem. **65**, 2139 (1985).
165. KOLASA, T., and A. CHIMIAK: Unambiguous Synthesis of N-Hydroxypeptides. Tetrahedron **33**, 3285 (1977).
166. ISOWA, Y., T. TAKASHIMA, M. OHMORI, H. KURITA, M. SATO, and K. MORI: Acylation of N⁶-Benzyloxyornithine. Bull. Chem. Soc. Japan **45**, 1464 (1972).

167. —————— Synthesis of N$^\delta$-Hydroxyornithine. Bull. Chem. Soc. Japan **45**, 1461 (1972).

168. —————— N$^\delta$-Hydroxyornithine. Japan Patent 7234314, **1972**; Chem. Abstr. **78**, 98013 q (1973).

169. Isowa, Y., and M. Ohmori: Synthesis of N$^\epsilon$-Hydroxy-L-lysine. Bull. Chem. Soc. Japan **47**, 2672 (1974).

170. Isowa, Y., H. Kurita, M. Ohmori, M. Sato, and K. Mori: Synthesis of Alanosine. Bull. Chem. Soc. Japan **46**, 1847 (1973).

171. Isowa, Y., and H. Kurita: A New Reagent for the Syntheses of N-Monoalkylated Hydroxylamines; N-Tosyl-O-2,4,6-trimethyl-benzylhydroxylamine. Bull. Chem. Soc. Japan **47**, 720 (1974).

172. Fujii, T., and Y. Hatanaka: A Synthesis of Rhodotorulic Acid. Tetrahedron **29**, 3825 (1973).

173. Isowa, Y., M. Ohmori, and H. Kurita: Total Synthesis of Ferrichrome. Bull. Chem. Soc. Japan **47**, 215 (1974).

174. Kamiya, T., K. Hemmi, H. Takeno, and M. Hashimoto: Studies on Phosphonic Acid Antibiotics. I. Structure and Synthesis of 3-(N-Acetyl-N-hydroxyamino)propyl-phosphonic Acid (FR-900098) and Its N-Formyl Analogue (FR-31564). Tetrahedron Letters **21**, 95 (1980).

175. Hashimoto, M., K. Hemmi, H. Takeno, and T. Kamiya: Studies on Phosphonic Acid Antibiotics. II. Synthesis of 3-(N-Acetyl-N-hydroxyamino)-2(R)-hydroxypropyl-phosphonic Acid (FR-33289) and 3-(N-Formyl-N-hydroxyamino)-1-trans-propenyl-phosphonic Acid (FR-32863). Tetrahedron Letters **21**, 99 (1980).

176. Hemmi, K., H. Takeno, M. Hashimoto, and T. Kamiya: Studies on Phosphonic Acid Antibiotics. III. Structure and Synthesis of 3-(N-Acetyl-N-hydroxyamino)propyl-phosphonic Acid (FR-900098) and 3-(N-Acetyl-N-hydroxyamino)-2(R)-hydroxypro-pylphosphonic Acid (FR-33289). Chem. Pharm. Bull. Japan **29**, 646 (1981).

177. Buehler, E.: Alkylation of *syn*- and *anti*-Benzaldoximes. J. Organ. Chem. (USA) **32**, 261 (1967).

178. Buehler, E., and G.B. Brown: A General Synthesis of N-Hydroxyamino Acids. J. Organ. Chem. (USA) **32**, 265 (1967).

179. Hurd, C.D., and J.M. Longfellow: Phenyl-N-(1-carboxyethyl)-nitrone. J. Amer. Chem. Soc. **73**, 2395 (1951).

180. Maehr, H., and M. Leach: Antimetabolites Produced by Microorganisms. IX. Chemical Synthesis of N^5-Hydroxyornithine and N^5-Hydroxyarginine. J. Organ. Chem. (USA) **39**, 1166 (1974).

181. Neunhoeffer, O., G. Lehmann, D. Haberer, and G. Steinle: Die Hydroxylamino-lyse N-substituierter Phthalimide, ein Verfahren zur Darstellung von Peptiden und N-Hydroxypeptiden. Liebigs Ann. Chem. **712**, 208 (1968).

182. Bellasio, E., F. Parravicini, T. La Noce, and E. Testa: α-Hydroxyamino Acids and Their Derivatives. Part II. Ann. chimie **58**, 407 (1968).

183. Połoński, T., and A. Chimiak: Nitrones as Intermediates in the Synthesis of N-Hydroxyamino Acid Esters. J. Organ. Chem. (USA) **41**, 2092 (1976).

184. Lau, H.H., and U. Schöllkopf: Synthesis of N-Hydroxy-α-amino Acids by Alkyla-tion of N-Benzylidene-α-amino Acid Methyl Ester N-Oxides. Liebigs Ann. Chem. **1981**, 1378.

185. Bentley, P.H., and G. Brooks: Semi-synthetic Penicillins and Cephalosporins Incor-porating a Hydroxyamino Group. Tetrahedron Letters **1976**, 3735.

186. Ross-Petersen, K.J., and H. Hjeds: Syntheses of Some Derivatives of α-Hydroxy-carboxylic Acids. Dansk. Tidsskr. Farm. **43**, 188 (1969).

187. Bellasio, E., F. Parravicini, A. Vigevani, and E. Testa: Synthesis and Properties of β-N-Hydroxyamino Acids. Part III. Gazz. chim. ital. **98**, 1014 (1968).

188. LIBEREK, B., and Z. PALACZ: Steric Course of Synthesis of N-Hydroxyamino Acids from Optically Active Amino Acids. Polish J. Chem. **45**, 1173 (1971).
189. SCHOENEWALDT, E.F., R.B. KINNEL, and P. DAVIS: Improved Synthesis of *anti*-Benzaldoxime. Concomitant Cleavage and Formylation of Nitrones. J. Organ. Chem. (USA) **33**, 4271 (1968).
190. EATON, C.N., G.H. DENNY JR., M.A. RYDER, M.G. LY, and R.D. BABSON: Improved Synthesis of DL-Alanosine. J. Med. Chem. **16**, 289 (1973).
191. CHIMIAK, A., and T. POŁOŃSKI: Application of Nitron Group for the Synthesis of N-Hydroxydipeptides and N,N'-Dihydroxyaminodicarboxylic Acids. Proceeding of Meeting of Polish Chem. Soc., Gliwice **1972**, 175.
192. POŁOŃSKI, T., and A. CHIMIAK: Oxaziridines as Intermediates in the Oxidation of Amino Acid Esters into N-Hydroxyamino Acid Derivatives. Bull. Acad. Polon. Sci. Ser. Chem. **27**, 459 (1979).
193. -- Oxidation of Amino Acid Esters into N-Hydroxyamino Acid Derivatives. Tetrahedron Letters **1974**, 2453.
194. WIDMER, J., and W. KELLER-SCHIERLEIN: Stoffwechselprodukte von Mikroorganismen. 130 Mitt. Synthese des δ-N-Hydroxy-L-arginins. Helv. Chim. Acta **57**, 657 (1974).
195. NAEGELI, H.U., and W. KELLER-SCHIERLEIN: Stoffwechselprodukte von Mikroorganismen. 174 Mitt. Eine neue Synthese des Ferrichroms; enantio-Ferrichrom. Helv. Chim. Acta **61**, 2088 (1978).
196. MAURER, P.J., and M.J. MILLER: Myobactins: Synthesis of (–)-Cobactin T from ε-Hydroxynorleucine. J. Organ. Chem. (USA) **13**, 2835 (1981).
197. MITSUNOBU, O.: The Use of Diethyl Azodicarboxylate and Triphenylphosphine in Synthesis and Transformation of Natural Products. Synthesis **1981**, 1.
198. OLSEN, R.K., K. RAMASAMY, and T. EMERY: Synthesis of N^α,N^δ-Hydroxy-L-ornithine from L-Glutamic Acid. J. Organ. Chem. (USA) **49**, 3527 (1984).
199. MAURER, P.J., and M.J. MILLER: Total Synthesis of a Mycobactin: Mycobactin S2. J. Amer. Chem. Soc. **105**, 240 (1983).
200. -- Microbial Iron Chelators: Total Synthesis of Aerobactin and Its Constituent Amino Acid, N^6-Acetyl-N^6-hydroxylysine. J. Amer. Chem. Soc. **104**, 3096 (1982).
201. KANAI, F., K. ISSHIKI, Y. UMEZAWA, H. MORISHIMA, H. NAGANAWA, T. TAKITA, T. TAKEUCHI, and H. UMEZAWA: Vanoxonin, a New Inhibitor of Thymidylate Synthetase. II. Structure Determination and Total Synthesis. J. Antibiotics **38**, 31 (1985).
202. LEE, B.H., G.J. GERFEN, and M.J. MILLER: Constituents of Microbial Iron Chelators. Alternate Syntheses of δ-N-Hydroxy-α-ornithine Derivatives and Applications to the Synthesis of Rhodotorulic Acid. J. Organ. Chem. (USA) **49**, 2418 (1984).
203. KOLASA, T., A. CHIMIAK, and A. KITOWSKA: Esters of N-Benzyloxyamino Acids. J. prakt. Chem. **317**, 252 (1975).
204. HERSCHEID, J.D.M., J.H. COLSTEE, and H.C.J. OTTENHEIJM: 1,4-Dihydro-2,5-dioxopiperazines from Activated N-Hydroxyamino Acids. J. Organ. Chem. (USA) **46**, 3346 (1981).
205. BENZ, G.: Albomycine, III. Synthese von N^5-Acetyl-N^5-hydroxy-L-ornithin aus L-Glutaminsäure. Liebigs Ann. Chem. **1984**, 1424.
206. NEILANDS, J.B., and P. AZARI: Synthesis and Reactions of the ω-N-Hydroxyamino Acids. Acta Chem. Scand. **17S**, 190 (1963).
207. SHIN, C., M. HAYAKAWA, T. SUZUKI, A. OHTSUKA, and J. YOSHIMURA: α,β-Unsaturated Carboxylic Acid Derivatives. XIII. The Synthesis and Configuration of Alkyl 2-Acylamino-2-alkenoates and Their cyclized 2,5-Piperazinedione Derivatives. Bull. Chem. Soc. Japan **51**, 550 (1978).
208. SHIN, C., K. NANJO, and J. YOSHIMURA: Cyclization Reaction of N-(Haloacyl)- or N-(Phthaloylglycyl)-hydroxyamino Acid Esters with Ammonia. Chem. Letters (Japan) **1973**, 1039.

209. SHIN, C., K. NANJO, M. KATO, and J. YOSHIMURA: α,β-Unsaturated Carboxylic Acid Derivatives. IX. The Cyclization of α-(N-Acyl-hydroxyamino) Acid Esters with Ammonia or Hydroxylamine. Bull. Chem. Soc. Japan **48**, 2584 (1975).
210. SHINMON, N., and M.P. CAVA: Total Synthesis of (±)-Mycelianamide. J. Chem. Soc. Chem. Commun. (London) **1980**, 1020.
211. HERSCHEID, J.D.M., R.J.F. NIVARD, M.W. TIJHUIS, M.P.H. SCHOLTEN, and H.C.J. OTTENHEIJM: α-Functionalized Amino Acid Derivatives. A Synthetic Approach of Possible Biogenetic Importance. J. Organ. Chem. (USA) **45**, 1880 (1980).
212. CHIMIAK, A., and T. POŁOŃSKI: The Use of o-Nitrophenylsulfenyl-N-carboxyanhydrides in N-Hydroxypeptide Synthesis. J. prakt. Chem. **322**, 669 (1980).
213. JENCKS, W.P.: The Reaction of Hydroxylamine with Activated Acyl Groups. II. Mechanism of the Reaction. J. Amer. Chem. Soc. **80**, 4585 (1958).
214. ZVILICHOVSKY, G., and L. HELLER: The synthesis of N-Hydroxy Peptides. Tetrahedron Letters **1969**, 1159.
215. SHIMIZU, K., M. HASEGAWA, and M. AKIYAMA: N-Hydroxy Amides. I. Synthesis of N-Benzyloxy and N-Hydroxy Peptides via Polymerization of N-Benzyloxy DL-α-Amino Acid N-Carboxy Anhydrides. Bull. Chem. Soc. Japan **57**, 495 (1984).
216. SHIMIZU, K., K. NAKAYAMA, and M. AKIYAMA: N-Hydroxy Amides. II. N-Benzyloxy-α-Amino Acid N-Hydroxysuccinimide Esters and Synthesis of a Hexapeptide Having an Alternating N-Hydroxy Amide – Amide Sequence. Bull. Chem. Soc. Japan **57**, 2456 (1984).
217. PALACZ, Z.: N-Benzylideneamino Acid N-Oxides and Their Use in the Synthesis of N-Benzylidenepeptide N-Oxide Esters and in the Synthesis of Optically Active N-Hydroxyamines. Ph. D. Thesis, University of Gdańsk 1978.
218. AKIYAMA, M., M. HASEGAWA, H. TAKEUCHI, and K. SHIMIZU: N-Hydroxypeptides. I. Preparation of N-Benzyloxy-α-amino Acid Anhydrides and Their Use in Peptide Synthesis. Tetrahedron Letters **1979**, 2599.
219. KOLASA, T.: An Effective Synthesis of α,β-Dehydro-α-amino Acid Derivatives from N-Acyl-N-hydroxy-α-amino Acid. Synthesis **1983**, 539.
220. OTTENHEIJM, H.C., and J.D.M. HERSCHEID: N-Hydroxy-α-amino Acids in Organic Chemistry. Chem. Rev. **86**, 697 (1986).
221. JALAL, M.A.F., J.L. GALLES, and D. VAN DER HELM: Structure of Des(diserylglycyl)-ferrirhodin, DDF, a Novel Siderophore from Aspergillus ochraceous. J. Organ. Chem. (USA) **50**, 5642 (1985).
222. FEENSTRA, R.W., E.H.M. STOKKINGREEF, R.J.F. NIVARD, and H.C.J. OTTENHEIJM: An Efficient Synthesis of N-Hydroxy-α-amino Acid Derivatives of High Optical Purity. Tetrahedron Letters **28**, 1215 (1987).
223. MILEWSKA, M.J., and A. CHIMIAK: Oxidation of Amino Acids. V. A Novel Synthesis of N^6-Acetyl-N^6-hydroxylysine from Lysine. Tetrahedron Letters **28**, 1817 (1987). – – Reaction of Dibenzoyl Peroxide with ω-Amino Acid Esters and Novel Synthesis of N^6-Acetyl-N^6-hydroxylysine and N^5-Acetyl-N^5-hydroxyornithine. 31st Int. Congress of Pure and Applied Chemistry, Sec. 6, Sofia, **1987**, 6.126.
224. MILEWSKA, M.J., T. KOLASA, and A. CHIMIAK: Oxidation of Amino Acid Esters with Acyl Peroxides. Polish J. Chem. **55**, 2215 (1981).
225. SHIMIZU, K., K. NAKAYAMA, and M. AKIYAMA: N-Hydroxy Amides. IV. Synthesis and Properties of a Trihydroxamic Acid Anilide as a Model for Ferrioxamines. Bull. Chem. Soc. Japan **59**, 2421 (1986).

(Received September 19, 1987)

Author Index

Page numbers printed in *italics* refer to References

Subject Index

Composition: Universitätsdruckerei H. Stürtz AG, D-8700 Würzburg

Fortschritte der Chemie organischer Naturstoffe
Progress in the Chemistry of Organic Natural Products

Volume 52:

1987. 65 figures. VIII, 224 pages. Cloth DM 210,–, öS 1470,–.
ISBN 3-211-81989-4
Contents: U. Weiss, L. Merlini, and G. Nasini: Naturally Occurring Perylene-quinones. – H. Achenbach: The Pigments of the Flexirubin-Type. A Novel Class of Natural Products. – T. Goto: Structure, Stability and Color Variation of Natural Anthocyanins. – P. Bhattacharyya and D. P. Chakraborty: Carbazole Alkaloids.

Volume 51:

1987. VII, 317 pages. Cloth DM 280,–, öS 1960,–.
ISBN 3-211-81972-X
Contents: M. Gill and W. Steglich: Pigments of Fungi (Macromycetes).

Volume 50:

1986. 71 figures. IX, 261 pages. Cloth DM 210,–, öS 1470,–.
ISBN 3-211-81969-X
Contents: L. Jaenicke and F.-J. Marner: The Irones and Their Precursors. – M. Lounasmaa and P. Somersalo: The Condylocarpine Group of Indole Alkaloids. – U. Séquin: The Antibiotics of the Pluramycin Group (4*H*-Anthra [1,2-*b*]pyran Antibiotics). – R. M. Wenger: Cyclosporine and Analogues – Isolation and Synthesis – Mechanism of Action and Structural Requirements for Pharmacological Activity. – H. Inouye and S. Uesato: Biosynthesis of Iridoids and Secoiridoids.

Volume 49:

1986. VIII, 400 pages. Cloth DM 290,–, öS 2030,–. ISBN 3-211-81910-X
Contents: R. A. Hill: Naturally Occurring Isocoumarins. – R. Wijnsma and R. Verpoorte: Anthraquinones in the Rubiaceae. – H. Chr. Krebs: Recent Developments in the Field of Marine Natural Products with Emphasis on Biologically Active Compounds.

Volume 48:

1985. 33 figures. IX, 285 pages. Cloth DM 220,–, öS 1540,–.
ISBN 3-211-81886-3
Contents: P. S. Steyn and R. Vleggaar: Tremorgenic Mycotoxins. – R. E. Moore: Structure of Palytoxin. – P. Crews and S. Naylor: Sesterterpenes: An Emerging Group of Metabolites from Marine and Terrestrial Organisms.

Volume 47:

1985. 16 figures. VIII, 290 pages. Cloth DM 198,–, öS 1390,–.
ISBN 3-211-81864-2
Contents: R. Southgate and S. Elson: Naturally Occurring β-Lactams. – I. Howe and M. Jarman: New Techniques for the Mass Spectrometry of Natural Products. – P. G. McDougal and N. R. Schmuff: Chemical Synthesis of the Trichothecenes. – J. Polonsky: Quassinoid Bitter Principles II.

Volume 46:

1984. 7 figures. IX, 253 pages. Cloth DM 178,–, öS 1250,–.
ISBN 3-211-81804-9
Contents: O. Tanaka and R. Kasai: Saponins of Ginseng and Related Plants. – E. Fujita and M. Node: Diterpenoids of *Rabdosia* Species. – S. Johne: The Quinazoline Alkaloids.

All Volumes and Cumulative Index 1–20 available

Price reduction for subscribers: 10%

Special reduced price (20% reduction) for the complete Series Vols. 1–52 incl. the Cumulative Index to Vols. 1–20

Springer-Verlag Wien New York

Mölkerbastei 5, A-1011 Wien
175 Fifth Avenue, New York, NY 10010, U.S.A.
Heidelberger Platz 3, D-1000 Berlin 33
37-3, Hongo 3-chome, Bunkyo-ku, Tokyo 113, Japan